# A LEVEL CHEMISTRY for OCR A

Primrose Kitten
Adam Robbins
Alyssa Fox-Charles
Mike Wooster
Josh Thomas

**3 steps to access your free online copy of this book**

1. Visit Education Bookshelf (http://bookshelf.oxfordsecondary.co.uk).
2. Create an account by clicking the *register* button or *sign in* if you already have an account.
3. Click on the *Activate Your Publication* button and enter your unique access code (printed on the inside cover).

OXFORD UNIVERSITY PRESS

# Contents

Shade in each level of the circle as you feel more confident and ready for your exam.

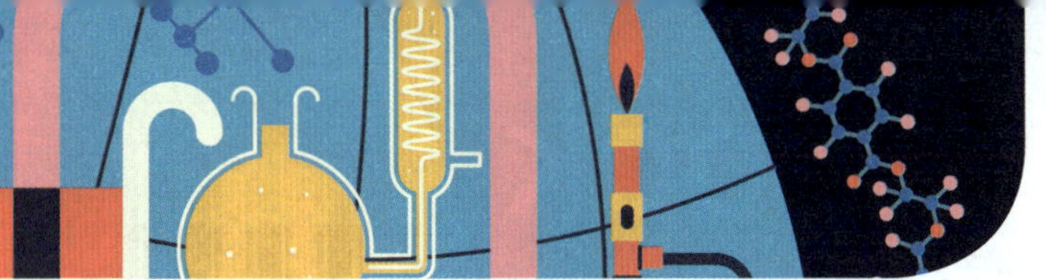

# How to use this book

This book uses a three-step approach to revision: **Knowledge**, **Retrieval**, and **Practice**. It is important that you do all three; they work together to make your revision effective.

## 1 Knowledge

**Knowledge** comes first. Each chapter starts with a **Knowledge Organiser**. These are clear, easy-to-understand, concise summaries of the content that you need to know for your exam. The information is organised to show how concepts relate to each other so you can understand how the knowledge fits together, rather than learning lots of disconnected facts.

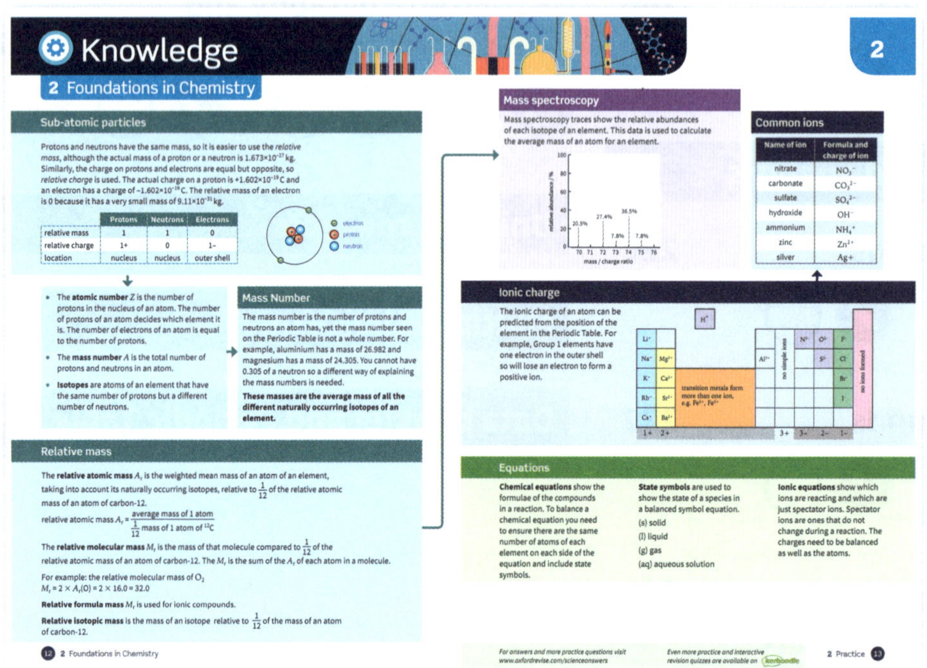


## 2 Retrieval

The **Retrieval questions** help you learn and quickly recall the information you've acquired. Memorise the short questions and answers about the content in the Knowledge Organiser, then cover the answers with some paper and write as many as you can from memory. Check back to the Knowledge Organiser for any you got wrong, then attempt the questions again until you can answer all the of them correctly.

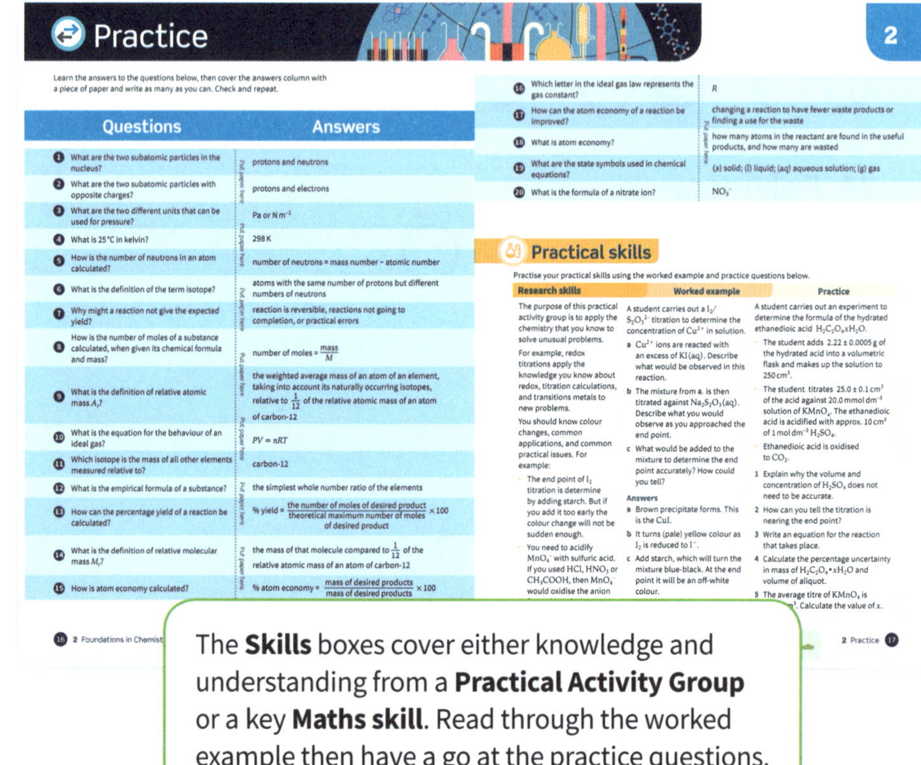


The **Skills** boxes cover either knowledge and understanding from a **Practical Activity Group** or a key **Maths skill**. Read through the worked example then have a go at the practice questions.

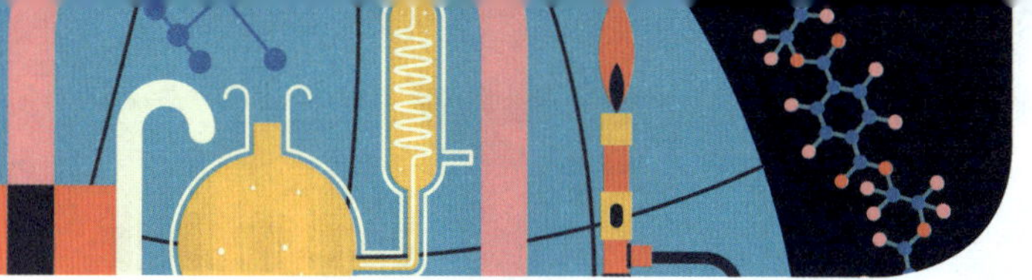

Storing these basic facts in your long-term memory through regular retrieval practice will mean you will find it easier to apply them to complex situations and difficult questions.

## 3 Practice

Once you think you know the Knowledge Organiser and Retrieval answers really well you can move on to the final stage: **Practice**.

Each chapter has lots of **exam-style questions**, that test your ability to:

- apply your knowledge and understanding, including from the Required Practicals
- analyse and evaluate information
- combine knowledge from different parts of the course (synoptic questions).

Questions with the link icon will have **synoptic links**. This means that they will assess the content from the current chapter along with knowledge from elsewhere in the course.

Practice

Exam-style questions

Questions with the apparatus icon test your **practical skills**. At least 15% of the marks in your exams will be about practical skills.

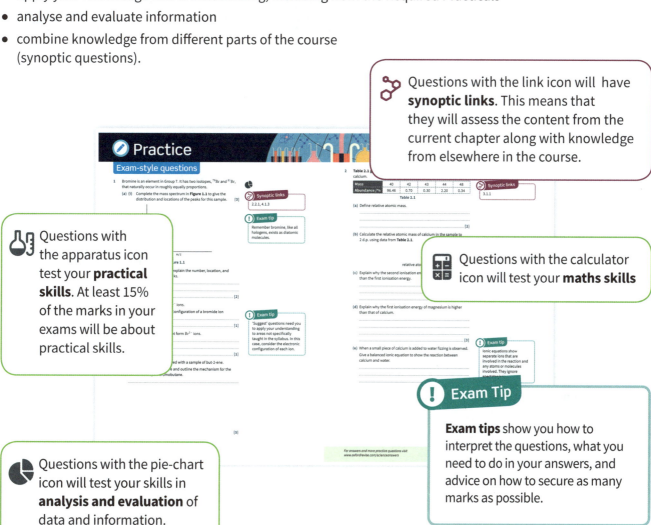

Questions with the calculator icon will test your **maths skills**

**Exam Tip**

**Exam tips** show you how to interpret the questions, what you need to do in your answers, and advice on how to secure as many marks as possible.

Questions with the pie-chart icon will test your skills in **analysis and evaluation** of data and information.

**kerboodle**

All the **answers** are on Kerboodle and the website. www.oxfordrevise.com/scienceanswers

# Knowledge

## Development of practical skills in Chemistry

### Assessment of practicals

Practical work is at the heart of all of the sciences. As such, written exams include questions on practical work in written exams. Over all of the exams, 15% of the marks will be associated with practical work. In order to answer these questions effectively, you will need to be confident that you know the details of the practicals you have carried out and you should be able to explain the steps that are taken.

### Practical skills and knowledge assessed in written papers

Keep track of your confidence in each skill that may be assessed in your exams.

| Check ✓ | Practical skill |
|---|---|
| | Solve problems in a practical context |
| | Apply scientific knowledge to practical contexts |
| | Comment on experiment design (what are the pros and/or cons) and evaluate scientific methods |
| | Present data in appropriate ways |
| | Evaluate results and draw conclusions. At A level you would be expected to comment on the uncertainties and errors |
| | Identify variables including controls |
| | Plot and interpret graphs – this includes calculating gradients |
| | Process and analyse data |
| | Consider margins of error, accuracy, and precision in data |
| | Know and understand the experimental procedures listed on the right. This includes understanding *why* steps are taken and not just *how* to do them. |

### Apparatus and techniques

You will have used a range of apparatus and techniques when carrying out your practical activity groups. In your written examinations, you might be asked about:

| | |
|---|---|
| a | Using appropriate apparatus to measure: mass, time, volume of liquids and gases, temperature |
| b | Using a water bath or electric heater or sand bath for heating |
| c | Measuring pH using pH charts, or a pH meter, or a pH probe on a data logger |
| d | Using laboratory apparatus for a variety of experimental techniques including:<br>• titration, using burette and pipette<br>• distillation and heating under reflux, including setting up glassware using retort stand and clamps<br>• qualitative tests for ions and organic functional groups<br>• filtration, including use of fluted filter paper, or filtration under reduced pressure |
| e | Using volumetric flask, including accurate technique for making up a standard solution |
| f | Using acid–base indicators in titrations of weak/strong acids with weak/strong alkalis |
| g | Purifying:<br>• a solid product by recrystallisation<br>• a liquid product, including use of a separating funnel |
| h | Using melting point apparatus |
| i | Using thin-layer or paper chromatography |
| j | Setting up electrochemical cells and measuring voltages |
| k | Safely and carefully handling solids and liquids, including corrosive, irritant, flammable, and toxic substances |
| l | Measuring rates of reaction by at least two different methods, for example:<br>• an initial rate method such as a clock reaction<br>• a continuous monitoring method. |

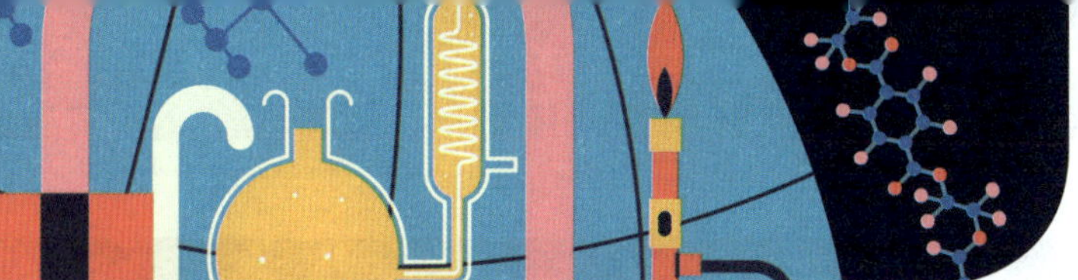



## PAG 1 – Moles determination

Titrations and mole calculations appear in many contexts and not just acid–base reactions.

**Key knowledge and skills**

- Use appropriate apparatus to record a range of measurements.

**Example practical**

Determination of the composition of copper(II) carbonate or determination of the $A_r$ of magnesium.

**Tips**

When measuring masses, remember to find the mass of empty clean containers before adding solids or liquids.

## PAG 2 – Acid–base titration

Titrations allow you to measure volumes very accurately.

**Key knowledge and skills**

- Use appropriate apparatus to record a range of measurements.
- Use laboratory apparatus for a variety of experimental techniques, including titration using a burette and a pipette.
- Use volumetric flask, including accurate technique for making up a standard solution.

**Example practical**

Titration of sodium hydrogencarbonate against hydrochloric acid or determination of the $M_r$ of an unknown acid.

**Tips**

When making a standard solution, if you overfill the volumetric flask you have to dispose of it all and start again, so be careful.

## PAG 3 – Enthalpy determination

When bonds are made, the chemical potential energy is transferred to thermal energy. A change in thermal energy can be measured and an enthalpy change determined.

**Key knowledge and skills**

- Use appropriate apparatus to record a range of measurements.
- Use laboratory apparatus for a variety of experimental techniques including titration, using a burette, and using a pipette.

**Example practical**

Determination of the enthalpy change of neutralisation or determination of a reaction using Hess' law.

**Tips**

Make sure the reaction vessel is well insulated; for example, by adding a lid. Often these experiments take place in a polystyrene cup.

## PAG 4 – Qualitative analysis of ions

Recording observations of test-tube reactions is an important skill for any chemist.

**Key knowledge and skills**

- Use laboratory apparatus for a variety of experimental techniques, including qualitative tests for ions and organic functional groups.

**Example practical**

Identification of the anions and cations present in a mixture of Group 2 salts.

**Tips**

Learn the tests, equations, and observations. Many are carried over from GCSE.

## PAG 5 – Synthesis of an organic liquid

After an organic reaction has completed there is likely to be a mixture of products, reagents, and impurities in the reaction flask. Distillation is a simple but effective technique to separate mixtures based on their boiling points.

**Key knowledge and skills**

- Use a water bath, electric heater, or sand bath for heating.
- Use laboratory apparatus for a variety of experimental techniques, including distillation and heating under reflux, and setting up glassware using a retort stand and clamps.
- Safely and carefully handle solids and liquids, including corrosive, irritant, flammable, and toxic substances.
- Purify a liquid product, including use of separating funnel.

**Example practical**

Synthesis of a haloalkane/cyclohexene or oxidation of ethanol.

**Tips**

When distilling, add anti-bumping granules and heat gently. Overheating is dangerous and can cause impurities in the distillate.

## PAG 6 – Synthesis of an organic solid

Synthesis and purification of an organic solid involves many different steps. After the reaction is complete you can purify by crystallisation and filtration.

**Key knowledge and skills**

- Use laboratory apparatus for a variety of experimental techniques including filtration, such as use of fluted filter paper, or filtration under reduced pressure.
- Purify a solid product by recrystallisation.
- Use melting point apparatus.

**Example practical**

Synthesis of aspirin or nitration of methylbenzene.

**Tips**

There are many steps in this experiment. Make sure you understand what is going on in each of them.

## PAG 7 – Qualitative analysis of organic functional groups

This is very similar to PAG 4, which tested inorganic ions. Here, organic functional groups are being tested instead. Some of these tests you would have come across before, at GCSE, but it is essential you memorise them and any observations that occur.

**Key knowledge and skills**

- Use a water bath, electric heater, or sand bath for heating.
- Use laboratory apparatus for a variety of experimental techniques including qualitative tests for ions and organic functional groups.
- Safely and carefully handle solids and liquids, including corrosive, irritant, flammable, and toxic substances.

**Example practical**

Identifying functional groups in a series of unknown organic compounds.

**Tips**

Learn the tests, equations, and observations. Many are carried over from GCSE.

## PAG 8 – Electrochemical cells

The electromotive force (EMF) is the potential difference of two cells. You can use a voltmeter to measure the potential difference between two cells.

**Key knowledge and skills**

- Set up electrochemical cells and measuring voltages.
- Safely and carefully handle solids and liquids, including corrosive, irritant, flammable, and toxic substances.

**Example practical**

Investigate the effect of concentration on the cell potential of an electrochemical cell or investigate how the concentration affects the potential of an electrochemical cell.

**Tips**

Remember concentrations need to be 1.00 mol $dm^{-3}$ with respect to the ion being investigated, and the temperature 298 K, to measure standard values.

## PAG 9 - Rates of reaction : continuous monitoring method

Measuring the rate of a reaction by a continuous monitoring method requires chemical reactions where there is an easily measurable change that happens with time.

**Key knowledge and skills**

- Use appropriate apparatus to record a range of measurements.
- Measure rates of reaction by an initial rate method such as a clock reaction.

**Example practical**

Finding the half-life of a reaction by investigating the decomposition of hydrogen peroxide or the reaction of dilute acid with metals/carbonates.

**Tips**

Graphs of concentration vs. time are straight for zero order, curved with constant half-life for first order and curved with not constant half-life for second order.

## PAG 10 - Rates of reaction : initial rate method

Another method to measure the rate of a reaction is the initial rate method. This involves a reaction where there is a clear colour change.

**Key knowledge and skills**

- Use appropriate apparatus to record a range of measurements.
- Measure rates of reaction by a continuous monitoring method.

**Example practical**

Finding the order and rate constant for a reaction or determining the activation energy of a reaction.

**Tips**

Rate–concentration graphs are flat, straight, and curved for zero, first, and second order respectively.

## PAG 11 - pH measurement

Titration curves can show the behaviour of strong acids, weak acids, strong bases, and weak bases.

**Key knowledge and skills**

- Use appropriate apparatus to record a range of measurements.
- Use acid–base indicators in titrations of weak/strong acids with weak/strong alkalis.

**Example practical**

Identifying unknown solutions via pH measurements or an investigation into buffers.

**Tips**

When you are getting nearer to the equivalence point, add a smaller volume between each pH measurement. This will result in more precise data.

## PAG 12 - Research skills

The purpose of these experiments is for you to do some research to help plan your experiment to solve a problem. Professional chemists seldom work in isolation, they research other people ideas and results and use them to develop their ideas and theories.

**Key knowledge and skills**

- Use appropriate apparatus to record a range of measurements.
- Safely and carefully handle solids and liquids, including corrosive, irritant, flammable, and toxic substances.

**Example practical**

Investigating how long it takes iron tablets to break down in the stomach, or the copper content of brass.

**Tips**

This practical activity group could include experiments from any part of the A level course. Its focus is on using your knowledge to solve problems.

# Knowledge

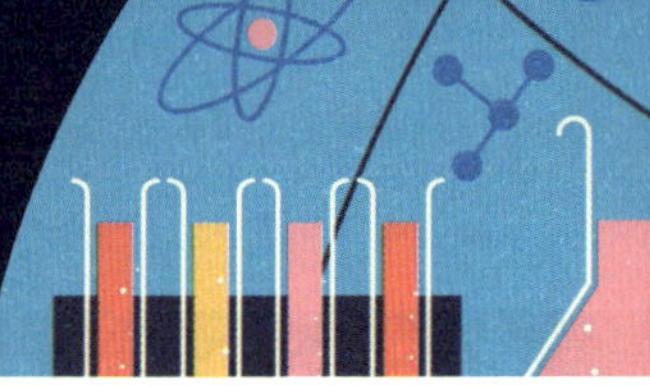

## 2 Foundations in Chemistry

### Sub-atomic particles

Protons and neutrons have the same mass, so it is easier to use the *relative mass*, although the actual mass of a proton or a neutron is $1.673 \times 10^{-27}$ kg. Similarly, the charge on protons and electrons is equal but opposite, so *relative charge* is used. The actual charge on a proton is $+1.602 \times 10^{-19}$ C and an electron has a charge of $-1.602 \times 10^{-19}$ C. The relative mass of an electron is 0 because it has a very small mass of $9.11 \times 10^{-31}$ kg.

| Properties | Protons | Neutrons | Electrons |
|---|---|---|---|
| relative mass | 1 | 1 | 0 |
| relative charge | 1+ | 0 | 1– |
| location | nucleus | nucleus | outer shell |

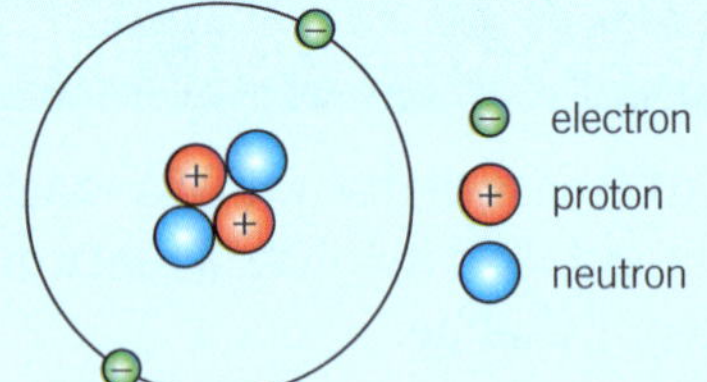

- The **atomic number**, $Z$, is the number of protons in the nucleus of an atom. The number of protons of an atom decides which element it is. The number of electrons of an atom is equal to the number of protons.
- The **mass number**, $A$, is the total number of protons and neutrons in an atom.
- **Isotopes** are atoms of an element that have the same number of protons but a different number of neutrons.

### Mass number

The mass number is the number of protons and neutrons an atom has, yet the mass number seen on the Periodic Table is not a whole number. For example, aluminium has a mass of 26.982 and magnesium has a mass of 24.305. You cannot have 0.305 of a neutron so a different way of explaining the mass numbers is needed.

**These masses are the average mass of all the different naturally occurring isotopes of an element.**

### Relative mass

The **relative atomic mass**, $A_r$, is the weighted mean mass of an atom of an element, taking into account its naturally occurring isotopes, relative to $\frac{1}{12}$ of the relative atomic mass of an atom of carbon-12.

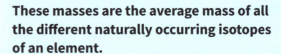

$$\text{relative atomic mass } A_r = \frac{\text{average mass of 1 atom}}{\frac{1}{12}\text{ mass of 1 atom of } ^{12}\text{C}}$$

The **relative molecular mass**, $M_r$, is the mass of that molecule compared to $\frac{1}{12}$ of the relative atomic mass of an atom of carbon-12. The $M_r$ is the sum of the $A_r$ of each atom in a molecule. For example, the relative molecular mass of $O_2$:

$$M_r = 2 \times A_r(\text{O}) = 2 \times 16.0 = 32.0$$

**Relative formula mass**, $M_r$, is used for ionic compounds.

**Relative isotopic mass** is the mass of an isotope relative to $\frac{1}{12}$ of the mass of an atom of carbon-12.

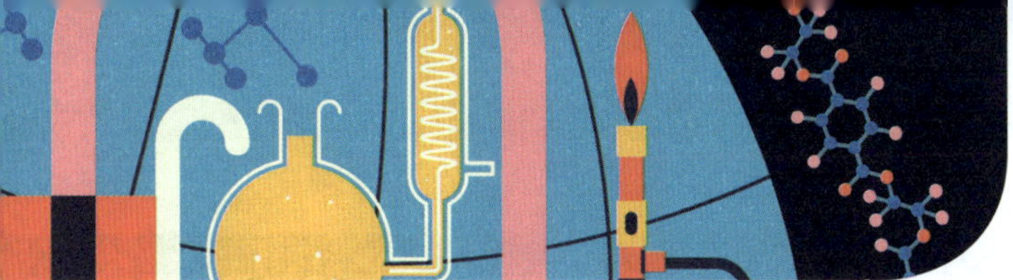



## Mass spectroscopy

Mass spectroscopy traces show the relative abundances of each isotope of an element. This data is used to calculate the average mass of an atom for an element.

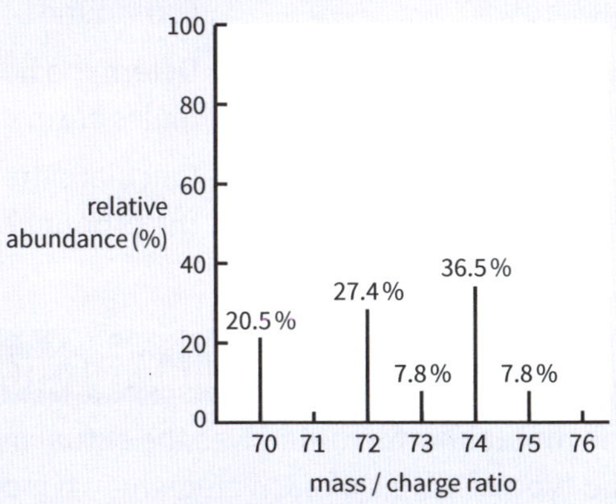


## Common ions

| Name of ion | Formula and charge of ion |
|---|---|
| nitrate | $NO_3^-$ |
| carbonate | $CO_3^{2-}$ |
| sulfate | $SO_4^{2-}$ |
| hydroxide | $OH^-$ |
| ammonium | $NH_4^+$ |
| zinc | $Zn^{2+}$ |
| silver | $Ag^+$ |

## Ionic charge

The ionic charge of an atom can be predicted from the position of the element in the Periodic Table. For example, Group 1 elements have one electron in the outer shell so will lose an electron to form a positive ion.

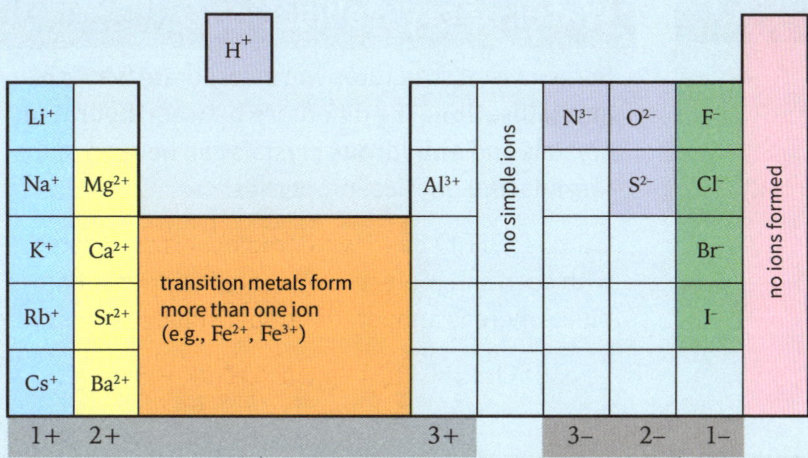


## Equations

**Chemical equations** show the formulae of the compounds in a reaction. To balance a chemical equation you need to ensure there is the same number of atoms of each element on each side of the equation and include state symbols.

**State symbols** are used to show the state of a species in a balanced symbol equation.

(s) = solid

(l) = liquid

(g) = gas

(aq) = aqueous solution

**Ionic equations** show which ions are reacting and which are just spectator ions. Spectator ions are ones that do not change during a reaction. The charges need to be balanced as well as the atoms.

## The mole

**Avogadro's number**, $N_A$, is the number of particles in one mole, $6.022 \times 10^{23}$ mol$^{-1}$, (or the number of atoms in 12 g of carbon-12). You are NOT expected to recall this value.

The **amount of substance** is measured in moles (mol).

$$\text{amount of substance } n = \frac{\text{mass } m}{\text{molar mass } M}$$

$$n = \frac{m}{M}$$

**Molar mass**, $M$ (g mol$^{-1}$), is the mass per mole.

## Determining mass

To determine the mass of products or reactants:

1 Balance the equation.
2 Determine the number of moles of each compound.
3 Determine how many moles are in the question.
4 Calculate the mass of the product or reactant.

## Determining empirical formula

The **empirical formula** for a compound is the simplest whole number ratio of the elements.

To work out the number of moles of each element:

$$\text{number of moles} = \frac{\text{mass of element}}{\text{relative atomic mass}}$$

Then work out the simplest ratio.

## Determining molecular formula

The **molecular formula** gives the actual number and type of atoms of each element in a molecule.

This can be determined from the empirical formula and the molecular mass.

$$\text{molecular formula ratio} = \frac{\text{molecular mass}}{\text{mass of empirical formula}}$$

## Calculating concentration

**Concentration** is measured in mol dm$^{-3}$ or in g cm$^{-3}$. This is the amount of solute in a known volume of a solution.

$$\text{concentration } c \text{ (mol dm}^{-3}) = \frac{\text{number of moles } n \text{ (mol)}}{\text{volume } V \text{ (dm}^{-3})}$$

## Water of crystallisation

Some crystal structures will incorporate **water of crystallisation**. The difference between **hydrated** crystals and **anhydrous** crystals can be seen in the formula and physical properties.

$CuSO4 \bullet 5H_2O$ has 5 water molecules associated with each salt molecule. Hydrated copper sulfate is blue, whereas anhydrous copper sulfate is white.

$$CuSO_4 \bullet 5H_2O(s) \rightarrow CuSO4(s) + 5H_2O(l)$$

## Ideal gas law

The volume of gas varies depending on the temperature of the gas and the pressure.

The **ideal gas law** is how an ideal gas would behave in a given situation.

$$PV = nRT$$

where:
$P$ = pressure / Pa
$V$ = volume / m$^3$
$n$ = number of moles / mol
$R$ = gas constant / J mol$^{-1}$ K$^{-1}$
$T$ = temperature / K

## Atom economy and percentage yield

No chemical reaction has a 100% yield.

When a process involves more than one step, then reactants and products are lost at each step. The more steps involved in the process, then generally the lower the yield.

One way of improving the atom economy is by reducing the number of stages so that there are fewer waste products. Improved atom economy will give a higher percentage yield.

## Percentage yield

**Percentage yield** is the efficiency of a process. This can be affected by the reaction being reversible, reactions not going to completion, or practical errors. Percentage yield of a reaction is worked out from this equation:

$$\%\ \text{yield} = \frac{\text{number of moles of desired product}}{\text{theoretical maximum number of moles of desires product}} \times 100$$

## Atom economy

**Atom economy** is how many atoms in the reactant are found in the useful products, compared to how many are wasted. Atom economy can be improved by changing a reaction to have fewer waste products or finding a use for the waste. Atom economy is worked out from this equation:

$$\%\ \text{atom economy} = \frac{\text{mass of desired products}}{\text{total mass of reactants}} \times 100$$

## Units

Always remember to check the units when carrying out calculations.

Convert all values to the correct units before you carry out the calculation.

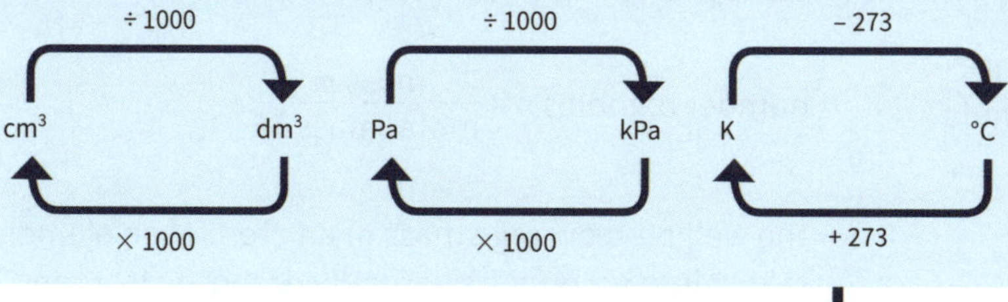


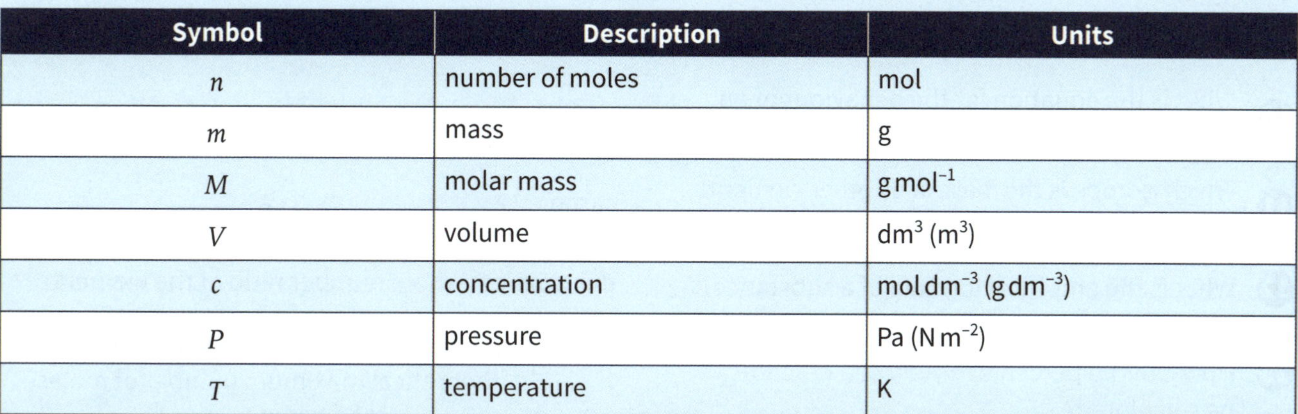

| Symbol | Description | Units |
|---|---|---|
| $n$ | number of moles | mol |
| $m$ | mass | g |
| $M$ | molar mass | $\text{g mol}^{-1}$ |
| $V$ | volume | $\text{dm}^3$ ($\text{m}^3$) |
| $c$ | concentration | $\text{mol dm}^{-3}$ ($\text{g dm}^{-3}$) |
| $P$ | pressure | Pa ($\text{N m}^{-2}$) |
| $T$ | temperature | K |

## Moles of a gas

At room temperature (25 °C) and pressure (101 kPa or 1 atm) the molar gas volume is $24.0\ \text{dm}^3$.

volume of gas = moles of gas × molar gas volume

**Molar gas volume** is the gas volume per mole in $\text{dm}^3\ \text{mol}^{-1}$.

# Retrieval

Learn the answers to the questions below, then cover the answers column with a piece of paper and write as many as you can. Check and repeat.

| | Questions | Answers |
|---|---|---|
| 1 | What are the two subatomic particles in the nucleus? | protons and neutrons |
| 2 | What are the two subatomic particles with opposite charges? | protons and electrons |
| 3 | What are the two different units that can be used for pressure? | Pa and $N\,m^{-2}$ |
| 4 | What is 25 °C in kelvin? | 298 K |
| 5 | How is the number of neutrons in an atom calculated? | number of neutrons = mass number − atomic number |
| 6 | What is the definition of the term isotope? | atoms with the same number of protons but different numbers of neutrons |
| 7 | Why might a reaction not give the expected yield? | reaction is reversible, reactions not going to completion, or practical errors |
| 8 | How is the number of moles of a substance calculated, when given its chemical formula and mass? | number of moles $n = \frac{\text{mass } m}{\text{molar mass } M}$ |
| 9 | What is the definition of relative atomic mass $A_r$? | the weighted average mass of an atom of an element, taking into account its naturally occurring isotopes, relative to $\frac{1}{12}$ of the relative atomic mass of an atom of carbon-12 |
| 10 | What is the equation for the behaviour of an ideal gas? | $PV = nRT$ |
| 11 | Which isotope is the mass of all other elements measured relative to? | carbon-12 |
| 12 | What is the empirical formula of a substance? | the simplest whole number ratio of the elements |
| 13 | How can the percentage yield of a reaction be calculated? | $\%\ \text{yield} = \frac{\text{the number of moles of desired product}}{\text{theoretical maximum number of moles of desired product}} \times 100$ |
| 14 | What is the definition of relative molecular mass $M_r$? | the mass of that molecule compared to $\frac{1}{12}$ of the relative atomic mass of an atom of carbon-12 |
| 15 | How is atom economy calculated? | $\%\ \text{atom economy} = \frac{\text{mass of desired products}}{\text{total mass of reactants}} \times 100$ |

Put paper here

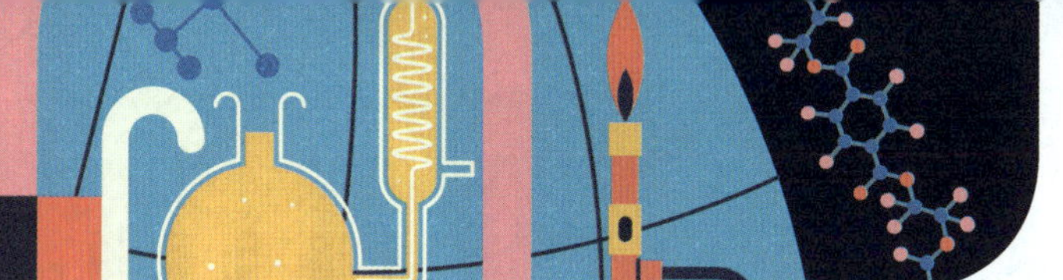



| | | |
|---|---|---|
| 16 | Which letter in the ideal gas law represents the gas constant? | $R$ |
| 17 | How can the atom economy of a reaction be improved? | changing a reaction to have fewer waste products or finding a use for the waste products |
| 18 | What is atom economy? | how many atoms in the reactant are found in the useful products, compared to how many are wasted |
| 19 | What are the state symbols used in chemical equations? | (s) solid; (l) liquid; (aq) aqueous solution; (g) gas |
| 20 | What is the formula of a nitrate ion? | $NO_3^-$ |

Put paper here

## Practical skills

Practise your practical skills using the worked example and practice questions below.

### Research skills

The purpose of this practical activity group is to apply the chemistry that you know to solve unusual problems.

For example, redox titrations apply the knowledge you know about redox, titration calculations, and transitions metals to new problems.

You should know colour changes, common applications, and common practical issues. For example:

- The end point of $I_2$ titration is determined by adding starch. But if you add it too early the colour change will not be sudden enough.
- You need to acidify $MnO_4^-$ with sulfuric acid. If you used HCl, $HNO_3$ or $CH_3COOH$, then $MnO_4^-$ would oxidise the anion formed in solution.

### Worked example

A student carries out a $I_2/S_2O_3^{2-}$ titration to determine the concentration of $Cu^{2+}$ in solution.

**Questions**

1 $Cu^{2+}$ ions are reacted with an excess of $KI(aq)$. Describe what would be observed in this reaction.

2 The mixture from **a** is then titrated against $Na_2S_2O_3(aq)$. Describe what you would observe as you approached the end point.

3 What would be added to the mixture to determine the end point accurately? How could you tell?

**Answers**

1 Brown precipitate forms. This is the CuI.

2 It turns (pale) yellow colour as $I_2$ is reduced to $I^-$.

3 Add starch, which will turn the mixture blue-black. At the end point it will be an off-white colour.

### Practice

A student carries out an experiment to determine the formula of the hydrated ethanedioic acid $H_2C_2O_4{\bullet}xH_2O$.

- The student adds $2.22 \pm 0.0005$ g of the hydrated acid into a volumetric flask and makes up the solution to 250 $cm^3$.
- The student titrates $25.0 \pm 0.1$ $cm^3$ of the acid against 20.0 mol $dm^{-3}$ solution of $KMnO_4$. The ethanedioic acid is acidified with approx. 10 $cm^3$ of 1 mol $dm^{-3}$ $H_2SO_4$.
- Ethanedioic acid is oxidised to $CO_2$.

1 Explain why the volume and concentration of $H_2SO_4$ does not need to be accurate.

2 How can you tell the titration is nearing the end point?

3 Write an equation for the reaction that takes place.

4 Calculate the percentage uncertainty in mass of $H_2C_2O_4{\bullet}xH_2O$ and volume of aliquot.

5 The average titre of $KMnO_4$ is 35.30 $cm^3$. Calculate the value of $x$.

For answers and more practice questions visit www.oxfordrevise.com/scienceanswers

Even more practice and interactive revision quizzes are available on kerboodle

# Practice

## Exam-style questions

1 Bromine is an element in Group 7. It has two isotopes, $^{79}Br$ and $^{81}Br$, that naturally occur in roughly equally proportions.

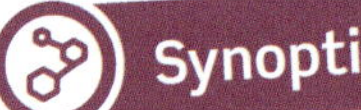

**Synoptic links**

2.2.1 4.1.3

**Exam tip**

Remember bromine, like all halogens, exists as diatomic molecules.

**(a) (i)** Complete the mass spectrum in **Figure 1.1** to give the distribution and locations of the peaks for this sample. **[3]**

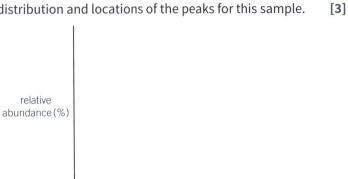

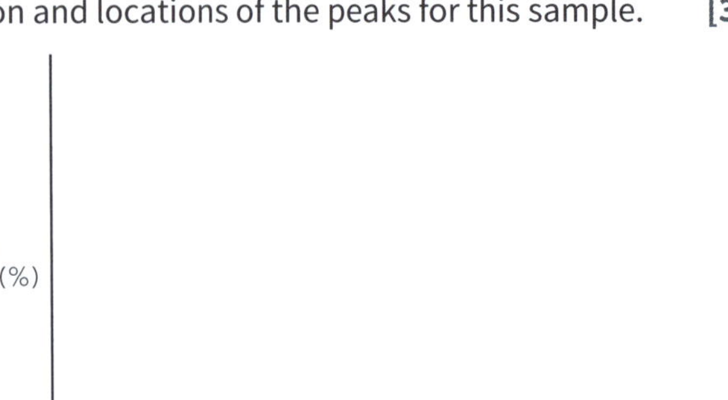


**Figure 1.1**

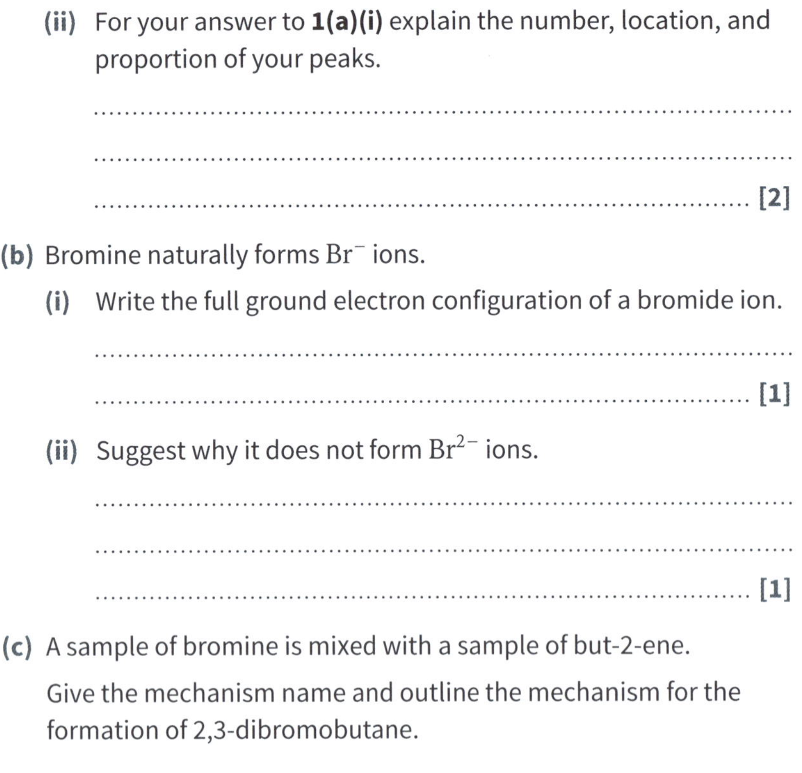

**(ii)** For your answer to **1(a)(i)** explain the number, location, and proportion of your peaks.

........................................................................................

........................................................................................

........................................................................................ **[2]**

**(b)** Bromine naturally forms $Br^-$ ions.

**(i)** Write the full ground electron configuration of a bromide ion.

........................................................................................

........................................................................................ **[1]**

**(ii)** Suggest why it does not form $Br^{2-}$ ions.

........................................................................................

........................................................................................

........................................................................................ **[1]**

**Exam tip**

'Suggest' questions need you to apply your understanding to areas not specifically taught in the syllabus. In this case, consider the electronic configuration of each ion.

**(c)** A sample of bromine is mixed with a sample of but-2-ene.

Give the mechanism name and outline the mechanism for the formation of 2,3-dibromobutane.

........................................................................................

**[5]**

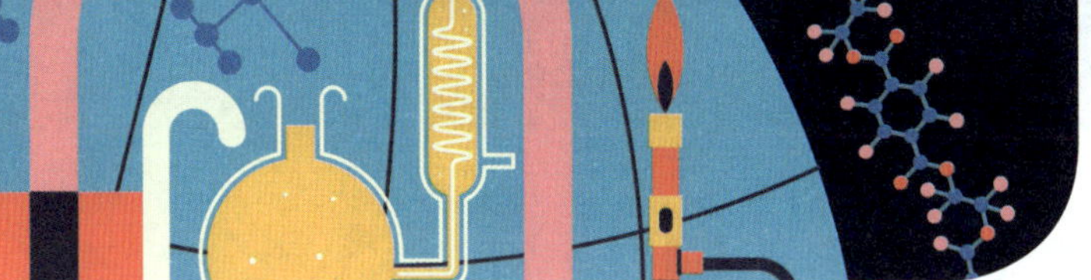



2 **Table 2.1** gives the atomic masses of the isotopes of a sample of calcium.

| **Mass** | 40 | 42 | 43 | 44 | 48 |
|---|---|---|---|---|---|
| **Abundance (%)** | 96.46 | 0.70 | 0.30 | 2.20 | 0.34 |

**Table 2.1**

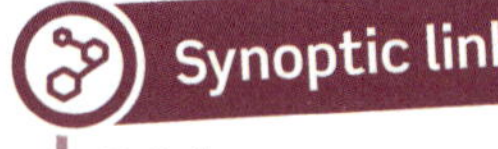

**Synoptic link**

3.1.1

**(a)** Define relative atomic mass.

..............................................................................................

..............................................................................................

..............................................................................................

.............................................................................................. **[2]**

**(b)** Calculate the relative atomic mass of calcium in the sample to 2 decimal places using data from **Table 2.1**.

relative atomic mass = ...................... **[2]**

**(c)** Explain why the second ionisation energy of calcium is greater than the first ionisation energy.

..............................................................................................

..............................................................................................

..............................................................................................

.............................................................................................. **[2]**

**(d)** Explain why the first ionisation energy of magnesium is higher than that of calcium.

..............................................................................................

..............................................................................................

..............................................................................................

..............................................................................................

..............................................................................................

.............................................................................................. **[3]**

**(e)** When a small piece of calcium is added to water, fizzing is observed.

Give a balanced ionic equation to show the reaction between calcium and water.

..............................................................................................

..............................................................................................

..............................................................................................

.............................................................................................. **[2]**

**Exam tip**

Ionic equations show separate ions that are involved in the reaction and any atoms or molecules involved. They ignore spectator ions.

3 An experiment was carried out to determine the relative molecular mass $M_r$ of a volatile hydrocarbon **X** that is a liquid at room temperature.

The mass of a sample of **X** was recorded, and heated until it evaporated. The gas flowed into a gas syringe and the volume was recorded. The temperature and air pressure were also recorded.

Data from this experiment are shown in **Table 3.1**.

| | |
|---|---|
| **Mass of X (mg)** | 184 |
| **Temperature (K)** | 361 |
| **Pressure (kPa)** | 102 |
| **Volume ($cm^3$)** | 53.0 |

**Table 3.1**

**(a)** Calculate the relative molecular mass of **X**.

Give your answer to an appropriate number of significant figures.
The gas constant $R = 8.31\ \text{J K}^{-1}\ \text{mol}^{-1}$

Take care to convert your units.

molecular mass of **X** = ..................... **[6]**

**(b) (i)** Analysis of a different hydrocarbon **Y** shows that it contains 90.0% by mass of carbon.

Calculate the empirical formula of **Y**.

empirical formula of **Y** = ..................... **[3]**

**(ii)** Use the empirical formula calculated in **3(b)(i)** and the relative molecular mass of **Y** to calculate the molecular formula of **Y**.

**Y** $M_r = 80.0$

molecular formula of **Y** = ..................... **[1]**

**(c)** A student repeated the experiment. After the experiment was complete, they noticed a small amount of liquid residue on the sides of the gas syringe.

Explain how this would affect the value of the $M_r$ calculated in **3(a)**.

..........................................................................................

..........................................................................................

.......................................................................................... **[2]**

**Exam tip**

In this question, you need to say how it will affect the value you obtained and why it will change.

4 Magnesium exists as three isotopes: $^{24}Mg$, $^{25}Mg$, and $^{26}Mg$.

(a) (i) State how the structures of the nuclei differ in the three isotopes.

.......................................................................................

....................................................................................... [1]

(ii) The three isotopes react the same way chemically. Explain why.

.......................................................................................

.......................................................................................

.......................................................................................

....................................................................................... [2]

(b) $^{24}Mg$ atoms make up 80.0% by mass in a sample of magnesium.

Use the Periodic Table and the information in the question to deduce the percentages of the other two magnesium isotopes present in the sample.

.......................................................................................

.......................................................................................

.......................................................................................

.......................................................................................

.......................................................................................

....................................................................................... [3]

**Exam tip**

Let the percentage of $^{25}Mg = a$ and the percentage of $^{26}Mg = 20 - a$.

5 A compound contains 36.5% of sodium and 25.5% of sulfur by mass, the rest being oxygen.

(a) Show that the empirical formula of the compound is $Na_2SO_3$. [2]

(b) Define relative molecular mass. [2]

(c) Calculate the relative molecular mass of $Na_2SO_3$. [1]

(d) (i) $Na_2SO_3$ reacts with hydrochloric acid according to the equation below:

$$Na_2SO_3 + HCl \rightarrow SO_2 + 2\ NaCl + H_2O$$

Complete the equation by balancing so that there is the same number of atoms on each side. [2]

(ii) Calculate the mass in g of $Na_2SO_3$ needed to produce 2.5 kg of $SO_2$. [4]

(iii) Calculate the temperature needed for the 2.5 kg of $SO_2$ to fill a 1500 $dm^3$ balloon at room temperature and 101 kPa. Include units in your answer. [3]

**Exam tip**

Calculate the percentage of oxygen first.

**Exam tip**

Always include state symbols in your equations.

**Exam tip**

Remember that $M_r$ is measured in $g\,mol^{-1}$.

6 Zinc forms many different salts including zinc sulfate, zinc chloride, and zinc fluoride.

**(a)** People who have a zinc deficiency can take hydrated zinc sulfate, $ZnSO_4.xH_2O$, in the form of pills as a dietary supplement.

The label on the bottle says that the hydrated zinc sulfate is 43.8% water by mass.

Calculate **X**, the number of moles of water surrounding 1 mole of zinc sulfate. **[4]**

**Synoptic links**

5.3.1 3.1.2

**(b) (i)** Zinc chloride can be prepared in the laboratory by the reaction between zinc oxide and hydrochloric acid. The equation for the reaction is:

**Reaction 1:** $ZnO + 2HCl \rightarrow ZnCl_2 + H_2O$

Zinc chloride can also be prepared in the laboratory by the reaction between zinc and hydrogen chloride gas:

**Reaction 2:** $Zn + 2HCl \rightarrow ZnC_2 + H_2$

Compare the atom economies of both reactions.

Use this information and other information from the equation to suggest which reaction a company should use to create a sustainable process. **[4]**

**Exam tip**

Consider the practical and safety implications of each reaction.

**(ii)** A 6.74 g sample of pure zinc oxide was added to 200 $cm^3$ of 1.20 mol $dm^{-3}$ hydrochloric acid.

Calculate the maximum mass of zinc chloride that could be obtained from the reaction.

Include units in your answer. **[5]**

**(iii)** An impure sample of zinc powder with a mass of 5.68 g was reacted with hydrogen chloride gas, as in the second reaction, until the reaction was complete. The zinc chloride produced had a mass of 10.7 g.

Calculate the percentage purity of the zinc metal.

Give your answer to **three** significant figures.

**[4]**

**(c)** Define transition element and explain why zinc does not fit that definition. **[2]**

**Exam tip**

When explaining the difference, ignore the general trends and focus on the key aspect of electron configuration.

**(d)** A student had two unlabelled solutions of zinc chloride and barium chloride. Both are colourless.

Using only one of the reagents below, how could the student deduce which solution was which?

- hydrochloric acid
- sulfuric acid
- nitric acid
- silver nitrate

State what would be observed with both solutions. **[3]**

7 Sodium azide, $NaN_3$, is a compound used in airbags. When heated, sodium azide decomposes as shown in the equation below. Nitrogen gas is used to inflate the air bag.

$$2NaN_3(s) \rightarrow 2Na(s) + 3N_2(g)$$

(a) Calculate the percentage atom economy of this reaction. [1]

**Exam tip**

Think carefully about the total number of moles of gas produced in this reaction.

(b) A driver's air bag is 56.0 $cm^3$ when inflated. To initiate the decomposition, a temperature of 300 K is required.

Calculate the mass in g of sodium azide needed to inflate the air bag at 101 kPa. [3]

(c) In reality, car manufacturers work on an assumption that the percentage yield of the reaction is 90%.

Calculate the actual mass in g of sodium azide placed into the air bag. [1]

8 Nitromethane, $CH_3NO_2$, is used as an 'energy rich' fuel for motor-racing. It burns in oxygen, forming three gases:

$$2CH_3NO_2 + 1\frac{1}{2}\ O_2 \rightarrow 2CO_2 + 3H_2O + N_2$$

(a) (i) 2 moles of nitromethane was reacted with 2 moles of oxygen at 100 kPa and 1000 K. Complete combustion occurred.

Complete **Table 8.1** by adding how many moles of each species was present at the end.

| Species | Amount of species at start / mol | Amount of species at the end / mol |
|---|---|---|
| $CH_3NO_3$ | 2 | |
| $O_2$ | 2 | |
| $CO_2$ | 2 | |
| $H_2O$ | 3 | |
| $N_2$ | 1 | |

**Table 8.1** [3]

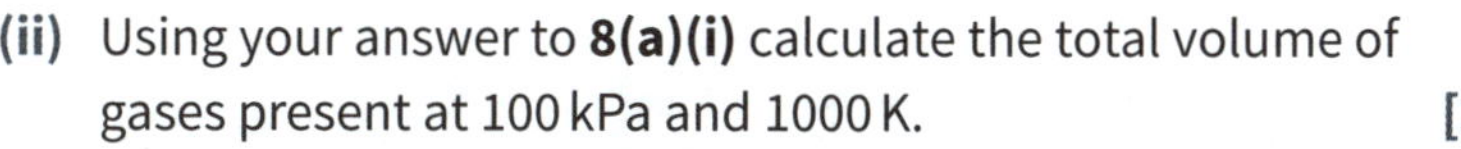

(ii) Using your answer to **8(a)(i)** calculate the total volume of gases present at 100 kPa and 1000 K. [2]

**Exam tip**

Assume the water is all gaseous.

(b) (i) Nitrogen forms several different oxides.

Calculate the empirical formula of an oxide of nitrogen that contains 46.7% of nitrogen by mass. [3]

(ii) Determine the molecular formula of the compound in **8(b)(i)** if it has a molecular mass of 60 g $mol^{-1}$. [1]

# Knowledge

## 3 Acid-base and redox reactions

### Acids

Acids contain hydrogen. When dissolved in water the hydrogen **dissociates** and becomes an aqueous hydrogen ion, $H^+$.

**Strong acids** fully dissociate, releasing all hydrogen ions into solution. An arrow → indicates this, for example:

$$HCl(aq) \rightarrow H^+(aq) + Cl^-(aq)$$

**Weak acids** only **partially dissociate**. Only some of the hydrogen is released into solution. These are incomplete reactions and a ⇌ sign shows this:

$$CH_3COOH(aq) \rightleftharpoons H^+(aq) + CH_3COO^-(aq)$$

#### Strong acids

| Name | Formula |
|---|---|
| hydrochloric acid | $HCl$ |
| sulfuric acid | $H_2SO_4$ |
| nitric acid | $HNO_3$ |

#### Weak acids

| Name | Formula |
|---|---|
| ethanoic acid | $CH_3COOH$ |
| phosphoric acid | $H_3PO_4$ |

### Bases

A **base** is a metal oxide, a metal hydroxide, a metal carbonate, or ammonia. These neutralise acids and form salts.

$$MgO(s) + 2HCl(aq) \rightarrow MgCl_2(aq) + H_2O(l)$$

An **alkali** is a base that dissolves in water and releases aqueous hydroxide ions, $OH^-$, for example, ammonia or metal hydroxides.

$$NaOH(s) + aq \rightarrow Na^+(aq) + OH^-(aq)$$

### Common bases

| Metal oxides | | Metal carbonate | | Alkali | |
|---|---|---|---|---|---|
| Name | Formula | Name | Formula | Name | Formula |
| magnesium oxide | $MgO$ | magnesium carbonate | $MgCO_3$ | ammonia | $NH_3$ |
| calcium oxide | $CaO$ | calcium carbonate | $CaCO_3$ | sodium hydroxide | $NaOH$ |
| copper(II) oxide | $CuO$ | beryllium carbonate | $BeCO_3$ | potassium hydroxide | $KOH$ |

3

## Neutralisation

**Neutralisation** is the reaction between hydrogen ions and a base to form a salt and water. The $H^+$ in the acid is replaced with the metal or the ammonium ions.

$$H^+(aq) + OH^-(aq) \rightleftharpoons H_2O(l)$$

## Neutralisation of acids with bases

acid + metal oxide → salt + water

acid + metal hydroxide → salt + water

acid + metal carbonate → salt + water + carbon dioxide

The name of the salt is linked to the acid used:

- **hydrochloric** acid makes **chloride** salts
- **sulfuric** acid makes **sulfate** salts
- **ethanoic** acid makes **ethanoate** salts.

## Titrations

**Titrations** are used to determine the concentration, purity, or identity of an unknown solution.

1 Fill a burette with acid of a known concentration.
2 Record the volume of acid in the burette.
3 Measure a known volume of alkali into a conical flask.
4 Add a suitable indicator.
5 Add acid from the burette to the conical flask slowly.
6 Stop adding acid when the end point is reached.
7 Record the volume of acid used.
8 Repeat until you get **concordant results**.
9 Calculate the concentration of the alkali.

Concordant results are when the titres are within 0.1 $cm^3$ of each other.

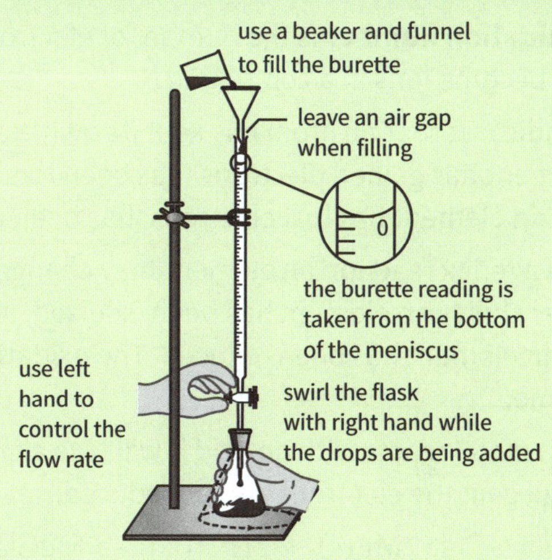
use a beaker and funnel to fill the burette
leave an air gap when filling
0
the burette reading is taken from the bottom of the meniscus
use left hand to control the flow rate
swirl the flask with right hand while the drops are being added

## Calculating unknown concentrations

Titrations can be used to find the concentration of an unknown solution. From titration results you know:

- the volume and concentration of a solution used to neutralise a solution with an unknown concentration
- the volume of the unknown solution.

At neutralisation: $H^+ + OH^- \rightleftharpoons H_2O$ there is an *equal* number of hydrogen ions and hydroxide ions.

1 Balance an equation for the reaction.
2 Determine the ratio of acid to alkali.
3 Calculate the amount ($n$) of hydrogen ions *OR* hydroxide ions in the known solution. This will be equal to the number of the opposite ions when neutralised.
4 Determine the concentration of the unknown solution.

When calculating the amount ($n$) of hydrogen ions *OR* hydroxide ions in the known solution, volume needs to be in $dm^3$. If volume is given in $cm^3$, divide by 1000.

$$\text{amount } n \text{ (mol)} = \frac{\text{concentration } c \text{ (mol dm}^3) \times \text{volume } V \text{ (cm}^3)}{1000}$$

## Redox reactions

**Oxidation** is the loss of electrons, the addition of oxygen, or the removal of hydrogen.

**Reduction** is the gain of electrons, the loss of oxygen, or the addition of hydrogen.

**Redox reactions** involve the transfer of electrons from one reagent to another. A redox reaction is when a reduction reaction and an oxidation reaction happen together.

**Oxidising agents** oxidise a reactant by removing electrons from other reactants, they gain these electrons, so they are reduced.

**Reducing agents** reduce other reactants by losing electrons, they donate these electrons to other reactants.

**Half equations** can be used to show the movement of electrons in redox reactions. The elements and the charges must both balance.

**Spectator ions**, that are not involved in the reaction, are omitted from half equations.

## Oxidation number

The **oxidation number** is the number of electrons lost or gained by an element when its atoms become ions in a compound.

Zero indicates that an atom has kept its own electrons. A negative oxidation number shows an element has gained electrons (has been reduced), whereas a positive oxidation number shows an element has lost electrons (has been oxidised).

During a redox reaction an element may change oxidation number. An increase in oxidation number shows an element has been oxidised, and a decrease in oxidation number will show an element has been reduced. The oxidation numbers of spectator ions will remain the same.

When a metal reacts with an acid it will lose electrons and be oxidised; the hydrogen in the acid will gain the electrons and be reduced:

$$Mg(s) + 2HCl(aq) \rightarrow MgCl_2(aq) + H_2(g)$$

In this reaction the oxidation state of magnesium changes from 0 to +2, and the the oxidation state of hydrogen changes from −1 to 0.

## Peroxides

Oxidation numbers shows which electrons are involved in the bonding.
The oxidation state of oxygen is normally −2 except in peroxides where it is −1.

A dot and cross diagram shows each oxygen has only gained one electron.
When chlorine is combined with fluorine or oxygen, both of which are more electronegative, these will attract the electron more.

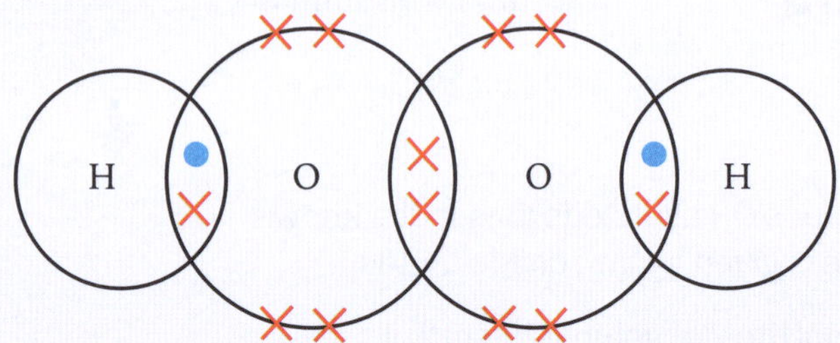

## How to assign oxidation numbers to elements and ions

These are some rules used to assign oxidation numbers to elements and ions:

- If an element is uncombined, its oxidation number is 0.
- Any diatomic element, for example, $H_2$, has an oxidation number of 0.
- An ion has the same oxidation number as its charge.
- Hydrogen has an oxidation number of +1, except in metal hydrides when it is –1.
- Group 1 and Group 2 ions have +1 or +2 oxidation numbers respectively.
- Oxygen is always –2, except when it is in a peroxide or bonded to fluorine.
- Fluorine is always –1.
- Chlorine is always –1, except when it is in a compound with oxygen or fluorine.

## How to calculate the oxidation numbers of compounds and ions

The overall charge on an ion or compound will equal the sum of the oxidation states.

To determine the oxidation number of sulfur($x$) in a sulfate ion $SO_4^{2-}$, assign the known oxidation numbers and use:

sum of oxidation numbers = total charge

$$x + (-2 \times 4) = -2$$

$$x + -8 = -2$$

$$x = +6$$

## Disproportionation reactions

A **disproportionation reaction** is a reaction where the same element is both oxidised and reduced:

$$\begin{array}{ccccccc} Cl_2 & + & H_2O & \rightarrow & HCl & + & HClO \\ 0 & + & +1\ -2 & & +1\ -1 & + & +1 +1 -2 \end{array}$$

**Half equations** show more detail. The elements and the charges must both be balanced. Electrons and hydrogen ions can be added to balance the charges and water can be added to balance the hydrogen ions.

The oxidation number of Cl in HCl has decreased from 0 to −1 so it has been reduced:

$$Cl_2 + 2e^- + 2H^+ \rightarrow 2HCl$$

The oxidation number of Cl in HClO has increased from 0 to +1 so it has been oxidised:

$$Cl_2 + 2H_2O \rightarrow 2HClO + 2e^- + 2H^+$$

The half equations are added together and duplicated species removed:

$$2Cl_2 + \cancel{2e^-} + \cancel{2H^+} + 2H_2O \rightarrow 2HCl + 2HClO + \cancel{2e^-} + \cancel{2H^+}$$

$$Cl_2 + H_2O \rightarrow HCl + HClO$$

## Elements with ions of different charges

Roman numerals are used to show the different charges an element has in a compound.

In iron(I) oxide, the iron ion has a +1 charge and the compound is $Fe_2O$, whereas in iron(II) oxide the iron ion has a +2 charge and the formula is FeO.

For nitrate $NO_3^-$, (nitrate(V)) and sulfate $SO_4^{2-}$ (sulfate(VI)), the oxidation numbers are usually not shown.

You need to know how to write formulae for chlorate(I) and chlorate(III) ions. For example, sodium chlorate(I) is NaOCl.

| Oxidation state | Number |
|---|---|
| I | 1 |
| II | 2 |
| III | 3 |
| IV | 4 |
| V | 5 |
| VI | 6 |
| VII | 7 |
| VIII | 8 |

# Retrieval

Learn the answers to the questions below, then cover the answers column with a piece of paper and write as many as you can. Check and repeat.

| | Questions | | Answers |
|---|---|---|---|
| 1 | What is the formula for ethanoic acid? | Put paper here | $CH_3COOH$ |
| 2 | What do reducing agents do? | | donate electrons |
| 3 | Is nitric acid a strong or weak acid? | | strong |
| 4 | What is the oxidation number of oxygen in a peroxide? | | −1 |
| 5 | What does the (II) in copper(II) oxide indicate? | Put paper here | the oxidation number for copper in this compound is +2 |
| 6 | What is a strong acid? | | an acid that fully dissociates releasing all its hydrogen ions |
| 7 | Which element always has an oxidation number of −1 in compounds? | Put paper here | fluorine |
| 8 | What is an alkali? | | a base that dissolves in water and releases aqueous hydroxide ions |
| 9 | What is a disproportionation reaction? | | a reaction where the same element is both oxidised and reduced |
| 10 | What are added to half equations to balance them? | Put paper here | electrons or hydrogen ions |
| 11 | Is ammonia an alkali? | | yes |
| 12 | What happens in oxidation? | | the loss of electrons, the addition of oxygen, or the removal of hydrogen |
| 13 | What is the equation for a neutralisation reaction? | Put paper here | $H^+(aq) + OH^-(aq) \rightleftharpoons H_2O(l)$ |
| 14 | What is the oxidation number of an uncombined element? | | 0 |
| 15 | What does an increase in oxidation number show? | Put paper here | an element has been oxidised |
| 16 | What salt is made from a reaction with ethanoic acid? | | ethanoate salts |
| 17 | What is the oxidation number of a Group 1 ion? | | +1 |
| 18 | What are spectator ions? | Put paper here | ions that are not changed in a reaction |
| 19 | What equation is used to calculate the number of moles of a solute in a solution with concentration $c$ mol dm$^{-3}$ and volume $V$ in cm$^3$? | | $n = \frac{c \times V}{1000}$ |
| 20 | When should you stop repeating a titration? | | when you get concordant results |

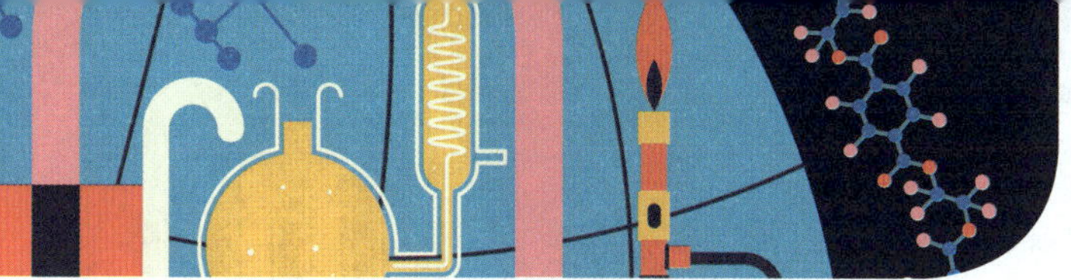



# Practical skills

Practise your practical skills using the worked example and practice questions below.

## Acid–base titration

Titrations are important skills to measure, as precisely as possible, the volume of a solution that reacts with another solution.

Common errors that can occur are:

- difficulty in finding the end point
- when transferring a solid that has been weighed and then dissolved into a volumetric flask, some of the washing may not have been transferred over
- averaging all the results and not just the concordant ones.

Examiners will allow any sensible suggestions to resolve these issues.

The percentage error for each measurement can be worked out by:

$$\%\text{ error} = \frac{\text{error}}{\text{measurement}} \times 100$$

You would need to double this if a measurement requires two judgements (e.g., a burette).

## Worked example

Some students are carrying out an experiment to measure the molecular mass of a solid acid.

- The students weighs a sample and dissolves it in distilled water.
- The solution is transferred to a volumetric flask and distilled water is added to the graduation mark.
- The students pipette out a sample into a conical flask and titrate against NaOH.

### Questions

1 The titre is 15.5 $cm^3$ and the maximum error in each reading is 0.05 $cm^3$. What is the percentage error in the titre volume?

2 State why the following have no effect on the titre volume:

a A student did not fill the burette with NaOH to the 0.00 $cm^3$ mark.

b A student dissolved the acid in distilled water at about 60 °C.

### Answers

1 $\%\text{error} = \frac{2 \times 0.05}{15.5} \times 100 = 0.06\%$

2 a The titre is the difference between two values on the burette, therefore it does not matter where you start.

b It does not change the number of moles of acid in the volumetric flask.

## Practice

Some students are carrying out an experiment to measure the mass of NaOH in a mixture of NaCl and NaOH.

- The students weigh out a sample and dissolve it in distilled water in a beaker.
- They transfer this to a volumetric flask and add distilled water to the graduation mark.
- They pipette out a sample into a conical flask and titrate against HCl.

| Equipment | Measurement | Error in measurement |
|---|---|---|
| balance | 5.00 g | ± 0.01 g |
| pipette | 25.0 $cm^3$ | ± 0.06 $cm^3$ |
| volumetric flask | 250 $cm^3$ | ± 0.3 $cm^3$ |
| burette | 11.0 $cm^3$ | ± 0.05 $cm^3$ |

1 Rank the measurements in order of accuracy, in terms of the percentage error.

2 State what effect, if any, each of the following has on the titre:

a A student does not rinse the beaker and stirring rod and transfer washings to the volumetric flask.

b A student over fills the volumetric flasks and then removes some of the liquid.

c A student washes out the conical flask with distilled water and pipettes the solution into a wet flask.

# Practice

## Exam-style questions

1 The equation for the reaction between magnesium carbonate and hydrochloric acid is: 

$$MgCO_3(s) + 2HCl(aq) \rightarrow MgCl_2(aq) + H_2O(l) + CO_2(g)$$

When 62.0 $cm^3$ of 0.600 mol $dm^{-3}$ hydrochloric acid was added to 1.25 g of impure $MgCO_3$ some of the acid did not react. This unreacted acid required 19.6 $cm^3$ of a 0.500 mol $dm^{-3}$ solution of sodium hydroxide for complete reaction.

**(a)** Calculate the number of moles of HCl in 62.0 $cm^3$ of 0.600 mol $dm^{-3}$ hydrochloric acid.

moles of HCl = .................... **[1]**

**(b)** Calculate the number of moles of NaOH used to neutralise the unreacted HCl.

moles of NaOH = .................... **[1]**

**(c)** Use your answers from **1(a)** and **1(b)** to calculate the number of moles of hydrochloric acid that reacted with the magnesium carbonate.

moles of HCl = .................... **[1]**

**(d)** Calculate the number of moles and the mass of $MgCO_3$ in the sample.
Deduce the percentage purity by mass of $MgCO_3$ in the sample.

Exam tip

Use your answer from **1(c)** and the equation at the start of the question to help you.

moles of $MgCO_3$ = ..............................

mass of $MgCO_3$ = ............................ g

percentage purity = ...................... % **[4]**

2 A student was asked to determine the concentration of ethanoic acid in a sample of white vinegar. They were given a standard solution of 0.100 mol dm$^{-3}$ NaOH solution, phenolphthalein indicator, and all the standard laboratory apparatus needed, including a burette and 25.0 cm$^3$ pipette.

**Synoptic link**

6.1.3

Ethanoic acid and sodium hydroxide react according to the following equation:

$$CH_3COOH(aq) + NaOH(aq) \rightarrow CH_3COONa(aq) + H_2O(l)$$

**(a)** Outline the practical method needed to obtain concordant titres for this reaction.

**Exam tip**

When outlining a method, be specific about which equipment you will use and how you will ensure accuracy.

.......................................................................................

.......................................................................................

.......................................................................................

.......................................................................................

.......................................................................................

.......................................................................................

....................................................................................... **[6]**

**(b) (i)** The student produced the results in **Table 2.1**.

| Trial | rough | 1 | 2 | 3 | 4 |
|---|---|---|---|---|---|
| Initial volume of NaOH (cm$^3$) | 23.55 | 46.60 | 23.65 | 47.25 | 23.75 |
| Final volume of NaOH (cm$^3$) | 0.00 | 23.55 | 0.00 | 23.65 | 0.00 |
| NaOH titre (cm$^3$) | 23.55 | 23.05 | 23.65 | 23.60 | 23.75 |

**Table 2.1**

Calculate the average titre.

average titre = .................... cm$^3$ **[1]**

**(ii)** Use your answer to **2(b)(i)** to calculate the concentration of ethanoic acid in the sample of white vinegar.

concentration = .................... mol dm$^{-3}$ **[3]**

**(c)** In one repeat of the titration, the student had an air bubble in their pipette.
Name this type of error.

....................................................................................... **[1]**

**(d)** Identify which titration result was caused by the error named in **2(c)**.
Give a reason for your answer.

..............................................................................................

..............................................................................................

.............................................................................................. **[2]**

**(e)** The pipette has a precision of ±0.3 cm$^3$.
Calculate the percentage error in titration 4.

percentage error = .................... % **[1]**

**(f)** Name the functional group that ethanoic acid contains.

.............................................................................................. **[1]**

**(g)** Draw the skeletal formula of the compound that can be formed from the reaction of ethanol with methanoic acid.

**[1]**

**3** 60 dm$^3$ of air containing a trace amount of chlorine was passed through an excess of potassium iodide solution at room temperature and pressure, forming iodine and chloride ions.

**Synoptic links**
2.1.2 2.1.3

The resulting solution was made up to 250 cm$^3$. A 25.0 cm$^3$ sample required 26.0 cm$^3$ of 0.10 mol dm$^{-3}$ sodium thiosulfate solution, $Na_2S_2O_3$, to fully react with the iodine present according to the following reaction:

$$2S_2O_3^{2-}(aq) + I_2(aq) \rightarrow 2I^-(aq) + S_4O_6^{2-}(aq)$$

**(a)** Give an ionic equation including state symbols to illustrate the reaction between chlorine and iodide ions.

..............................................................................................

.............................................................................................. **[2]**

**(b)** State the role of chlorine in the reaction.

..............................................................................................

.............................................................................................. **[1]**

**Exam tip**
When asked the role of a reactant, think about the purpose that it plays in the reaction. What does it do to the other reactant?

**(c)** Calculate the number of moles of chlorine present in the original sample.

number of moles = .................... **[4]**

**(d)** Calculate the volume in $m^3$ of chlorine present in the 60 $dm^3$ sample of air.

volume = .................... $m^3$ **[3]**

**(e)** Calculate the percentage by volume of the chlorine in the air sample.

percentage by volume = .................... % **[3]**

**(f)** The safe limit of chlorine in the atmosphere is 2.9 mg $m^{-3}$. Show that this sample is beyond safety limits.

..........................................................................................

..........................................................................................

.................................................................................... **[2]**

**4** A sample of nitrogen dioxide gas, $NO_2$, was prepared by the reaction of copper with concentrated nitric acid.

**Synoptic links**

3.2.3 5.1.2

**(a) (i)** Give the balanced equation for the reaction of copper with concentrated nitric acid:

$Cu + \ldots\ldots\ldots HNO_3 \rightarrow Cu(NO_3)_2 + \ldots\ldots\ldots NO_2 + \ldots\ldots\ldots H_2O$ **[1]**

**(ii)** Give the oxidation states of nitrogen in $HNO_3$ and $NO_2$. **[2]**

**(iii)** Give the half-equation for the conversion of $HNO_3$ into $NO_2$ in this reaction. **[1]**

**(b)** The following equilibrium is established between colourless dinitrogen tetraoxide gas, $N_2O_4$, and dark brown nitrogen dioxide gas.

$$N_2O_4(g) \rightleftharpoons 2NO_2(g) \qquad \Delta H = +58\ kJ\ mol^{-1}$$

**(i)** Give **two** features of a reaction at equilibrium. **[2]**

**(ii)** Explain why the mixture of gases becomes darker in colour when the mixture is heated at constant pressure. **[2]**

**(iii)** Explain the change in the amount of $NO_2$ in the mixture when the pressure is increased at constant temperature. **[3]**

**Exam tip**

It is acceptable to use 'shift to the left/right' in your answer, but it will be interpreted by the examiner in the context of how the equation is written in the question.

**(c)** 2.50 moles of $N_2O_4$ were placed in a sealed vessel and allowed to reach equilibrium. At equilibrium, the container was found to contain 1.00 moles of $N_2O_4$. The total pressure of all the gases in the container was 200 kPa.

Calculate the value of the equilibrium constant $K_p$ and state its units. **[6]**

5 Vanadium is a metal. Pure vanadium is used in nuclear reactors.

One method of producing pure vanadium is the reaction between $V_2O_5$ and calcium at high temperature:

$$5Ca(s) + V_2O_5(s) \rightarrow 2V(s) + 5CaO(s)$$

**Synoptic links**

2.1.2 3.2.1

**Table 5.1** gives some enthalpy of formation data.

| **Compound** | $V_2O_5(s)$ | $CaO(s)$ |
|---|---|---|
| $\Delta_f H^\ominus$ / kJ mol$^{-1}$ | −1560 | −635 |

**Table 5.1**

**(a)** State the oxidation number of vanadium in $V_2O_5$. **[1]**

**(b)** State the role calcium plays in this reaction. **[1]**

**(c)** Define enthalpy of formation. **[3]**

**(d)** Use the data in **Table 5.1** to calculate the enthalpy change in kJ mol$^{-1}$ for this reaction. **[3]**

**(e)** Vanadium can also be extracted from $VCl_2$ by reacting it with hydrogen gas.

Give a balanced symbol equation to show the reaction that produces vanadium and HCl gas only. **[1]**

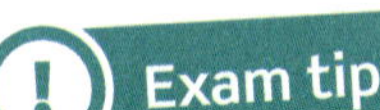

**Exam tip**

Consider the states of the reaction products.

**(f)** Suggest why the vanadium formed by the equation in **5(e)** is pure and contains no HCl. **[1]**

6 Propanoic acid can be prepared from propan-1-ol. The reaction uses acidified potassium chromate(VII), $K_2Cr_2O_7$, which produces chromium(III) ions.

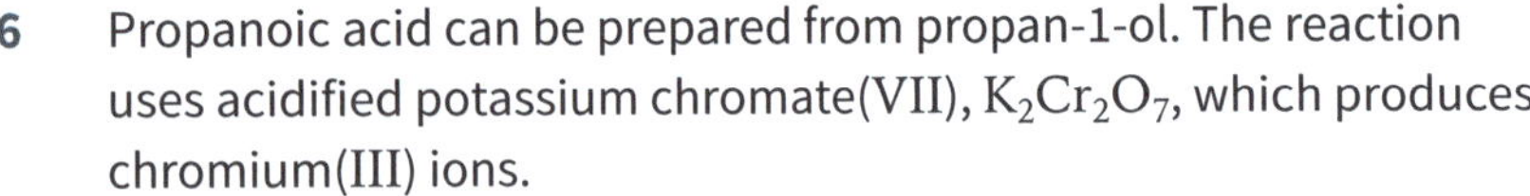

**Synoptic links**

3.1.4 4.2.1 6.1.2 6.3.1

**(a)** Outline the practical procedure, including specific pieces of lab equipment, used to produce a sample of propanoic acid from propan-1-ol. **[5]**

**(b)** Give the half-equation for the conversion of $Cr_2O_7^{2-}$ ions into $Cr^{3+}$ ions in acidic conditions. **[1]**

**Exam tip**

Make sure you balance charge and atoms.

**(c)** Give the colour change observed during the reaction that confirms oxidation has taken place. **[1]**

**(d)** After a class experiment a teacher had an unlabelled sample that could have contained propan-1-ol, propanal, or propanoic acid.

Outline a simple series of laboratory tests that could be used to prove the identity of the sample. **[6]**

7 Chlorine is a useful chemical with many industrial and commercial uses. Chlorine can be produced by reacting $KMnO_4$ with HCl.

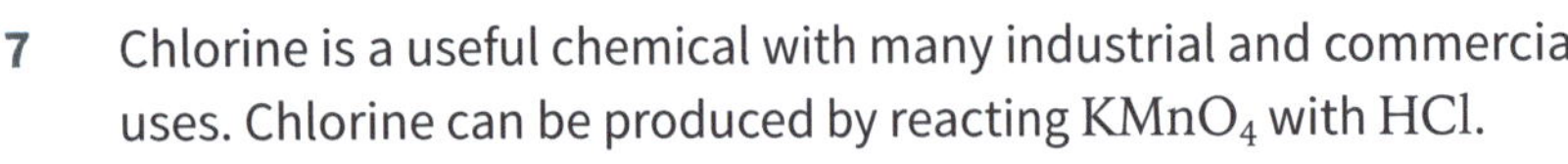

The half equations involved are:

$$Cl_2(g) + 2e^- \rightarrow 2Cl^-(aq)$$

$$MnO_4^-(aq) + 8H^+(aq) + 5e^- \rightarrow Mn^{2+}(aq) + 4H_2O(l)$$

**Synoptic links**

2.1.2 2.2.2

**(a) (i)** State the oxidation number of Mn in the $MnO_4^-$ ion. **[1]**

**(ii)** Give the balanced equation for the overall reaction. **[2]**

**(b)** Chlorine is used to extract bromine from seawater. Chlorine gas is bubbled through a solution of bromide ions.

**(i)** Give the simplest ionic equation for the reaction of chlorine with bromide ions. **[1]**

**(ii)** Name the species behaving as an oxidising agent in the reaction. **[1]**

**(iii)** Define oxidising agent in terms of electrons. **[1]**

**Exam tip**

Remember, in the overall equation, the electrons must cancel out.

**(c)** Chlorine has a boiling point of −34 °C. Bromine has a boiling point of 59 °C.

Explain the difference in boiling points of these two molecules. **[3]**

**8** A student carried out an experiment to find the percentage purity of an impure sample of iron(II) sulfate, $FeSO_4 \cdot 7H_2O$.

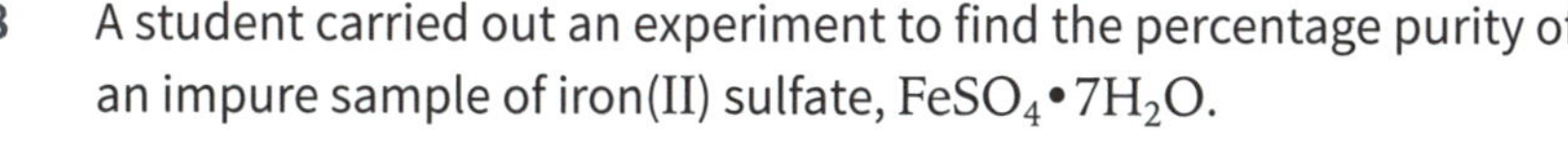

The student dissolved 5.78 g of iron(II) sulfate in water and made the solution up to 250 cm$^3$.

The student found that, after an excess of acid had been added, 25.0 cm$^3$ of this solution reacted with 21.0 cm$^3$ of a 0.0160 mol dm$^{-3}$ solution of potassium dichromate, $K_2Cr_2O_7$.

**(a) (i)** Give the half-equation including state symbols for the oxidation of iron(II) ions to iron(III) ions in solution. **[1]**

**(ii)** Give the half-equation including state symbols for the reduction of dichromate ions to chromium(III) ions in acidic conditions. **[1]**

**(iii)** Use your answers to **8(a)(i)** and **8(a)(ii)** to write the overall equation for the reaction, including state symbols. **[1]**

**(b)** Calculate the mass in g of iron(II) sulfate in the impure sample. **[4]**

**(c)** Determine the percentage purity of the sample. **[1]**

# Knowledge

## 4 Electrons, bonding, and structure

### Electrons and shells

Electrons can be thought of as being a cloud of negative charge in **orbitals**. An orbital is a way of defining the energy of that electron: each orbital can hold two electrons of opposite spin.

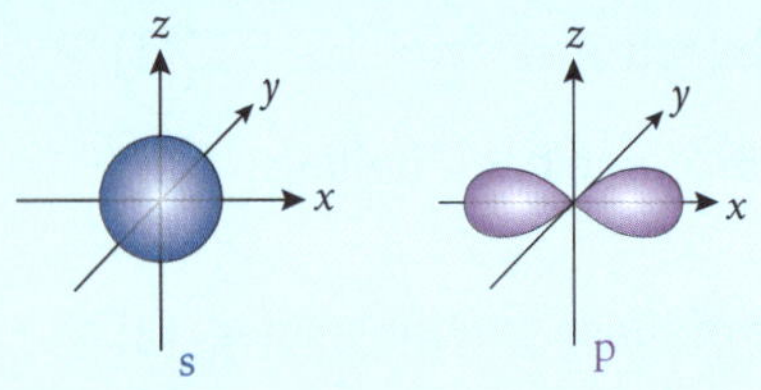


There are different types of orbitals: s-, p-, d-, and f-orbitals. The s-orbital is spherical in shape and the p-orbital is a dumb-bell shape. Each shell ($n$) contains $2n^2$ electrons. Orbitals of the same type are grouped together in **sub-shells**.

| Shell, $n$ | Number of electrons $= 2n^2$ | s | p | d | f | Sub-shells | Number of electrons / each sub-shell |
|---|---|---|---|---|---|---|---|
| 1 | 2 | 1 | | | | 1s | 2 |
| 2 | 8 | 1 | 3 | | | 2s 2p | 2, 6 |
| 3 | 18 | 1 | 3 | 5 | | 3s 3p 3d | 2, 6, 10 |
| 4 | 32 | 1 | 3 | 5 | 7 | 4s 4p 4d 4f | 2, 6, 10, 14 |

### Filling orbitals

1. Fill in order of increasing energy, from the bottom.
2. Each orbital has a pair of *opposite spin* electrons.
3. Fill each orbital with a single electron first, before pairing, electrons repel.

Remember, the 4s subshell has a *lower* energy than the 3d, so is filled first.

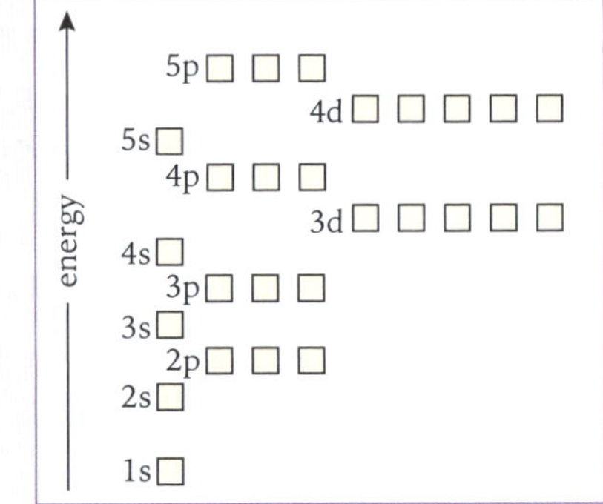


### Ionic bonding

**Ionic bonding** is the transfer of outer shell electrons from a metal atom to a non-metal atom.

This creates oppositely charged ions and between these ions there is an **electrostatic attraction**.

### Electron arrangements

The electron arrangement of an atom can be written using sub-shells.

For example, the electron arrangement of phosphorus is $1s^2 2s^2 2p^6 3s^2 3p^3$.

### Properties of ionic compounds

The physical properties are due to the strong electrostatic attraction between the ions in the **giant ionic lattice**.

Most ionic compounds:

- have high melting and boiling points
- in the *solid state*, do not conduct electricity because ions are in a fixed position
- conduct electricity when *melted* or *dissolved* in water because lattice breaks down and ions are free to move
- are soluble in water.

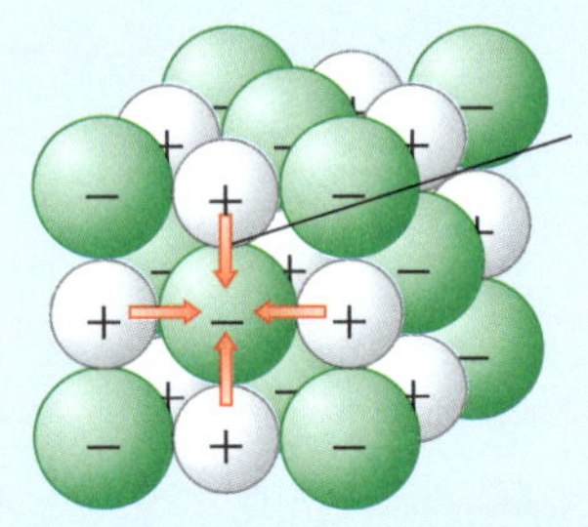

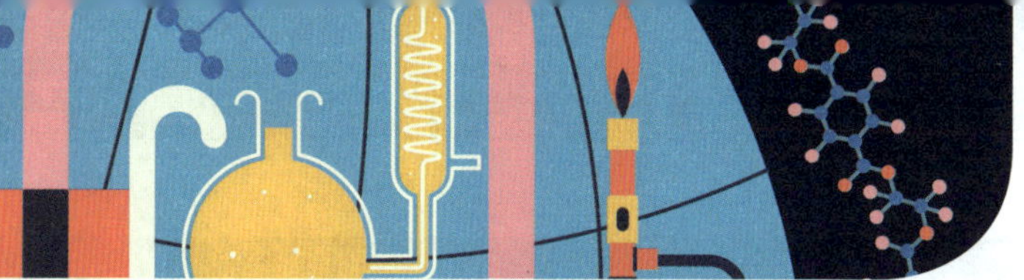



## Covalent bonding

**Covalent bonding** is the strong electrostatic attraction between a shared pair of electrons and the nuclei of the bonded atoms. A single bond is one pair of electrons and double bonds are two pairs of electrons. Covalent bonding occurs between non-metals.

**Average bond enthalpy** is a measure of the strength of a covalent bond. The larger the value, the stronger the bond.

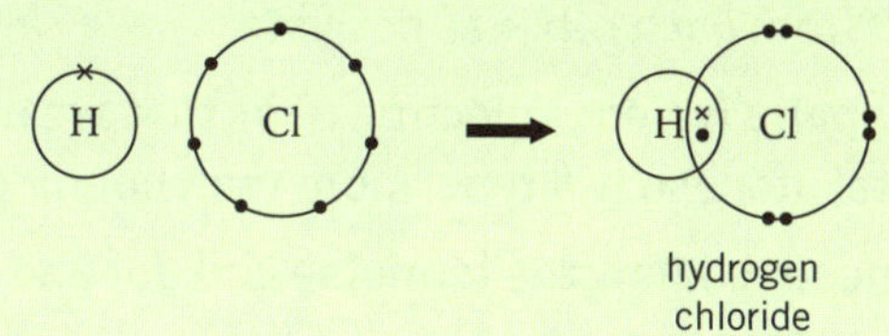


## Properties of simple covalent compounds

**Simple covalent compounds** have weak intermolecular forces of attraction, so need little energy to overcome these forces of attraction.

This means that they:

- have low melting and boiling points
- generally exist as liquids or gases at room temperature
- do not conduct electricity.

## Properties of giant covalent compounds

**Giant covalent compounds** (e.g., silicon dioxide, diamond, and graphite):

- have large lattices
- have very strong covalent bonds, so require large amounts of energy to break bonds
- have high melting and boiling points
- do not conduct electricity except for graphite.

## Dative covalent bonding

**Dative covalent bonding** (a co-ordinate bond) is where the bonding electrons in a covalent bond are from one of the atoms. This is represented using an arrow instead of a line.

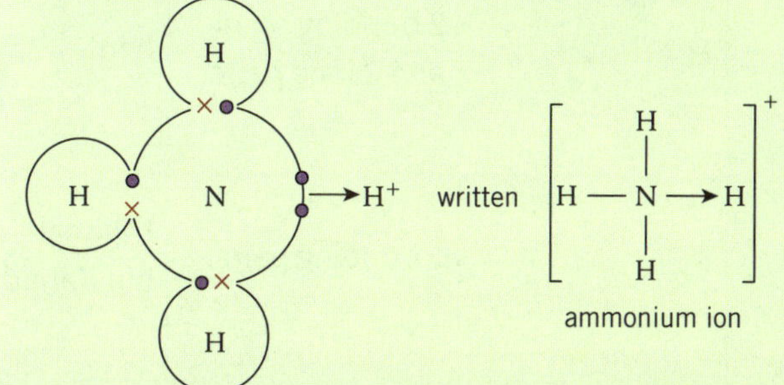


## Metallic bonding

**Metallic bonding** is a lattice of positive metal ions within a 'sea' of delocalised outer electrons.

The positive ions repel each other but they are held together by their strong electrostatic attraction to the delocalised electrons.

This means that metals:

- have high melting and boiling points
- have high heat and electrical conductivity
- are strong, malleable, and ductile.

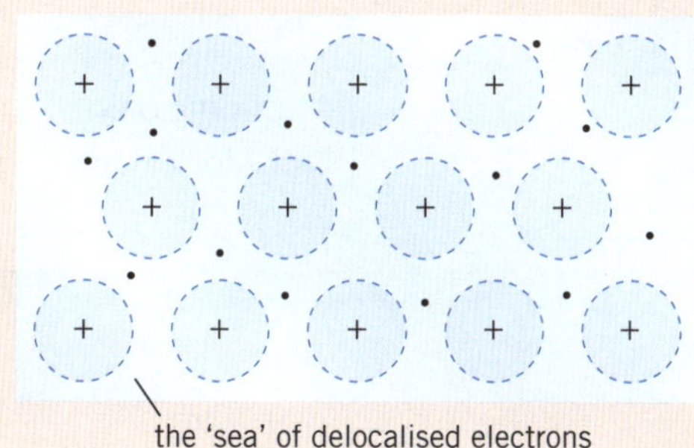

the 'sea' of delocalised electrons

For answers and more practice questions visit www.oxfordrevise.com/scienceanswers

Even more practice and interactive revision quizzes are available on kerboodle

## The shapes of simple molecules and ions

**Electron-pair repulsion theory** explains the shapes of molecules.

Electron pairs repel each other, forcing the pairs of electrons to be as far apart from each other as possible.

**Lone pairs** (valence electron pairs that are not shared) sit closer to the central atom so will repel more than bonding pairs of electrons.

You need to know the formulae and dot and cross diagrams for common covalent compounds: $H_2O$, $CO_2$, $O_2$, $NH_3$, and $NH_4^+$.

| Example | Number of pairs of electrons | Shape | Bond angles | Diagram |
|---|---|---|---|---|
| $CO_2$ or $BeCl_2$ | 2 bonding pairs | linear | 180° | Cl—Be—Cl 180° |
| $BF_3$ (outer shell not full) | 3 bonding pairs | trigonal planar | 120° | $BF_3$ 120° |
| $CH_4$ or $NH_4^+$ | 4 bonding pairs | tetrahedral | 109.5° | $CH_4$ 109.5° |
| $NH_3$ | 3 bonding pairs and 1 lone pair | trigonal pyramidal | 107° | $NH_3$ |
| $H_2O$ | 2 bonding pairs and 2 lone pairs | bent | 104.5° | $H_2O$ 104.5° |
| $PCl_5$ | 5 bonding pairs | trigonal bipyramidal | 90° and 120° | $PCl_5$ 120° |
| $SF_6$ (expanded octet) | 6 bonding pairs | octahedral | 90° | $SF_6$ 90° |
| $ClF_4^-$ | 4 bonding pairs and 2 lone pairs | square planar | 90° | $[ClF_4]^-$ |

## Electronegativity

**Electronegativity** is the ability of an atom to attract the bonding electrons in a covalent bond. The **Pauling scale** is a measure of electronegativity: the more electronegative (higher number on the Pauling scale), the more it will attract electrons.

Electronegativity depends on:

1. the nuclear charge
2. the distance between the outer electrons shells and the nucleus
3. the level of shielding by the other electron shells.

- Across a period the electronegativity increases, because the nuclear charge increases.
- As you go down a group the electronegativity decreases, as the shielding increases.

## Dipoles

As electrons move towards more electronegative atoms this creates a **dipole** (the separation of opposite charges) within a bond. In symmetrical molecules these dipoles cancel out so the molecule is non-polar.

$\overset{\delta+}{H}-\overset{\delta-}{F}$

hydrogen fluoride – the molecule is unsymmetrical so dipoles do not cancel

$Cl^{\delta-}$ (top), $\overset{\delta-}{Cl}-\overset{4\delta+}{C}-\overset{\delta-}{Cl}$, $Cl^{\delta-}$ (bottom)

carbon tetrachloride – the molecule is symmetrial so the dipoles cancel

## Intermolecular forces

strongest intermolecular force

weakest intermolecular force

Intermolecular forces are weak interactions between the dipoles of different molecules. There are three main ones: **hydrogen bonding, permanent dipole–dipole,** and **induced dipole–dipole** interactions.

### Hydrogen bonding

Hydrogen bonding occurs between the H atom that is bonded to either an N, O, or F atom and another electronegative atom. The electronegativity of the O, F, or N atom pulls an electron away from the H atom, giving it a small positive charge. This then forms a hydrogen bond with the lone pair of electrons on the N, O, or F atom.

### Permanent dipoles

Permanent dipole–dipole bonding occurs between permanent δ− ends and δ+ ends of molecules.

$\overset{\delta+}{H}-\overset{\delta-}{Cl} \cdots\cdots \overset{\delta+}{H}-\overset{\delta-}{Cl}$

covalent bond

permanent dipole–dipole interaction

### Induced dipoles

Induced dipole–dipole interactions (London dispersion forces) exist between all molecules.

1. The random movement of electrons introduces an **instantaneous dipole**.
2. This dipole will induce a dipole in a neighbouring molecule.
3. The induced dipole induces more dipoles, which attract one another.

# Retrieval

Learn the answers to the questions below, then cover the answers column with a piece of paper and write as many as you can. Check and repeat.

| | Questions | Answers |
|---|---|---|
| 1 | What bond angles can be found in $PCl_5$? | 90° and 120° |
| 2 | Which types of compound have high melting points? | ionic, metallic, and giant covalent |
| 3 | Which subshell comes after 3p? | 4s |
| 4 | How many electrons are involved in a double bond? | 4 |
| 5 | Which is the weakest type of intermolecular force? | induced dipoles (London/dispersion forces) |
| 6 | Which covalent lattice can conduct electricity? | graphite |
| 7 | What is the shape of $SF_6$? | octahedral |
| 8 | Is carbon dioxide polar? Explain why. | no, it is symmetrical, so the dipoles cancel each other out |
| 9 | Which end of H—Cl is the negative dipole? | Cl |
| 10 | What are the structures of diamond and graphite? Relate their properties to this. | diamond is a lattice and is very hard, graphite has a lattice structure but forms layers that can slide over each other making it soft |
| 11 | Which element is most electronegative? | fluorine |
| 12 | Why do metals conduct electricity? | they have a sea of delocalised electrons that can move |
| 13 | How many bonding pairs and lone pairs of electrons surround the oxygen atom in water? | 2 bonding pairs and 2 lone pairs |
| 14 | Which pair of electrons is more repulsive: bonding pairs or lone pairs? | lone pairs |
| 15 | What is the bond angle of water? | 104.5° |
| 16 | Which elements must hydrogen be bonded to for hydrogen bonding to occur? | oxygen, fluorine, or nitrogen |
| 17 | What is the order of the orbitals? | s p d f |
| 18 | How is the bonding in diamond and graphite similar and different? | Both have C—C bonds, but diamond makes 4 C—C bonds and graphite makes 3 C—C bonds. |
| 19 | Which type of intermolecular force is the strongest? | hydrogen bonding |
| 20 | What effect does hydrogen bonding have on Group 6 hydrides' boiling points? | water forms hydrogen bonds so has a higher than expected boiling point |

Put paper here

# Maths skills

Practise your maths skills using the worked example and practice questions below.

## VSEPR theory

You can use valence-shell electron-pair repulsion (VSEPR) theory to predict the shapes and bond angles of ions and molecules.

**Step 1:** Count the outer electrons of the central atom (check its group in the Periodic Table).

**Step 2:** Add an electron for each bond. Halogens contribute one electron to make a single bond, oxygen contributes one for a single bond or two for a double bond, and so on.

**Step 3:** For ions, add (if negatively charged) or remove (if positively charged) a number of electrons equal to the magnitude of the charge on the ion.

**Step 4:** Divide your total by two to get the number of electron pairs.

**Step 5:** For each double bond (e.g., $=$O) remove one pair of electrons from your count; for each triple bond (e.g., $\equiv$N) remove two.

**Step 6:** Count the lone pairs.

**Step 7:** Predict (or recall) the shape with that many bonding and lone pairs of electrons.

*OR* you can draw a dot-and-cross diagram and count the lone and bonding pairs, allowing for double and triple bonds.

## Worked example

**Questions**

Consider the molecule $ClF_3$. Draw its shape, name the shape, and state the bond angles.

**Answers**

**Steps 1–3:** Count the number of electrons

| $ClF_3$ | Number of electrons |
|---|---|
| central atom: Cl | 7 |
| other atoms: $3 \times$ F | $3 \times 1$ |
| charge | 0 |
| total | 10 |

**Steps 4–6:** There are five electron pairs. Three of the pairs are bonding pairs.

**Step 7:** Species with five electron pairs are based on a trigonal bipyramidal shape with bond angles of 90° and 120°.

Lone pairs repel more than bonding pairs, so they occupy two of the equatorial positions. This leaves one of the *F* atoms in the third equatorial position, and the other two axial.

## Practice

For each of the following molecules or ions:

1. Draw the shape of the molecule (include any lone pairs).
2. Name the shape
3. Label any bond angles.

**a** $AlCl_4^-$

**b** $PCl_4^+$

**c** $PCl_6^-$

**d** $IF_5$

**e** $NH_4^+$

**f** $OF_2$

**g** $IF_3$

**h** $NH_2^-$

**i** $SO_2^{4-}$

# Practice

## Exam-style questions

1 Potassium bromide is commonly found in medicines that prevent seizures. Potassium bromide is a white crystalline solid with a high melting point and is soluble in water.

**(a) (i)** Explain in terms of bonding why potassium bromide has a high melting point.

........................................................................................

........................................................................................

........................................................................................

........................................................................................ **[2]**

**(ii)** Suggest in terms of bonding why potassium bromide is soluble in water.

........................................................................................

........................................................................................ **[1]**

> **Synoptic link**
>
> 2.1.2

**(b)** Potassium is a Group 1 metal.

Give a balanced ionic equation to show the reaction between potassium and water.

**[1]**

> **Exam tip**
>
> Remember, ionic equations need to show the charges for each ion and the state symbols.

**(c)** Give the full ground state electron configuration of the $K^+$ ion.

........................................................................................

........................................................................................ **[1]**

**(d)** Use your knowledge of structure and bonding to explain why potassium bromide has a melting point that is higher than both potassium and bromine.

........................................................................................

........................................................................................

........................................................................................

........................................................................................

........................................................................................

........................................................................................

........................................................................................

........................................................................................

........................................................................................

........................................................................................ **[6]**

> **Exam tip**
>
> The key to getting this correct is identifying the different types of structure and bonding correctly.

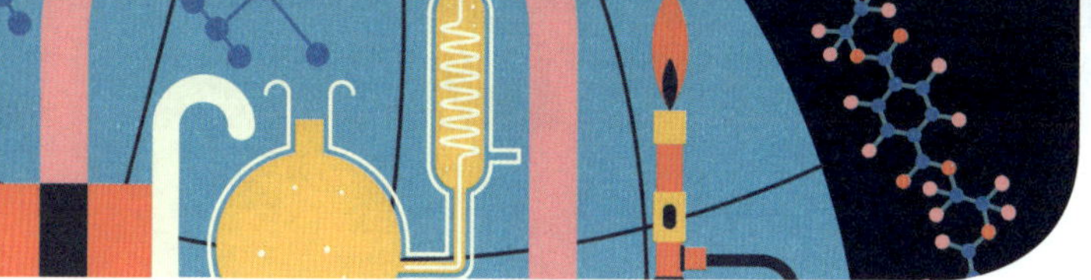



2 This question is about compounds containing hydrogen.

The boiling points of three common hydrogen-based compounds are presented in **Table 2.1**.

| Compound | Boiling point (°C) |
|---|---|
| water | 100 |
| ammonia | −33 |
| methane | −162 |

**Table 2.1**

Synoptic link

2.1.3

**(a)** Use your understanding of structure and bonding to explain this difference in boiling points.

........................................................................................................

........................................................................................................

........................................................................................................

........................................................................................................

........................................................................................................

........................................................................................................

........................................................................................................

........................................................................................................**[6]**

Exam tip

Remember, physical changes are always explained by intermolecular forces.

**(b)** Explain why ammonia is soluble in water but methane is not.

........................................................................................................

........................................................................................................**[1]**

**(c)** **(i)** Ammonia is a base. It reacts with $BH_3$ to form $[NH_3BH_3]$.

Illustrate the shapes of ammonia and $BH_3$ by drawing.

Name the shapes and give the approximate bond angles.

**[6]**

**(ii)** Name the type of bond that forms between ammonia and $BH_3$. Give a reason for your answer.

........................................................................................................

........................................................................................................

........................................................................................................

........................................................................................................**[2]**

**(d)** Ammonia also reacts with hydrochloric acid to form a white ionic solid, ammonium chloride, $NH_4Cl$, which is used in fertilisers.

The balanced equation for this reaction is:

$$NH_3(g) + HCl(g) \rightarrow NH_4Cl(s)$$

Calculate the mass of ammonia needed to create 25 kg of ammonium chloride. **[4]**

**(e)** Hydrochloric acid with a concentration of 2.50 mol dm$^{-3}$ was used to produce ammonium chloride.

Calculate the volume of acid needed to completely react with the ammonia from **2(d)**.

If you were unable to obtain a value for **2(d)**, you should assume that 14.2 kg of ammonia was used. This is *not* the correct answer.

volume of acid = .................... dm$^{-3}$ **[2]**

**(f)** Use your knowledge of structure and bonding to explain why ammonium chloride is an ionic solid at room temperature but ammonia is a gas.

.....................................................................................

.....................................................................................

.....................................................................................

..................................................................................... **[5]**

**3** This question is about intermolecular forces.

**(a)** Define the term electronegativity.

.....................................................................................

..................................................................................... **[1]**

**(b)** Explain how permanent dipole–dipole forces arise between hydrogen chloride molecules.

.....................................................................................

.....................................................................................

..................................................................................... **[2]**

**(c) (i)** Illustrate the shapes by drawing and give the bond angles of $BeCl_2$ and $CH_3Cl$.

**Exam tip**

Make sure you clearly show the number of bonds and the lone pairs.

**[4]**

**(ii)** The Be—Cl bond is polar yet beryllium chloride is not a polar molecule. Explain why.

..........................................................................................

..........................................................................................**[1]**

**4** Group 6 elements react with hydrogen to form hydrides.

**Figure 4.1** shows the boiling points of the Group 6 hydrides.

Synoptic link

3.1.1

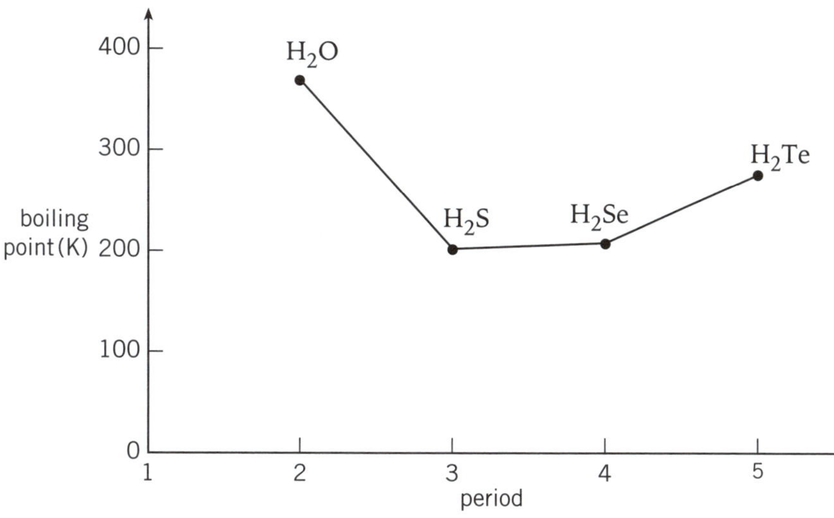


**Figure 4.1**

**(a)** Explain why water has a higher boiling point than dihydrogen sulfide. **[3]**

**(b)** Explain why dihydrogen selenide has a higher boiling point than dihydrogen sulfide. **[2]**

**(c)** Explain why the density of water gets lower when the temperature drops below 273 K. **[1]**

**(d)** Ilustrate the shape of $H_2Se$ by drawing.

State the name of the shape and its bond angles. **[3]**

**(e) (i)** Carbon exists as two distinct allotropes: graphite and diamond.

Name the structure in each allotropes **[1]**

**(ii)** Explain why diamond is hard but graphite is soft. **[2]**

**(iii)** Explain why graphite can conduct electricity but diamond cannot. **[2]**

**(iv)** Explain why both diamond and graphite have high melting points. **[2]**

**(v)** Give the shape of the repeated structures of both diamond and graphite.

In each case, identify the bond angle and any lone pairs.

Draw at least four carbon atoms in each diagram.

**[6]**

Exam tip

Remember, physical changes are always explained by intermolecular forces.

5 Ethanol and propane are two organic molecules with similar relative molecular masses. Their displayed formula are shown in **Figure 5.1**.

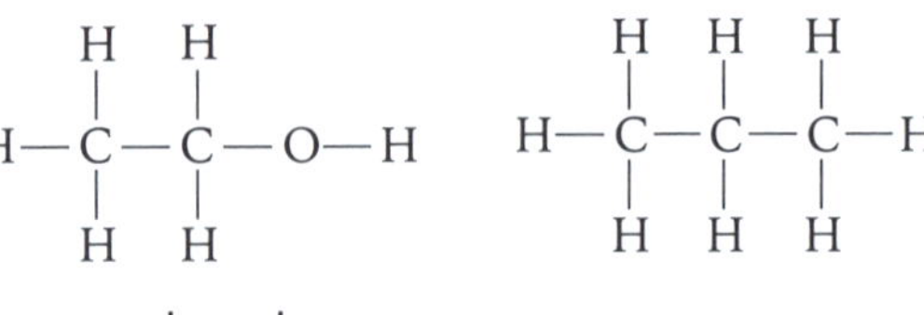


**Figure 5.1**

**(a)** Name the type of bonding found within these molecules and explain how it occurs. **[2]**

**(b)** Explain why ethanol has a boiling point of 78 °C whereas the boiling point of propane is −42 °C. **[3]**

**(c)** Ethanol is soluble in water. Draw a diagram to show the bonding between one molecule of water and one molecule of ethanol. Show any partial charges and lone pairs. **[3]**

**(d)** Ethanol and propane are both able to undergo combustion. Give a balanced chemical equation for the complete combustion of ethanol to produce only carbon dioxide and water vapour. **[1]**

**(e)** Ethanol can be used as a fuel in car engines.

Calculate the volume of exhaust gases produced when 25.0 g of ethanol is burnt at 2000 K and 101 kPa.

Give your answer to 3 significant figures. **[5]**

**Synoptic links**

2.1.2 2.1.3

**Exam tip**

Make sure you draw the water molecules as 'V-shapes' and include their lone pairs. Use a dotted or dashed line to show the intermolecular forces.

6 Aluminium is a Group 3 metal that is often found in the Earth's crust as aluminium oxide, $Al_2O_3$. Unlike sodium oxide and magnesium oxide, aluminium oxide does not react with water.

**(a)** Explain aluminium oxide's lack of reactivity in terms of the relative strength of the ionic bonds. **[3]**

**(b)** Explain the trend in ionic radii as you move from Na to Al in Period 3. **[3]**

**(c)** Aluminium oxide will react with hot dilute hydrochloric acid to form aluminium chloride and water.

Give a balanced equation to show this reaction. **[1]**

**(d)** Unusually for a metal compound, aluminium chloride is a covalent molecule. **Figure 6.1** shows the structure of aluminium chloride.

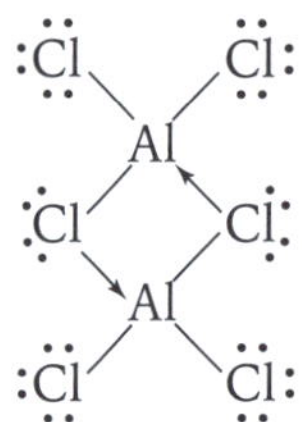

**Figure 6.1**

Name the type of bonding shown by the arrows in **Figure 6.1** and explain how it forms. **[2]**

**Synoptic links**

2.1.2 2.1.3 3.1.1

**Exam tip**

Make sure you mention the size, charge, and amount of each ion.

**(e)** Aluminium chloride has a melting point of 192 °C. Aluminium oxide has a melting point of 2072 °C.

Explain how the bonding in each substance shows this difference. **[5]**

**(f)** Explain why aluminium chloride does not conduct electricity in its molten form. **[2]**

**7** Hydrogen iodide can be produced from the reaction of hydrogen and iodine.

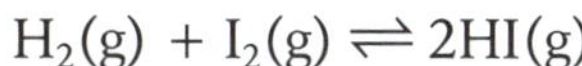

$$H_2(g) + I_2(g) \rightleftharpoons 2HI(g) \qquad \Delta H^{\ominus} = +50\ \text{kJ mol}^{-1}$$

**(a)** Explain why iodine has a much higher boiling point than hydrogen. **[3]**

**(b) (i)** Explain the effect, if any, on the concentration of HI if the reaction is heated. **[2]**

**(ii)** Explain the effect, if any, on the position of equilibrium if the pressure is increased. **[2]**

**(c)** Determine if the reaction is feasible at 300 K. **[4]**

**(d)** State the role of iodine, with reference to redox reactions, in this reaction. **[1]**

**(e)** The HI produced in the reaction is dissolved in water to form hydroiodic acid, HI(aq), a strong acid, with a concentration of 0.015 mol $dm^{-3}$.

Give the ionic equation for the formation of hydroiodic acid. **[1]**

**(f)** Calculate the pH for the 0.015 mol $dm^{-3}$ hydroiodic acid solution. **[2]**

**(g)** State what would be observed if chlorine gas was bubbled through the hydroiodic acid.

Give the two half-equations for the reduction and oxidation occurring and the overall equation. **[5]**

**Synoptic links**

3.2.3 5.2.2 2.1.5
2.1.2 5.1.3

**Exam tip**

Any question about physical properties is a bonding question. Remember, covalent bonds do not break when molecular covalent substances change state.

**Exam tip**

Take care! This question has four parts to it.

**8** Fritz Haber, a German chemist, first manufactured ammonia in 1908. Ammonia is very soluble in water.

**(a)** State the strongest type of intermolecular force between one molecule of ammonia and one molecule of water. **[1]**

**(b)** Give a diagram that shows how one molecule of ammonia is attracted to one molecule of water.

Include all partial charges and all lone pairs of electrons in your diagram. **[3]**

**(c)** Phosphine ($PH_3$) has a structure that is similar to ammonia.

In terms of intermolecular forces, suggest the main reason why phosphine is almost insoluble in water. **[1]**

**(d)** Draw and name the shape of one molecule of phosphine.

State the bond angles. **[3]**

# Knowledge

## 5 The Periodic Table and periodicity

### Structure of the Periodic Table

Elements are arranged in the Periodic Table by increasing proton (or atomic) number. Their positions on the Periodic Table are related to their physical and chemical properties. Metals are found on the left of the red line and non-metals are on the right.

**Periods** show repeating trends in physical and chemical properties. **Groups** show trends and similar chemical properties.

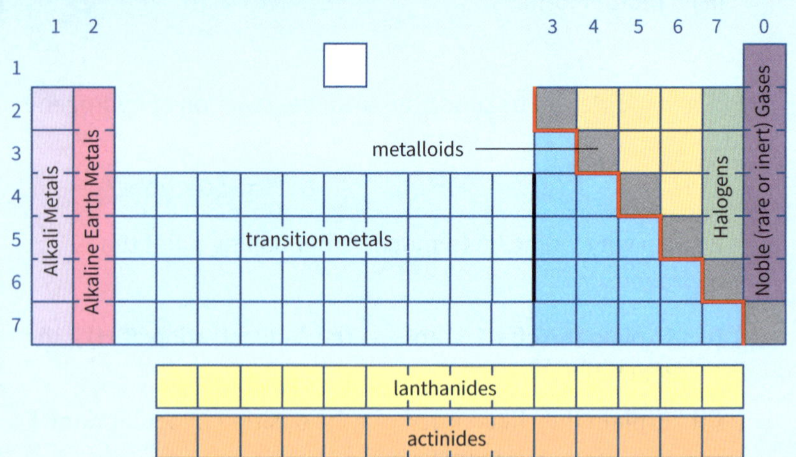


### The development of atomic models

| 1661 | 1803 | 1897 | 1911 | 1913 | 1926 | 1932 |
|---|---|---|---|---|---|---|
| Elements defined | Dalton defined elements as atoms of the same mass | Thomson discovered electrons, development of the plum pudding model | Rutherford performed the gold foil experiment and proposed the idea of the atomic nucleus | Bohr developed the nuclear model of the atom | Schrödinger discovered that electrons can act as waves or particles | Chadwick discovered neutrons |

### Ionisation energies

The **first ionisation energy** is the value of energy that is required to convert one mole of gaseous atoms into one mole of positively charged gaseous ions.

$$X(g) \rightarrow X^+(g) + e^-$$

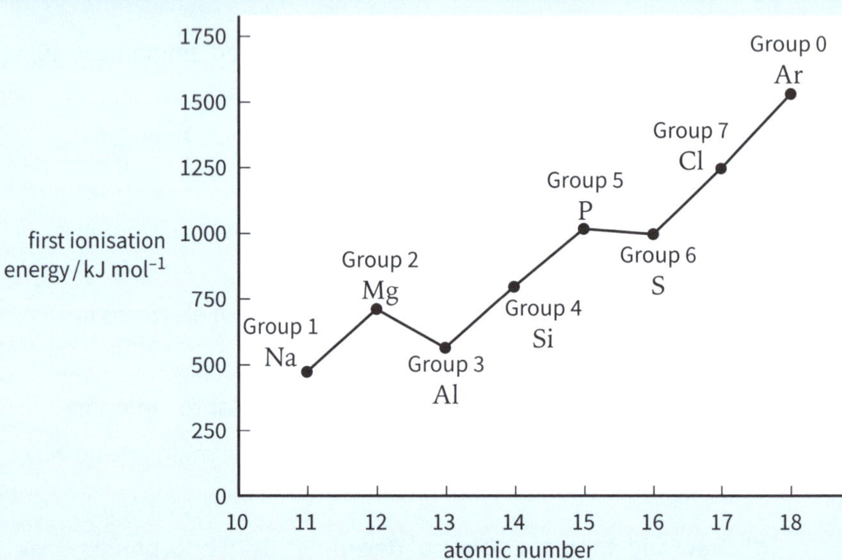

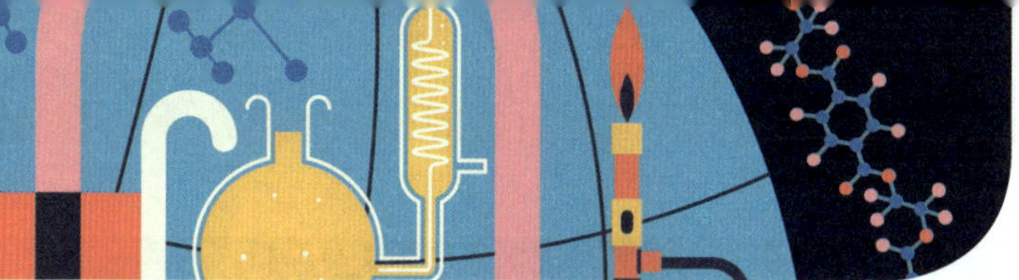



## Link between trends in first ionisation energy and period

There is a general *increase* in first ionisation energy across a period. More energy is needed to remove the first electron because across a period:

- nuclear charge *increases*
- shielding stays the same as number of shells *stays the same*
- nuclear force of attraction *increases*
- atomic radius *decreases*.

## First ionisation energy and groups

First ionisation energy *decreases* down a group. Moving down a group, the number of filled energy levels increases shielding and atomic radius *increases*, so the outer electron is easier to remove as it is further away from the nucleus. Metallic elements in Groups 1 and 2 have the lowest first ionisation energy as the first electron from their outer shell is lost easily.

## Trends in first ionisation energy and sub-shells across Period 2

Between Groups 2 and 3 the drop occurs when the 2s-sub-shell gets filled up and the 2p-sub-shell is started. The 2p-sub-shell in B has a higher energy than the 2s-sub-shell in Be. The move to a new shell makes the first electron in the p-sub-shells easier to remove than the electron in the s-sub-shell. The dip occurs between Groups 5 and 6 due to the pairing of electrons. The pairing makes the 8th electron slightly easier to remove due to the repulsion between the paired electrons in the p-orbital.

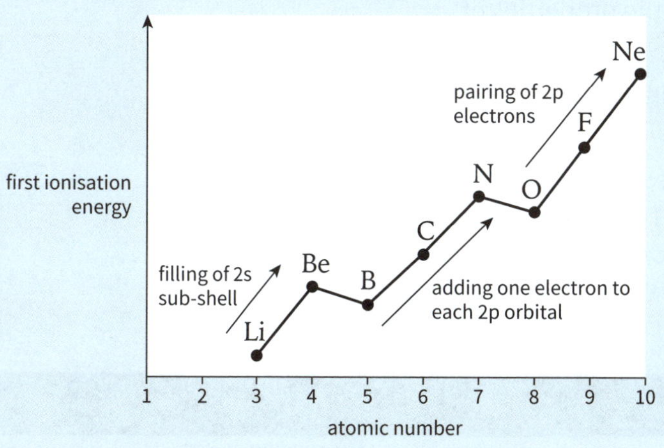


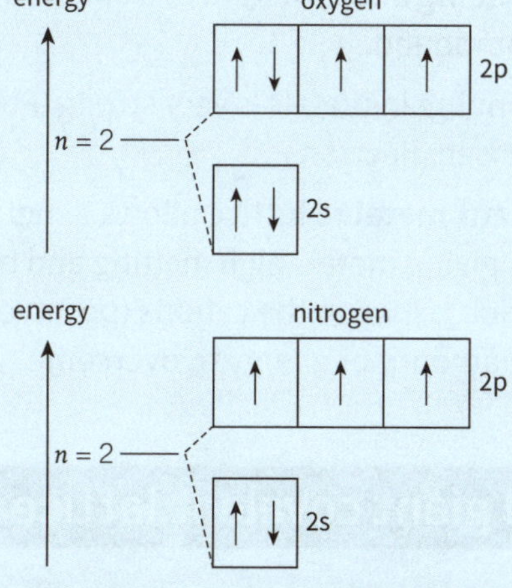


## Making predictions from ionisation energies

Successive ionisation energies can predict:

- the number of electrons in each shell
- the group for the element in the Periodic Table
- the identity of an element.

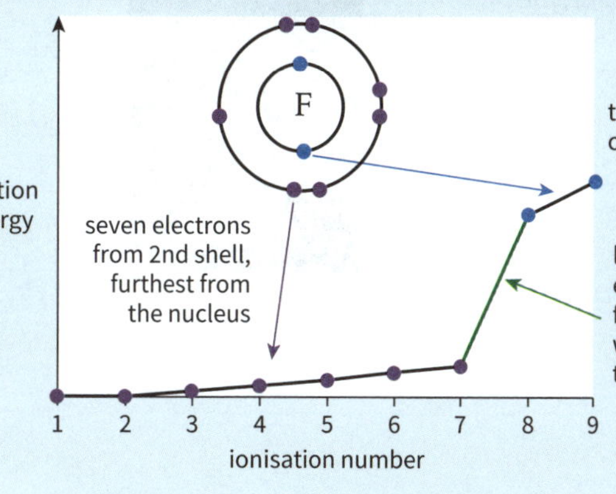

## Periodic trends in bonding and structure of Period 3

The bonding and structure of elements across a period show a trend from metals on the left to non-metals on the right. For example, the elements in Period 3 have the following properties:

- Sodium, magnesium, and aluminium are metals. The electrons in the outer shell are delocalised so they form giant metallic structures.
- Silicon is a semi-metal and will form giant covalent structures.
- Phosphorus, sulfur, and chlorine are non-metals and will form simple molecular covalent compounds.
- Argon is a gas, it is monoatomic and unreactive.

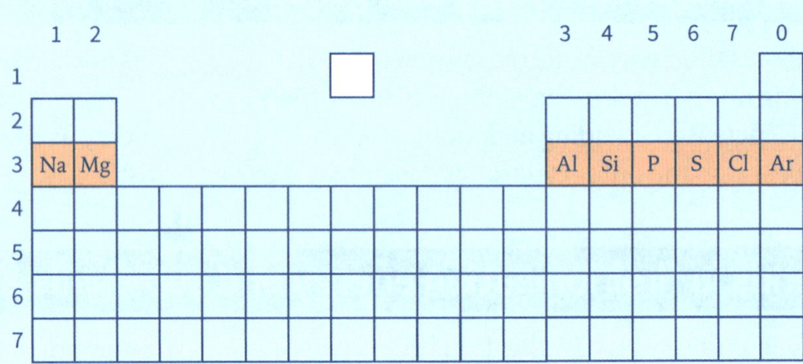


The physical and chemical properties of the elements vary across the period because there is:

- a decrease in atomic radius
- an increase in first ionisation energy
- a variation in melting points.

## Metallic bonding

The bonding and structure of a compound determine the properties of that compound.

The bonding in metals is very strong. Positive ions sit in a sea of delocalised electrons.

This **giant metallic lattice** allows a metal to conduct electricity and heat, as well as giving metals high melting and boiling points. The strong electrostatic attraction between the cations (positive ions) and delocalised electrons requires a large amount of energy to overcome.

## Solid giant covalent lattice

Diamond, graphite, and graphene are **solid giant covalent lattices** made up of pure carbon, but they have very different properties.

| Lattice | diamond | graphite | graphene |
|---|---|---|---|
| Number of C—C bonds | 4 | 3 | 3 |
| Structure | lattice | layers | layers |
| Hard/soft | hard | soft (layers can slide over each other) | hard |
| Conducts | no | yes (free electron between layers can carry charge) | yes |

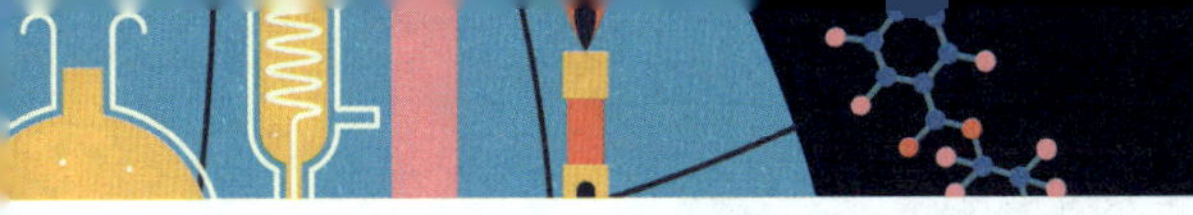



## Properties of diamond and graphite

Diamond makes 4 carbon–carbon bonds and forms a **tetrahedral structure**, graphite makes 3 carbon–carbon bonds and has layers. The layers mean that graphite is soft (whilst diamond is hard) as the layers can slide over each other.

Both of these structures have high melting and boiling points as the covalent bonds are very strong and require large amounts of energy to break.

They are both insoluble as the covalent bonds are too strong to be broken by solvents.

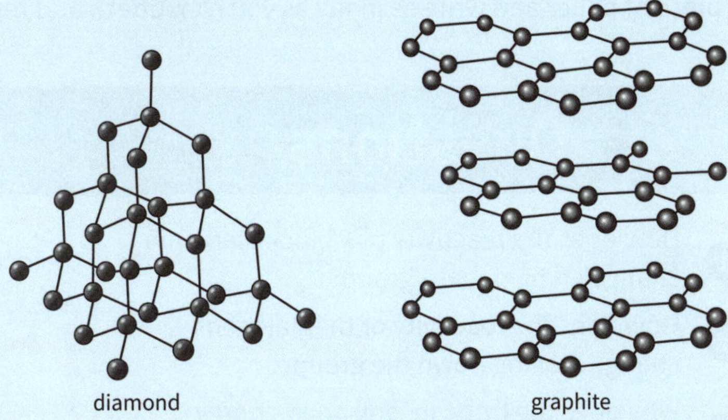


## Graphite and graphene

The 3 carbon–carbon bonding in graphite means that there is an extra electron that is not involved in the bonding. This extra electron is between layers and gives graphite the ability to conduct electricity.

Graphene is a single layer of graphite. It has the same properties, making it incredibly useful as it is thin, flexible, and can conduct electricity. It is used in electrical circuits in mobile phones, and in sensors to detect explosives.

## Silicon dioxide

Silicon dioxide is commonly known as sand and has a giant covalent lattice. It consists of a network of atoms that are bonded together by strong covalent bonds. It has similar properties to diamond: a high melting and boiling point, and it does not conduct electricity.

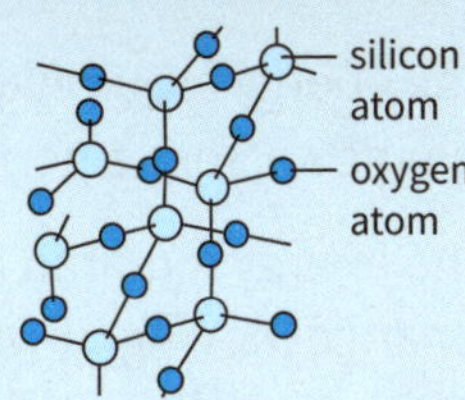


## Trends in melting points and boiling points in Period 2 and Period 3

Elements that tend to form giant structures (on the left of the period) have high melting points and boiling points, which increase from Groups 1 to 4. Groups 1 to 3 elements have giant metallic structures. Group 4 elements have a giant covalent structure.

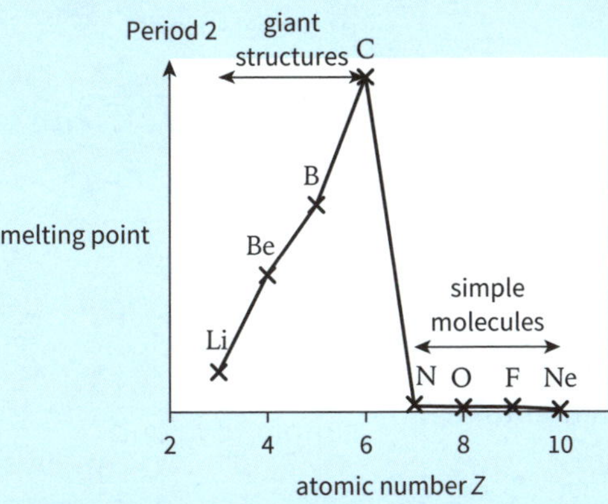


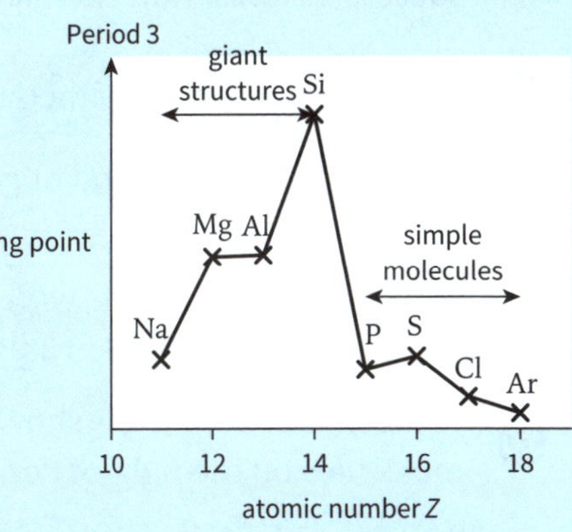


For those with simple molecular structures the size of the van der Waal forces between molecules will determine melting and boiling points. Those with more electrons will have stronger intermolecular forces.

# Retrieval

Learn the answers to the questions below, then cover the answers column with a piece of paper and write as many as you can. Check and repeat.

| | Questions | Answers |
|---|---|---|
| 1 | How does the reactivity of s-block elements change going down a group? | increases |
| 2 | How does the reactivity of the Halogens change moving down the group? | decreases |
| 3 | Why does the trend in ionisation energy change between Groups 2 and 3? | in Group 3 the outermost electron is now in a p-sub-shell |
| 4 | Why does the trend in ionisation energy change between Groups 5 and 6? | pairing of electrons starts in the p-sub-shell in Group 6 |
| 5 | What is metallic bonding? | positive ions in a sea of delocalised electrons leading to strong electrostatic attraction between the cations and electrons |
| 6 | How are graphite and graphene linked? | graphene is a single layer of graphite |
| 7 | How many C—C bonds are there between atoms in diamond? | 4 |
| 8 | How many C—C bonds are there between atoms in graphite? | 3 |
| 9 | How does the energy required for the first ionisation change across a period? | increases |
| 10 | How does the energy required for the first ionisation change moving down a group? | decreases |
| 11 | What is the general equation to show the first ionisation of an element? | $X(g) \rightarrow X^{+}(g) + e^{-}$ |
| 12 | What can be predicted about an element from successive ionisation energies? | the number of electrons in each shell; the group the element is in on the Periodic Table, and the identity of the element |
| 13 | Are the boiling points of metals high or low? | high |
| 14 | Which of these: diamond or graphite, conducts electricity? | graphite |
| 15 | Where will the elements with high melting points be found in Period 3 (left or right)? | left |
| 16 | What type of bonding is shown by elemental molecules on the right of Period 2? | simple covalent |
| 17 | Where are the d-block elements found in the Periodic Table? | in the middle |
| 18 | What three trends are seen across Period 3? | a decrease in atomic radius; an increase in first ionisation energy; a variation in melting point |

Put paper here

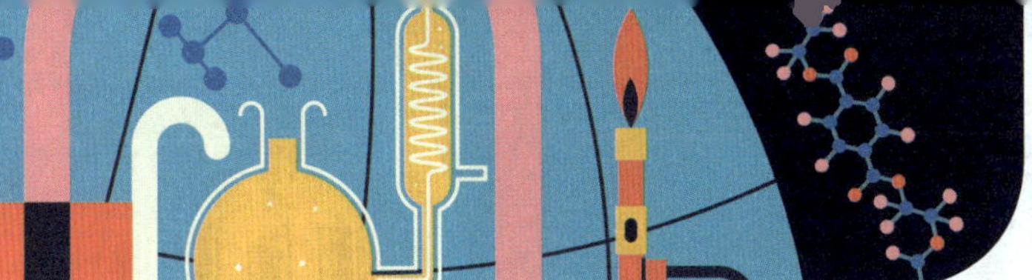



| | | |
|---|---|---|
| 19 | How does increasing nuclear charge affect the ease of removing electrons across a period? | as charge increases it gets more difficult to remove electrons |
| 20 | What is the trend in successive ionisation energies as atomic radius decreases? | increases |

Put paper here

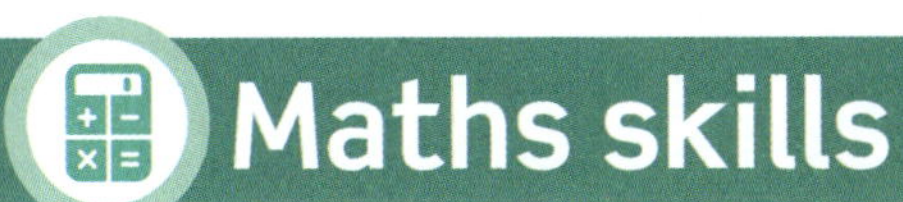

## Maths skills

Practise your maths skills using the worked example and practice questions below.

### Significant figures

Significant figures are essential to reporting calculated values to appropriate accuracy. For example, if you had 4.0 g of water then the number of moles of water is:

$\frac{4.0}{18.0} = 0.222222222$ mol

You cannot claim to know the amount of substance to that many significant figures. The rule is, when carrying out a calculation the final answer must have the same number of significant figures as the input with the **lowest** number of significant figures.

If you are given the value to two significant figures (i.e. 4.0 g), the answer should be given to two significant figures also. Therefore, the number of moles of water is 0.22 mol.

### Worked example

**Questions**

Given the following values:

$x = 27.3$

$y = 16.9351$

$z = 4$

calculate the following to an appropriate number of significant figures.

1 $x \div y$

2 $\log_{10}(3z)$

**Answers**

1 $\frac{x}{y} = \frac{27.3}{16.9351} = 1.6120365....$
1.61 (to 3 s.f.)

Explanation: $x$ is given to 3 significant figures, so the answer has to be given to 3 significant figures too.

2 $\log_{10}(3 \times 4) = 1.07918....$
$= 1$ (to 1 s.f.)

Explanation: $z$ is given to one significant figure, so the answer has to be given to one significant figure too.

### Practice

Given the following values:

$a = 1.538 \times 10^4$

$b = 15.98704$

$c = 19$

$d = 3 \times 10^{-6}$

calculate the following to an appropriate number of significant figures.

1 $ab^2$

2 $a+b+c$

3 $d \times (b+c)$

4 $\log_{10}b$

5 $10^d$

For answers and more practice questions visit www.oxfordrevise.com/scienceanswers

Even more practice and interactive revision quizzes are available on kerboodle

# Practice

## Exam-style questions

1 Ionisation energies allow us to determine electronic structure. **Table 1.1** shows the first eight successive ionisation energies of a Period 3 element, **Y**.

| Ionisation energy | Value ($kJ\,mol^{-1}$) |
|---|---|
| 1st | 577.54 |
| 2nd | 1816.8 |
| 3rd | 2744.8 |
| 4th | 11 577 |
| 5th | 14 841 |
| 6th | 18 379 |
| 7th | 23 326 |
| 8th | 27 464 |

**Table 1.1**

**(a)** Define first ionisation energy.

..............................................................................................

..............................................................................................

..............................................................................................**[2]**

**(b)** Identify the element using the ionisation energies in **Table 1.1**.

Explain your choice.

..............................................................................................

..............................................................................................

..............................................................................................**[2]**

**Exam tip**

Larger jumps in ionisation energy come from removing an electron from a closer shell.

**(c)** The first ionisation energy is given to a different number of decimal places than the second ionisation energy.

State why this is appropriate.

..............................................................................................

..............................................................................................**[1]**

2 The following table shows the masses of the subatomic particles in kg.

**Synoptic link**

2.1.1

| Subatomic particle | Mass (kg) |
|---|---|
| proton | $1.660\,540 \times 10^{-27}$ |
| neutron | $1.674\,929 \times 10^{-27}$ |
| electron | $9.109\,390 \times 10^{-31}$ |

**Table 2.1**

**(a)** Calculate the mass of an atom of beryllium-9.

Give your answer to an appropriate number of significant figures.

mass of atom = .................... kg **[2]**

**Exam tip**

Remember your answer should have the same number of significant figures as the least precise data you have been given.

**(b)** Give an equation to show the second ionisation energy of beryllium.

........................................................................

........................................................................

........................................................................ **[2]**

**(c)** Explain why there is a large jump between the second and third ionisation energies of beryllium.

........................................................................

........................................................................

........................................................................

........................................................................ **[3]**

**(d)** Beryllium has a smaller atomic radius than lithium. Explain why.

........................................................................

........................................................................

........................................................................

........................................................................ **[3]**

**Exam tip**

Remember to mention nuclear charge and shielding in your answer even if they do not change.

**3** Copper is a transition element that has many uses. Copper exists in two stable isotopes. **Figure 3.1** shows a mass spectrum of a sample of copper from a meteorite.

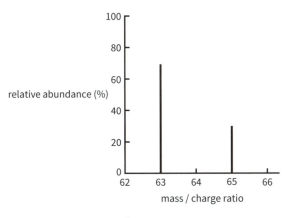

**Synoptic links**

2.1.1 5.3.1

relative abundance (%)

100, 80, 60, 40, 20, 0

62, 63, 64, 65, 66

mass / charge ratio

**Figure 3.1**

**(a)** Give the full electron configuration of Cu.

........................................................................

........................................................................ **[1]**

**(b)** Give the full electron configuration of $Cu^{+}$.

........................................................................

........................................................................ **[1]**

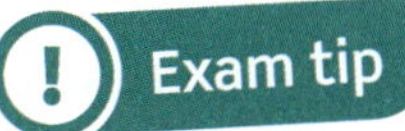

**Exam tip**

Think carefully about which shell empties first.

**(c)** Define the term transition element.

........................................................................

........................................................................ **[2]**

For answers and more practice questions visit www.oxfordrevise.com/scienceanswers

Even more practice and interactive revision quizzes are available on kerboodle

**(d)** Define the term isotope.

........................................................................................................

........................................................................................................

........................................................................................................ **[2]**

**(e)** On the mass spectrum of this sample of copper, the $^{63}Cu$ isotope was found to have an abundance of 71.23%.

Calculate the $A_r$ from this sample of copper.

Give your answer to an appropriate number of significant figures.

........................................................................................................

........................................................................................................

........................................................................................................

........................................................................................................ **[3]**

**Exam tip**

Check the mass spectrum to find any missing information you need.

**(f) (i)** Copper forms a soluble complex cation that has a distinctive blue colour.

Give the molecular formula for the complex ion and state its name.

........................................................................................................

........................................................................................................

........................................................................................................ **[2]**

**(ii)** Give equations to show how the ion in **3(f)(i)** reacts in the presence of dilute sodium hydroxide solution.

State any observations you would see.

........................................................................................................

........................................................................................................

........................................................................................................ **[2]**

**4** Beryllium and boron are both elements in Period 2.

**Synoptic link**

2.2.2

**(a)** State why can boron be classified as a p-block element.

........................................................................................................

........................................................................................................ **[1]**

**(b)** Give the full ground state electron configuration of boron.

........................................................................................................

........................................................................................................ **[1]**

**(c)** State why beryllium's first ionisation energy is higher than that of boron.

........................................................................................................

........................................................................................................

........................................................................................................ **[2]**

**(d)** Beryllium and boron both react with chlorine. Give the structures of $BeCl_2$ and $BCl_3$.

Name the shapes and give the bond angles.

.......................................................................................... **[6]**

> **Exam tip**
> The exceptions to the overall trends are important to learn.

**5** Iron is a transition metal and has typical metallic properties.

**(a) (i)** Describe the bonding in iron.

.......................................................................................... **[2]**

> **Synoptic link**
> 2.2.2

**(ii)** Explain why iron is ductile.

.......................................................................................... **[1]**

> **Exam tip**
> Ductile means 'can be pulled into wires'.

**(b)** Iron forms $Fe^{3+}$, which forms an oxide in air. Give the chemical formula of this oxide.

.......................................................................................... **[1]**

**(c)** Write the full ground state electronic configuration of $Fe^{3+}$.

.......................................................................................... **[1]**

**(d)** Explain why metallic iron is able to conduct electricity as a solid but iron(III) oxide cannot.

.......................................................................................... **[4]**

> **Exam tip**
> Make sure you are clear about the type of bonding and how it is linked to the property being discussed.

6 Period 3 contains elements that display characteristic properties of both metals and non-metals.

(a) Define the term periodicity.

..........................................................................................

..........................................................................................

..........................................................................................

.......................................................................................... [1]

(b) Explain the trend in first ionisation energy from sodium to argon.

..........................................................................................

..........................................................................................

..........................................................................................

..........................................................................................

..........................................................................................

..........................................................................................

..........................................................................................

..........................................................................................

..........................................................................................

..........................................................................................

..........................................................................................

.......................................................................................... [6]

**Exam tip**

Make sure you talk about each change and the general trend and explain each section.

(c) **Table 6.1** gives the successive ionisation energies of an element in Period 3.

| Ionisation number | Ionisation energy (kJ $mol^{-1}$) |
|---|---|
| 1 | 1000 |
| 2 | 2260 |
| 3 | 3390 |
| 4 | 4540 |
| 5 | 6990 |
| 6 | 8490 |
| 7 | 27 100 |
| 8 | 31 700 |

**Table 6.1**

Identify the element and explain your answer using the data in **Table 6.1**.

..........................................................................................

..........................................................................................

..........................................................................................

..........................................................................................

.......................................................................................... [2]

7 **Table 7.1** shows the melting point of some elements in Period 3.

| **Name of element (symbol)** | sodium (Na) | magnesium (Mg) | aluminium (Al) | silicon (Si) | phosphorus (P) | sulfur (S) | chlorine (Cl) | argon (Ar) |
|---|---|---|---|---|---|---|---|---|
| **Melting point (°C)** | 98 | 639 | 660 | 1410 | 44 | 113 | −101 | −189 |

**Table 7.1**

**(a)** Explain why the melting point of magnesium is higher than sodium. **[2]**

**(b)** Explain why the melting point of silicon is higher than aluminium. **[3]**

**(c)** Explain why sulfur has a higher melting point than phosphorus. **[3]**

> **Exam tip**
> Be specific with the type and number of bonds and link this to the physical property.

8 Diamond, graphite, and graphene are useful allotropes of carbon.

**(a)** Diamond is hard whereas graphite is a lubricant. Both have very high melting points.

Explain each of these properties using your knowledge of structure and bonding. **[8]**

**(b)** **Figure 8.1** shows the structure of graphene and its bonding.

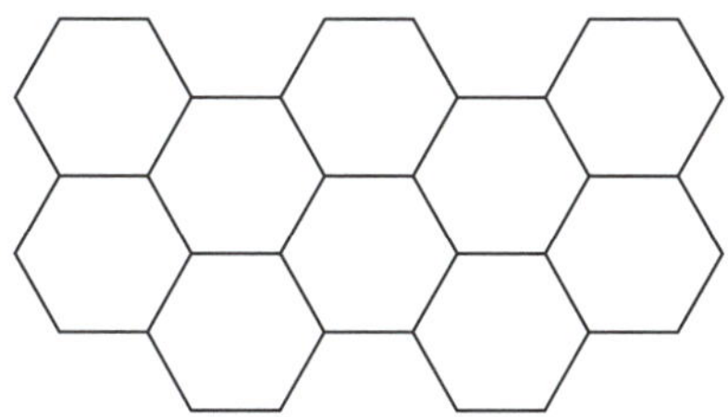

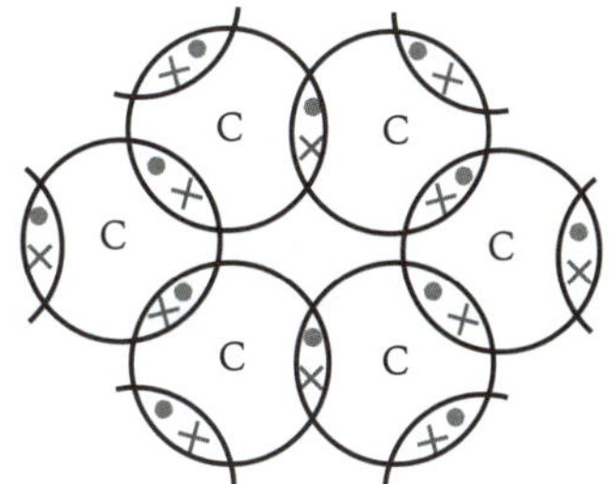


**Figure 8.1**

**(i)** Explain why graphene can conduct electricity. **[2]**

**(ii)** State the type of crystal structure shown in graphene. **[1]**

**(iii)** Explain in terms of bonding why graphene has a high strength per unit mass. **[2]**

> **Exam tip**
> Think about the other allotropes of carbon and apply your knowledge.

# Knowledge

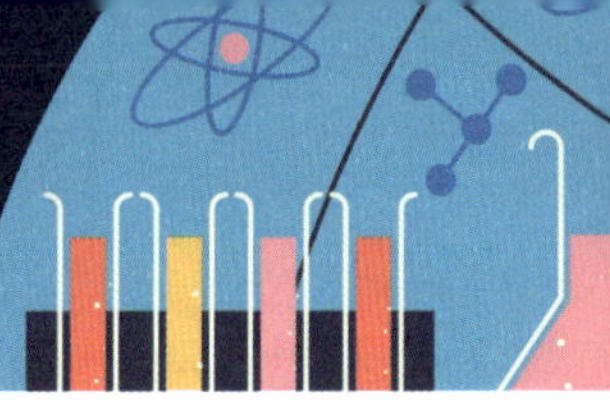

## 6 Group 2 and the Halogens

### Melting points and ionisation energy in Group 2

Group 2 elements are also known as the Alkaline Earth Metals. They *increase* in atomic radius down the group, as the 2 electrons in the s-orbital, get further away from the nucleus.

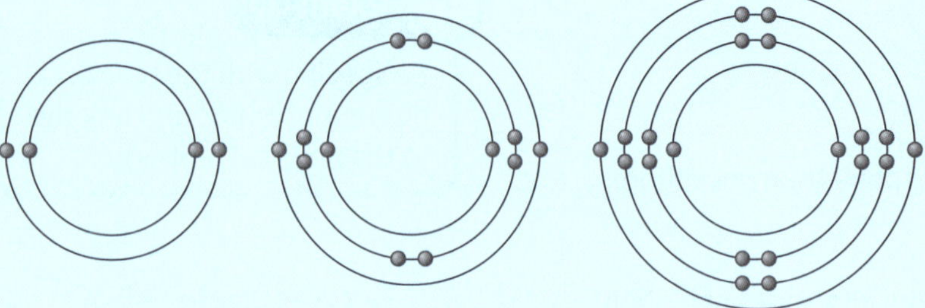

Going down Group 2:

- the first ionisation energy *decreases*
- less energy is needed to remove the first electron.

All Group 2 elements will tend towards losing 2 electrons and forming 2+ ions.

$M(g) \rightarrow M^+(g) + e^-$ first ionisation energy

$M^+(g) \rightarrow M^{2+}(g) + e^-$ second ionisation energy

All elements in Group 2 form metallic structures and so have high melting points. Going down the group:

- the electrons that are available to be delocalised are further away from the nucleus
- the strength of the bond *decreases*
- the melting point *decreases*.

### Trends in Group 2

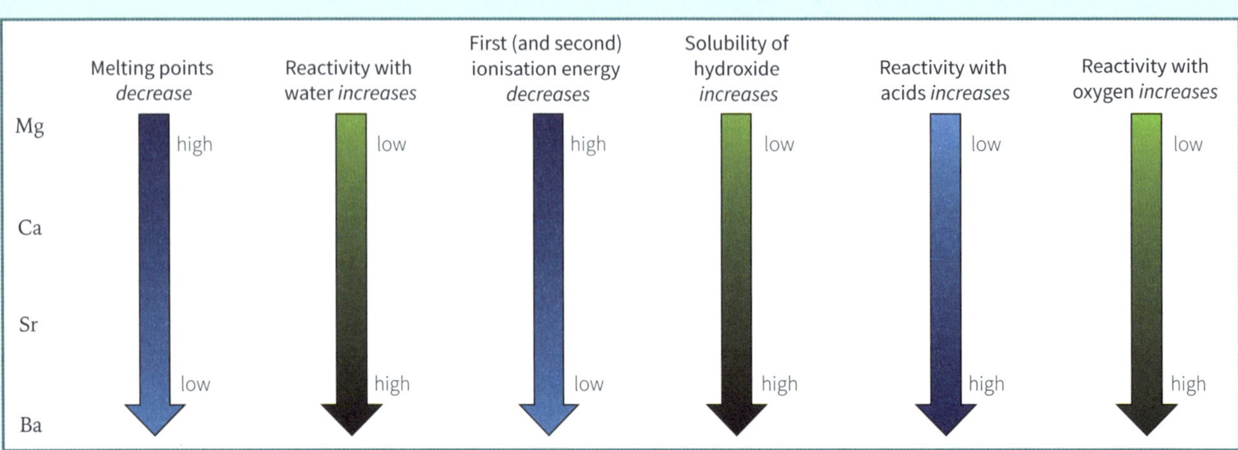

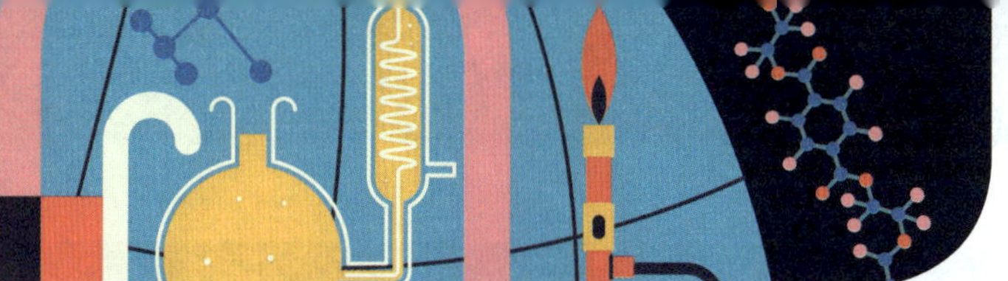



## Redox reactions

Group 2 elements have 2 electrons in their outer shell, so in reactions will tend towards losing these two electrons. This is an **oxidation reaction** and the **oxidation state** will change from 0 to +2.

The Group 2 element will be the **reducing agent** as another element will be reduced after receiving these electrons.

## Reaction with oxygen

The elements in Group 2 will react in an oxidation reaction with oxygen to give a metal oxide. The reactivity *increases* down the group.

$$2X(s) + O_2(g) \rightarrow 2XO(s)$$

## Reaction with acids

Elements in Group 2 will react in a **redox reaction** with dilute acids – forming a salt and hydrogen gas. Reactivity *increases* down the group.

$$X(s) + 2HCl(aq) \rightarrow XCl_2(aq) + H_2(g)$$

## Reaction with water

When Group 2 elements react with water, a hydroxide is formed. The reactivity with water *increases* down the group. X is any Group 2 metal.

$$X(s) + H_2O(l) \rightarrow X(OH)_2(aq) + H_2(g)$$

Magnesium reacts slowly with cold water but vigorously with steam. The other elements react in a similar but increasingly vigorous way.

## Uses of Group 2 compounds as bases

The solubility of hydroxides of Group 2 elements *increases* down the group.

- Magnesium hydroxide, $Mg(OH)_2$, is sparingly soluble and is sold as an indigestion remedy as a suspension rather than a solution.
- Calcium hydroxide, $Ca(OH)_2$, (lime water) is slightly more soluble, and can be used in agriculture to treat acidic soil.
- Barium hydroxide readily forms an alkaline solution.
- Calcium carbonate, $CaCO_3$ can also be used as an indigestion remedy.

## Reactions with water

Oxides of Group 2 elements can react with water to give an alkaline solution.

$$XO(s) + H_2O(l) \rightarrow X^{2+}(aq) + 2OH^-(aq)$$

The alkalinity of the solution *increases* down the group.

## Physical properties of the Halogens

The Halogens, Group 7 of the Periodic Table, tend to be found as diatomic molecules.

These diatomic Halogens have instantaneous dipole–induced dipole interactions. The boiling point of the Halogens *increases* down the group, because:

- the more electrons the atom, has the stronger the London forces between atoms
- more energy is needed to overcome these forces.

The **electronegativity** of the Halogens *decreases* down the group. Fluorine is the most electronegative element in the Periodic Table.

Going down Group 7:

- shielding *increases*, so will attract electrons less,
- electronegativity *decreases*.

## Trends of the Halogens

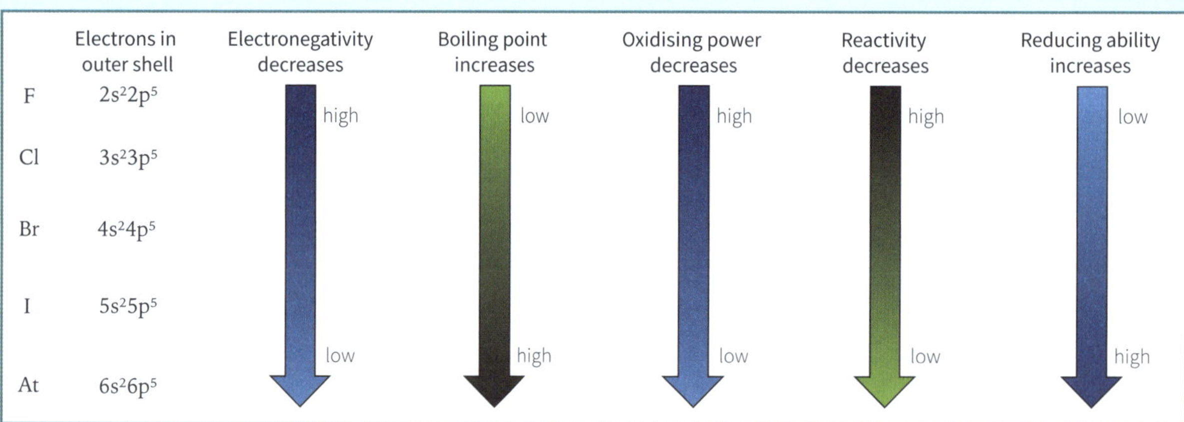


## Reactions of Halogens with halides

The Halogens have 7 electrons on their outer shell, so tend to react by gaining one electron and becoming −1 ions. This gain of electrons is **reduction** and the **oxidation state** of an atom will decrease from 0 to −1. Fluorine will readily gain electrons and is one the most powerful oxidising agents.

Going down Group 7:

- shielding *increases*, so will attract electrons less
- this oxidising power *decreases*.

Metal halides will react with other Halogens, so that if the less reactive Halogen is in the metal halide, it will be displaced by the more reactive Halogen.

Chlorine is more reactive than bromine so will displace it in a reaction.

$$Cl_2(aq) + 2KBr(aq) \rightarrow 2KCl(aq) + Br_2(aq)$$

Chlorine is acting as the oxidising agent and is oxidising bromine.

In a **displacement reaction** there will be a colour change. In water, chlorine will form a pale green solution, bromine will form an orange–brown solution, and iodine will form a brown solution.

Bromine and iodine solutions are both similar in colour, so to tell them apart an organic solvent such as cyclohexane can be added. If iodine is present it will give a deep violet colour.

## Reaction of chlorine with water

Chlorine and fluorine are added to water supplies to make them safer and to give health benefits. The health benefits of this outweighs the toxic effects of chlorine gas.

When chlorine reacts with water it is both reduced and oxidised. This is an example of a **disproportionation reaction**. The oxidation state of chlorine goes from 0 in $Cl_2$ to +1 in HClO and −1 in HCl. Chloride ions and chlorate ions are produced. Chloric(I) acid can be used as bleach or to kill bacteria due to its oxidising properties.

$$Cl_2(g) + H_2O(l) \rightleftharpoons HClO(aq) + HCl(aq)$$

Chlorine can be used to treat swimming pools to keep them clean. However, the chlorine is rapidly lost. In the presence of sunlight, chlorine can also react with water to form oxygen:

$$2Cl_2(g) + 2H_2O(l) \rightleftharpoons 4HCl(aq) + O_2(g)$$

## Reaction of sodium chlorate and water

Alternatively sodium chlorate(I) can be added to form chloric(I) acid. This is a reversible reaction so the position of the equilibrium needs to be carefully controlled.

$$NaClO(s) + H_2O(l) \rightleftharpoons Na^+(aq) + OH^-(aq) + HClO(aq)$$

Chlorine reacts with cold dilute aqueous NaOH in a disproportionation reaction to form a range of products. Sodium chlorate (I) is an oxidising agent and the main ingredient in household bleach.

$$Cl_2(g) + 2NaOH(aq) \rightleftharpoons NaClO(aq) + NaCl(aq) + H_2O(l)$$

## Reducing powers of halide ions

The larger the ion, the more shielding between the outer electrons and the nucleus, making it more likely for an electron to be lost. The electron can then reduce a substance. The halide ion acts as a reducing agent.

- Reducing power *increases* down the Halogen group.
- The halide ions lose electrons and become Halogens.

Halogens are *oxidising agents*.

Halide ions are *reducing agents*.

## Tests for halide ions

Most metal halides (except for metal fluorides) can be identified by using silver ions.

1 *Add acidified solution (nitric acid) to an aqueous solution of a silver halide*. Nitric acid is used to remove any carbonate ions as carbon dioxide, because any contaminating carbonate ions will react with the silver ions to form precipitate of silver carbonate or silver hydroxide.

2 *Add aqueous silver nitrate and a precipitate will form*. The colour of any precipitate will give an indication of the halide ion present.

3 *Add aqueous ammonia*. The colours of the silver halides can be very hard to distinguish from each other; the addition of ammonia solution tests the solubility of the precipitate, which will help identify the halide.

| **Halide** | silver chloride | silver bromide | silver iodide |
|---|---|---|---|
| **Colour of precipitate** | white | cream | pale yellow |
| **Does it dissolve in ammonia?** | in dilute ammonia | in concentrated ammonia | does not dissolve |

# Retrieval

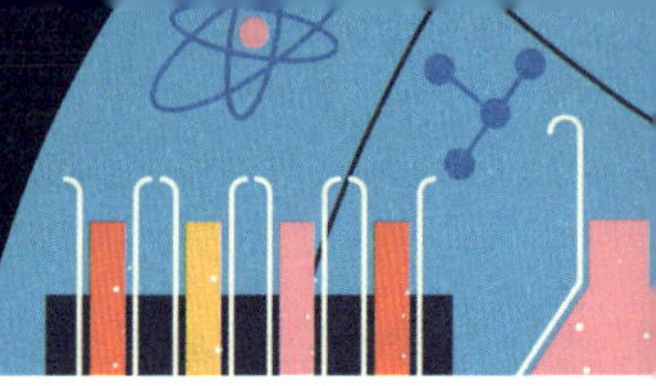

Learn the answers to the questions below, then cover the answers column with a piece of paper and write as many as you can. Check and repeat.

| | Questions | Answers |
|---|---|---|
| 1 | How does the reactivity of Group 2 elements with water change down the group? | it decreases |
| 2 | How does the electronegativity of Halogens change down the group? | it decreases |
| 3 | How does atomic radius change down a group? | it increases |
| 4 | When testing for halide ions, why must the solution first be acidified? | to remove any carbonate ions |
| 5 | How does chlorine react with water in the presence of sunlight? | $Cl_2(g) + H_2O(l) \rightleftharpoons 4HCl(aq) + O_2(g)$ |
| 6 | How does the oxidising power of the Halogens change down the group? | it decreases |
| 7 | What is a disproportionation reaction? | where the same element is both oxidised and reduced |
| 8 | One of the products of the reaction between chlorine and sodium hydroxide is bleach. What is the formula of this product? | $NaClO$ |
| 9 | Which metal halide does not react with silver ions to form a precipitate? | metal fluorides |
| 10 | How soluble is magnesium hydroxide? | sparingly |
| 11 | What is the test for halide ions? | addition of silver nitrate to acidified sample, in a positive test a precipitate will form |
| 12 | When will one Halogen displace another? | if it is more reactive |
| 13 | What is the trend in first ionisation energy down a group? | it decreases |
| 14 | What is the electronic structure of magnesium? | $[Ne]\ 3s^2$ |
| 15 | What is the colour of the precipitate of silver bromide? | cream |
| 16 | What electrons are in the outer shell of astatine? | $6s^26p^5$ |
| 17 | What halides are added to water for health benefits? | chlorides and fluorides |
| 18 | How does the melting point change down Group 2? | it decreases |
| 19 | How does the boiling point change down the Halogen group? | it increases |

Put paper here

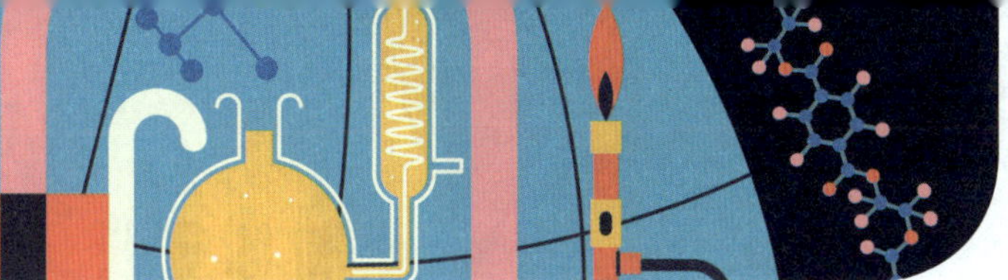



# Maths skills

Practise your maths skills using the worked example and practice questions below.

## Finding the arithmetic mean

Finding the mean is not a new concept. At GCSE, you will remember how to find a mean of a set of numbers: 'add them all together and divide by the number you've got'. At A level it is a little more complicated.

Firstly, not all the data should be included. For example, when calculating titres you do *not* include all your data in the calculation. You ignore anomalous results and the rough titre.

Secondly, calculating a *weighted mean* requires you to multiply the value by the percentage or fraction of that value first.

## Worked example

**Questions**

**a** Calculate the average titre from the following results:

| Trial | Titre (cm$^3$) |
|---|---|
| rough | 24.00 |
| 1$^{st}$ | 25.50 |
| 2$^{nd}$ | 25.50 |
| 3$^{rd}$ | 27.50 |
| 4$^{th}$ | 25.50 |

**b** Calculate the relative atomic mass of Ga from this sample of isotopes. Give your answer to 3 significant figures.

| Isotope | Relative abundance (%) |
|---|---|
| $^{69}$Ga | 60.11 |
| $^{71}$Ga | 39.89 |

**Answers**

**a** When calculating the mean, you ignore the rough and the anomalous results. So, ignore the rough and the 3$^{rd}$ result. Therefore, the average is:

$$\frac{25.50 + 25.50 + 25.\ 50}{3} = 25.5\,\text{cm}^3$$

**b** You need to multiply the mass by the percentage and then divide by 100.

$$\frac{(69 \times 60.11) + (71 \times 39.89)}{100} = 69.8$$

## Practice

**1** Calculate the average titre from the following results.

| Trial | Titre (cm$^3$) |
|---|---|
| rough | 32.50 |
| 1$^{st}$ | 31.30 |
| 2$^{nd}$ | 32.00 |
| 3$^{rd}$ | 32.55 |
| 4$^{th}$ | 32.50 |

**2** Calculate the relative atomic mass of Ge from the following isotopic data. Give your answer to 3 significant figures.

| Isotope | Relative abundance (%) |
|---|---|
| $^{70}$Ge | 20.52 |
| $^{72}$Ge | 27.45 |
| $^{73}$Ge | 7.76 |
| $^{74}$Ge | 36.52 |
| $^{76}$Ge | 7.75 |

**3** Calculate the average titre from the following results.

| Trial | Titre (cm$^3$) |
|---|---|
| rough | 12.00 |
| 1$^{st}$ | 14.35 |
| 2$^{nd}$ | 14.45 |
| 3$^{rd}$ | 14.50 |
| 4$^{th}$ | 12.20 |

# Practice

## Exam-style questions

**1** This question is about atomic structure.

**(a)** Give the full ground state electron configuration for $Al^{3+}$ and $Cl^{-}$.

................................................................................

................................................................................

................................................................................ **[2]**

**(b)** State why both aluminium and chlorine are considered p-block elements.

................................................................................

................................................................................ **[1]**

**(c)** Aluminium and silicon are both elements in Period 3.

State which has the larger atomic radius.

Explain your answer.

................................................................................

................................................................................

................................................................................

................................................................................ **[3]**

**(d)** A student had two unlabelled solutions. One was potassium bromide and one was potassium chloride.

Outline a series of practical steps and the observations that would allow you to determine the identity of these two unlabelled solutions.

................................................................................

................................................................................

................................................................................

................................................................................

................................................................................

................................................................................

................................................................................ **[6]**

**Synoptic links**

2.2.1 3.1.1

**Exam tip**

Make sure you are specific with the reagents you will use and the observations at each stage.

**2** Fluorine is a highly reactive element in Group 7.

**(a)** Explain the trend in first ionisation energy as you move down Group 7.

................................................................................

................................................................................

................................................................................

................................................................................ **[3]**

**(b)** Explain why iodine has a higher melting point than fluorine.

................................................................................

................................................................................

................................................................................ **[2]**

**Synoptic links**

2.2.2 2.1.2
2.1.3 3.1.1

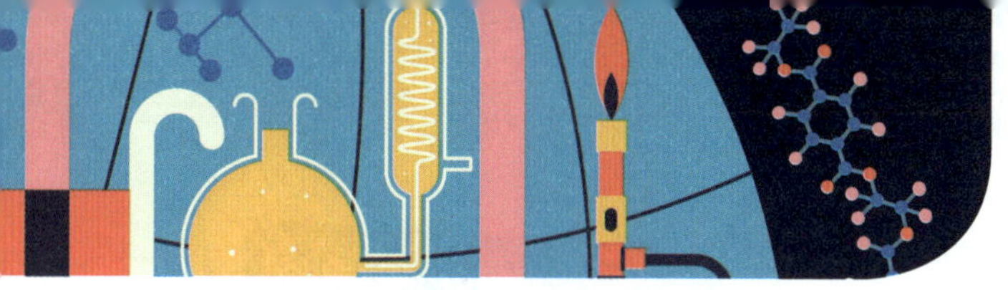



**(c)** Fluorine gas is labelled as toxic and corrosive, yet compounds containing fluoride ions are readily added to toothpaste. Explain why.

.......................................................................................................

.......................................................................................................

.......................................................................................................**[2]**

**(d)** Fluorine is reactive enough to react with xenon to form the solid xenon tetrafluoride ($XeF_4$).

Give a balanced symbol equation for the formation of xenon tetrafluoride from xenon and fluorine in their standard states.

.......................................................................................................

.......................................................................................................**[1]**

**(e)** Fluorine also reacts with krypton to form krypton difluoride ($KrF_2$).

Give the shapes of $XeF_4$ and $KrF_2$, label the bond angles, and name the shapes of the molecules.

**Exam tip**

With all 'shapes of molecules' questions, remember to identify the number of bonding pairs and lone pairs before you decide on a shape.

**[6]**

**(f)** Xenon also can form a similar difluoride, $XeF_2$, which is a solid at room temperature.

Determine the type of intermolecular force in $XeF_2$ crystals.

.......................................................................................................

.......................................................................................................**[1]**

**(g)** Calculate the mass of fluorine needed to make 4.5 kg of $XeF_2$.

mass = .................... **[3]**

3 This question is about the characteristics of Period 3 and Group 7 elements.

**(a)** State and explain which Period 3 element has the highest melting point.

..........................................................................................

..........................................................................................

..........................................................................................

.......................................................................................... **[3]**

**(b)** State and explain which Period 3 element has the highest first ionisation energy.

..........................................................................................

..........................................................................................

..........................................................................................

.......................................................................................... **[3]**

**(c)** State and explain the trend in boiling points of the elements in Group 7 from chlorine to iodine.

..........................................................................................

..........................................................................................

..........................................................................................

.......................................................................................... **[3]**

**(d)** State which element in Group 7 is the most electronegative. Give a reason for your answer.

..........................................................................................

..........................................................................................

..........................................................................................

.......................................................................................... **[3]**

**Synoptic links**

2.2.2 3.1.1

**Exam tip**

All questions about physical changes are bonding questions. Explicitly state the bonding of each substance.

4 This question is about chlorine.

**(a) (i)** Give the balanced ionic equation for the reaction between chlorine and sodium iodide.

**[1]**

**(ii)** Name the species in the reaction written in **4(a)(i)** that is the strongest reducing agent.

..........................................................................................

.......................................................................................... **[1]**

**Synoptic links**

2.1.2 3.2.3

**Exam tip**

Reducing agents are always species that have been oxidised.

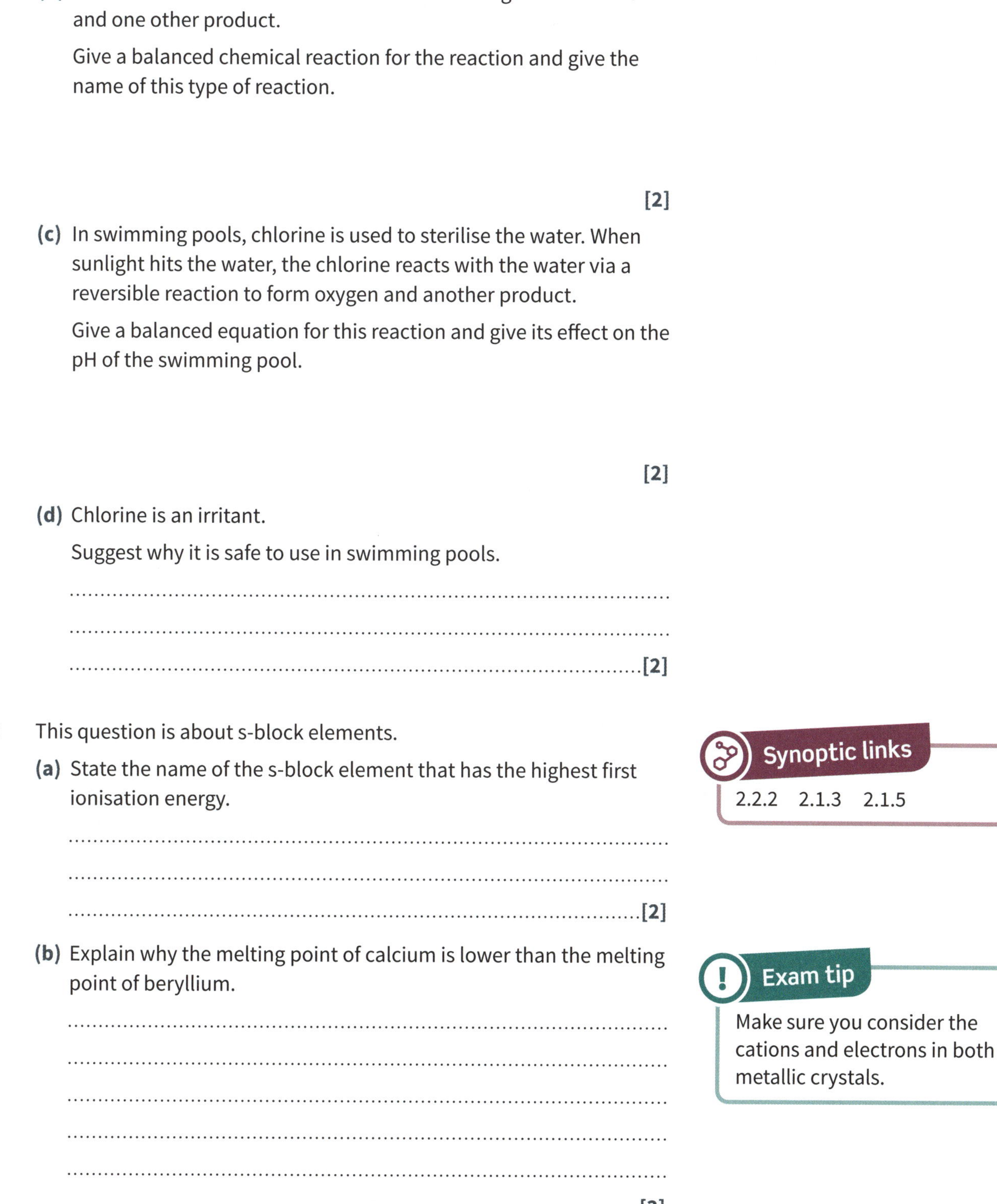

**(b)** Chlorine and water react in the absence of sunlight to form HCl and one other product.

Give a balanced chemical reaction for the reaction and give the name of this type of reaction.

**[2]**

**(c)** In swimming pools, chlorine is used to sterilise the water. When sunlight hits the water, the chlorine reacts with the water via a reversible reaction to form oxygen and another product.

Give a balanced equation for this reaction and give its effect on the pH of the swimming pool.

**[2]**

**(d)** Chlorine is an irritant.

Suggest why it is safe to use in swimming pools.

..........................................................................................

..........................................................................................

..........................................................................................**[2]**

**5** This question is about s-block elements.

**(a)** State the name of the s-block element that has the highest first ionisation energy.

..........................................................................................

..........................................................................................

..........................................................................................**[2]**

**(b)** Explain why the melting point of calcium is lower than the melting point of beryllium.

..........................................................................................

..........................................................................................

..........................................................................................

..........................................................................................

..........................................................................................

..........................................................................................**[3]**

**(c)** Give the ionic formula for the least soluble hydroxide of Group 2 from Mg to Ba.

..........................................................................................

..........................................................................................**[1]**

**Synoptic links**

2.2.2 2.1.3 2.1.5

**Exam tip**

Make sure you consider the cations and electrons in both metallic crystals.

**(d)** A student added 11 $cm^3$ of 0.25 mol $dm^{-3}$ barium chloride solution to 6 $cm^3$ of 0.35 mol $dm^{-3}$ sodium sulfate solution.

The student filtered off the precipitate and collected the filtrate.

**(i)** Give an ionic equation for the formation of the precipitate.

**[1]**

**(ii)** Illustrate by calculation which reagent is in excess.

reagent in excess = .................... **[2]**

**(iii)** Calculate the total volume of the other reagent that should be used by the student so that the excess reagent would be completely reacted.

volume = .................... $cm^3$ **[2]**

**6** This question is about Group 7 chemistry. Seawater is a major source of iodine.

The iodine extracted from seawater is impure. It is purified in a two-stage process.

**Stage 1** $I_2 + 2H_2O + SO_2 \rightarrow 2HI + H_2SO_4$

**Stage 2** $2HI + Cl_2 \rightarrow I_2 + 2HCl$

**(a)** State the role of $Cl_2$ in **Stage 2** of the process.

..........................................................................................

..........................................................................................

..........................................................................................

..........................................................................................**[1]**

**(b)** Calculate the volume, in $dm^3$, of sulfur dioxide needed to purify 1 kg of iodine under standard conditions.

The gas constant $R$ is 8.31 J $mol^{-1}$.

volume = .................... $dm^3$ **[5]**

Synoptic links

2.1.3 2.1.5 3.1.4

**Exam tip**

Any time you see $R$ stated, it is a clue to use the ideal gas equation. The first two marks are always for stating the formula and converting units, so even if you are not sure what to do next, you can score two marks.

7 This question is about the reactivity of Group 2.

**(a)** **(i)** Give an ionic equation, including state symbols, for the reaction between calcium and water.

State the role of water.

............................................................................................

............................................................................................

............................................................................................

............................................................................................

............................................................................................

............................................................................................

............................................................................................ **[3]**

**(ii)** State a possible pH for the reaction outlined in **7(a)(i)**.

............................................................................................

............................................................................................ **[1]**

**(b)** Outline a simple test tube experiment a student could perform to distinguish between unlabelled samples of solid barium sulfate and barium nitrate.

Give the ionic equation for the decisive reaction.

............................................................................................

............................................................................................

............................................................................................

............................................................................................

............................................................................................

............................................................................................

............................................................................................

............................................................................................

............................................................................................ **[4]**

**(c)** State the trend in solubility of Group 2 hydroxides as you descend the group.

............................................................................................

............................................................................................

............................................................................................

............................................................................................ **[1]**

**(d)** Hydrochloric acid is found in the stomach of humans.

Give a balanced equation that demonstrates the action of a Group 2 hydroxide as an antacid.

............................................................................................

............................................................................................

............................................................................................

............................................................................................ **[1]**

**Exam tip**

When asked for the role of an ion or molecule, you should consider whether it is acid or base, oxidising or reducing agent, or a catalyst.

**Exam tip**

It is important to state the observations for each reactant even if this is 'no visible change'.

# Knowledge

## 7 Qualitative analysis

### Tests for anions

**Carbonate ions**, $CO_3^{2-}$, can be identified after addition of dilute nitric acid.

If the carbonate ion is present, bubbles of carbon dioxide will be formed:

$$Na_2CO_3(aq) + 2HNO_3(aq) \rightarrow 2NaNO_3(aq) + CO_2(g) + H_2O(l)$$

The carbon dioxide can be identified by limewater turning cloudy.

The cloudiness is the solid calcium carbonate precipitating out:

$$CO_2(g) + Ca(OH)_2(aq) \rightarrow CaCO_3(s) + H_2O(l)$$

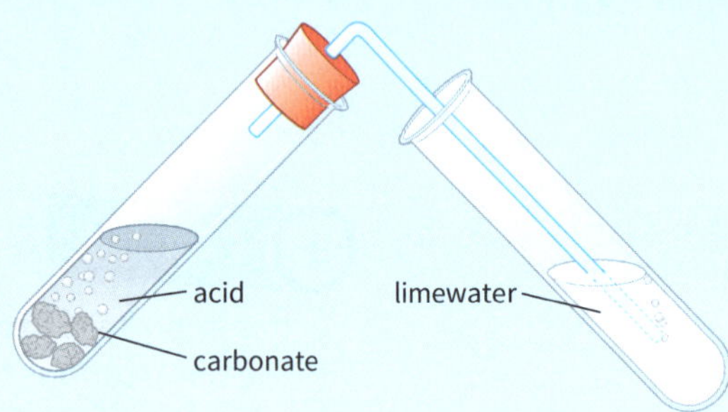


The test for **sulfate ions**, $SO_4^{2-}$, involves the addition of barium ions to give a white precipitate of $BaSO_4$.

This can be done using barium chloride or barium nitrate. Barium sulfate is one of the few sulfates that is insoluble in water:

$$Ba^{2+}(aq) + SO_4^{2-}(aq) \rightarrow BaSO_4(s)$$

The **halide ions**, $Cl^-$, $Br^-$, and $I^-$, can be identified by the addition of silver nitrate.

Chloride ions will give a *white precipitate* (AgCl) that is soluble in dilute ammonia; bromide ions will give a *cream precipitate* (AgBr); and iodide ions will give a *yellow precipitate* (AgI):

$$Ag^+(aq) + X^- \rightarrow AgX(s)$$

As these colours are so similar it can be hard to differentiate them individually.

It can be a good idea to have known reference samples for comparison.

The solubilities of the halide precipitates can be tested by the addition of ammonia.

| Halide | Colour of precipitate | Solubility in $NH_3$ |
|---|---|---|
| chloride, $Cl^-$ | white | soluble in dilute $NH_3(aq)$ |
| bromide, $Br^-$ | cream | soluble in concentrated $NH_3(aq)$ |
| iodide, $I^-$ | yellow | insoluble in dilute and concentrated $NH_3(aq)$ |

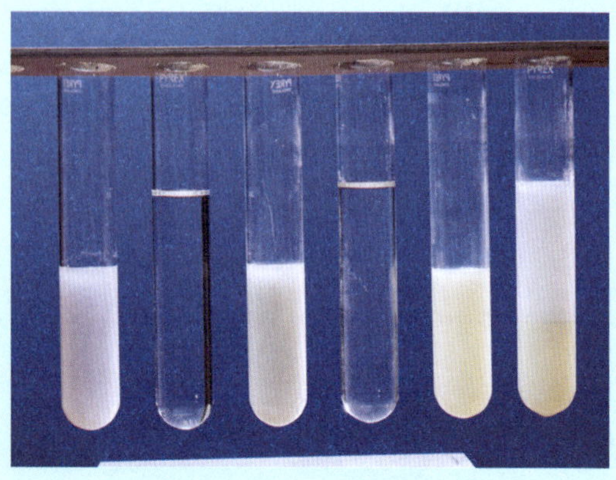

## Tests for ammonium ions

Ammonium ions, $NH_4^+$, can be identified by adding hydroxide ions (e.g. NaOH) and heating the solution. A positive result will release ammonia gas, which you will be able to smell and can be tested for with damp red litmus paper turning blue.

$$NH_4^+(aq) + OH^-(aq) \rightarrow NH_3(g) + H_2O(l)$$

## Sequence of tests

To analyse a mixture of chemicals you must carry out tests *in the correct sequence* with the same solution.

- Tests for both halide ions and sulfate ions should be carried out in acidified conditions to ensure removal of carbonate ions.
- Barium carbonate is a white precipitate so all carbonate ions need to be removed before barium is added or a false positive will be seen.
- The test for sulfate ions should be done before the test for halide ions, and barium nitrate should be used.
- The silver from the silver nitrate in the halide test will form silver sulfate, which is insoluble and gives a false positive for the sulfate test, whereas the use of barium chloride may introduce a false positive in the halide test.

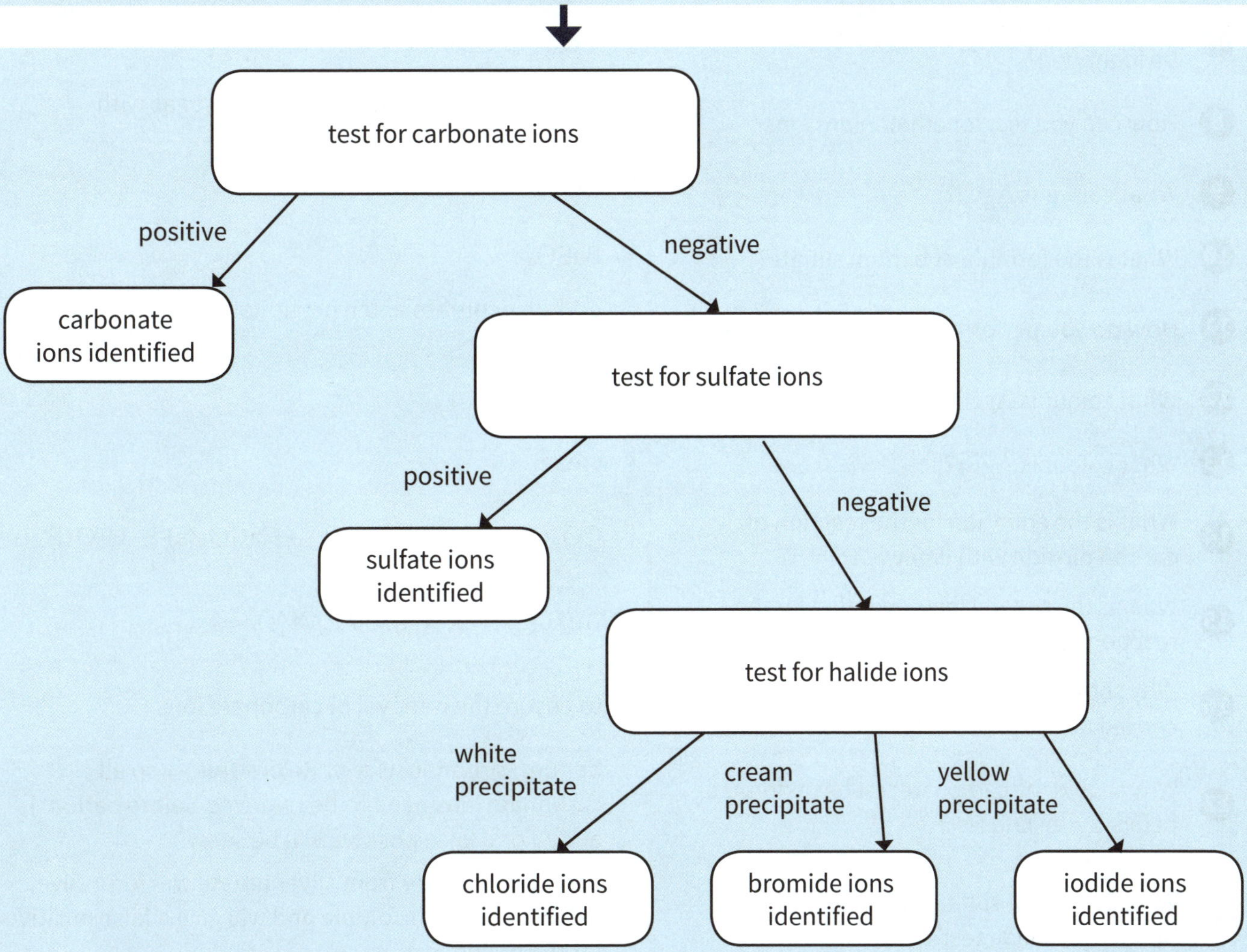

test for carbonate ions

- positive → carbonate ions identified
- negative → test for sulfate ions
  - positive → sulfate ions identified
  - negative → test for halide ions
    - white precipitate → chloride ions identified
    - cream precipitate → bromide ions identified
    - yellow precipitate → iodide ions identified

# Retrieval

Learn the answers to the questions below, then cover the answers column with a piece of paper and write as many as you can. Check and repeat.

| | Questions | | Answers |
|---|---|---|---|
| 1 | How can you test for carbonate ions? | Put paper here | add acid then bubble the gas produced through limewater |
| 2 | Which silver halide does not dissolve in concentrated ammonia? | | AgI |
| 3 | What is the test for ammonia gas? | | it turns damp red litmus paper blue |
| 4 | What is the correct order of tests for identifying cations in a solution? | Put paper here | carbonate, sulfate, halide |
| 5 | What is a positive result for sulfate ions? | | a white precipitate on addition of barium nitrate or barium chloride |
| 6 | Why should you *not* use barium chloride to test for the sulfate ion if you are then going to test for a halide ion? | | because the chloride ions will be identified in the halide test |
| 7 | How can bubbles of carbon dioxide be identified? | Put paper here | limewater becomes cloudy |
| 8 | How can you test for ammonium ions? | | heating with hydroxide ions and test gas with litmus paper |
| 9 | What colour is AgCl? | Put paper here | white |
| 10 | What is the formula of barium sulfate? | | $BaSO_4$ |
| 11 | How do you perform a test for halide ions? | | add silver nitrate and a precipitate will form for a positive test |
| 12 | What colour is AgI? | Put paper here | yellow |
| 13 | What colour is AgBr? | | cream |
| 14 | What is the equation for the reaction of carbon dioxide with limewater? | | $CO_2(g) + Ca(OH)_2(aq) \rightarrow CaCO_3(s) + H_2O(l)$ |
| 15 | What is the general ionic equation for the reaction of sliver nitrate with halide salts? | Put paper here | $Ag^+(aq) + X^-(aq) \rightarrow AgX(s)$ |
| 16 | Why should the test for sulfate ions be carried out in acidified conditions? | | to ensure the removal of carbonate ions |
| 17 | Why should the carbonate test be performed before the sulfate test? | Put paper here | barium carbonate is a white precipitate so all carbonate ions need to be removed before barium is added or a false positive will be seen |
| 18 | Why should the sulfate test be performed before the halide test? | | because the silver from silver nitrate will form silver sulfate which is insoluble and will give a false positive for the halide test |

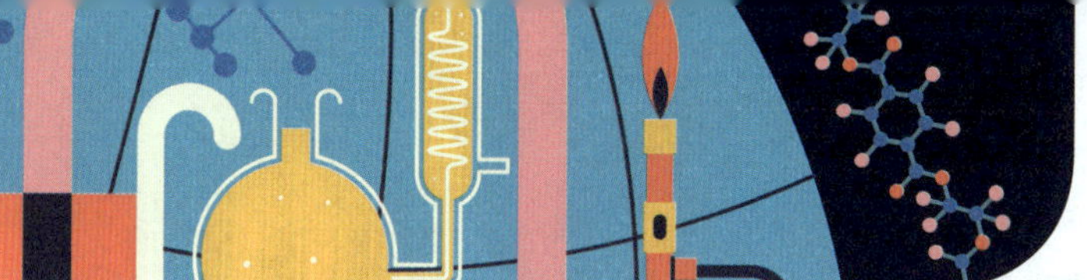



| | | |
|---|---|---|
| 19 | How could you determine the identity of a precipitate in the halide test? | you could compare the the precipitate to known reference samples and test the solubility upon addition of ammonia |
| 20 | What is the ionic equation for the reaction performed in the test for ammonium ions? | $NH_4^+(aq) + OH^-(aq) \rightarrow NH_3(g) + H_2O(l)$ |

Put paper here

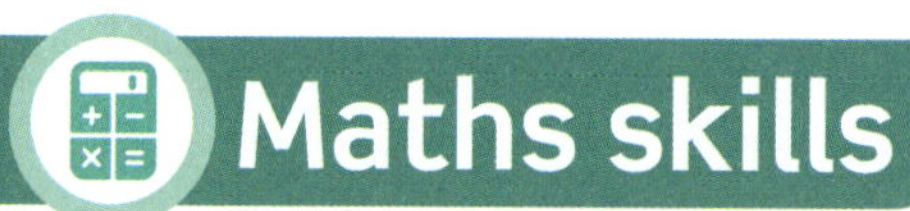

## Maths skills

Practise your maths skills using the worked example and practice questions below.

### Identifying and determining uncertainties

Every measurement has an uncertainty associated with it. For example, a 200 cm$^3$ volumetric flask might have an uncertainty of 0.15 cm$^3$: this is the *absolute* uncertainty. This means if you measure the volume of a solution in that flask, its actual volume is somewhere between 199.85 cm$^3$ and 200.15 cm$^3$. Unless stated otherwise, the uncertainty is the resolution of the equipment you are using.

Percentage uncertainty is the ratio of the absolute uncertainty to the measured value. So, in the above example the percentage uncertainty is:

$$\frac{0.15}{200} \times 100 = 0.075\%$$

When measurements are added or subtracted the *absolute* uncertainty of the measurement is the sum of the absolute uncertainties of each measurement.

Some measurements involve measuring a change, for example, $\Delta T$ (temperature change). In this case, the uncertainty is multiplied by two because there is uncertainty in both the initial and the final measurement.

### Worked example

**Question**

A student measures a temperature change from 10.0 °C to 24.0 °C using a thermometer that is accurate to the nearest 0.5 °C. What is the percentage uncertainty in their measurement?

**Answer**

The temperature change is:

24.0 – 10.0 = 14.0 °C

The overall uncertainty is 2 × ±0.5. This is because there is a ± ±0.5 °C error at 10.0 °C and 24.0 °C.

Therefore, the overall percentage error is:

$$\frac{2 \times 0.5}{14.0} \times 100 = 7.1\%$$

Remember, you only multiply the uncertainty by two if there is a change.

### Practice

1 What is the percentage uncertainty in a measurement of mass of 12 g on a balance that is accurate to 0.1 g?

2 What is the percentage uncertainty in a measurement of mass of 45 g on a balance that is accurate to 0.1 g?

3 A student is using a burette to measure a volume of a liquid. The initial burette reading is 12.50 cm$^3$ and the final burette reading is 36.75 cm$^3$. The uncertainty in each measurement is 0.05 cm$^3$. What is the percentage uncertainty in the volume of liquid?

4 In an exothermic reaction the temperature increases from 22.5 °C to 45.0 °C. The uncertainty on the thermometer is 0.5 °C. What is the percentage uncertainty for the temperature change?

# Practice

## Exam-style questions

1 Strontium is a Group 2 metal that was commonly used in the screens of the first colour televisions.

**(a)** Give an equation for the third ionisation energy of strontium.

................................................................................

................................................................................ **[1]**

**(b)** Both strontium carbonate and strontium sulfate are white solids that are insoluble in water. Strontium carbonate reacts with hydrochloric acid to produce a solution of strontium chloride. Strontium sulfate does not react with hydrochloric acid.

Describe how you would obtain strontium sulfate from a mixture of strontium carbonate and strontium sulfate.

................................................................................

................................................................................

................................................................................ **[2]**

**(c)** Describe how you would obtain a pure sample of magnesium sulfate from a mixture of strontium sulfate and magnesium sulfate.

................................................................................

................................................................................

................................................................................

................................................................................ **[3]**

**(d)** A student has two unlabelled containers; one is for strontium sulfate and the other is for strontium carbonate. Both are white powders that are insoluble in water.

Describe the simple chemical test that could be performed to determine the identity of each sample.

State clearly the observation you would expect.

................................................................................

................................................................................

................................................................................ **[2]**

**Synoptic links**

3.1.1 3.1.2

**Exam tip**

To ensure you score full marks, clearly state which will react and which will not.

2 Sodium is a Group 1 metal that readily forms ionic compounds.

**(a)** Explain, in terms of the structure and bonding involved, why sodium metal has a lower melting point than sodium chloride.

................................................................................

................................................................................

................................................................................

................................................................................

................................................................................

................................................................................ **[5]**

**Synoptic link**

2.2.2

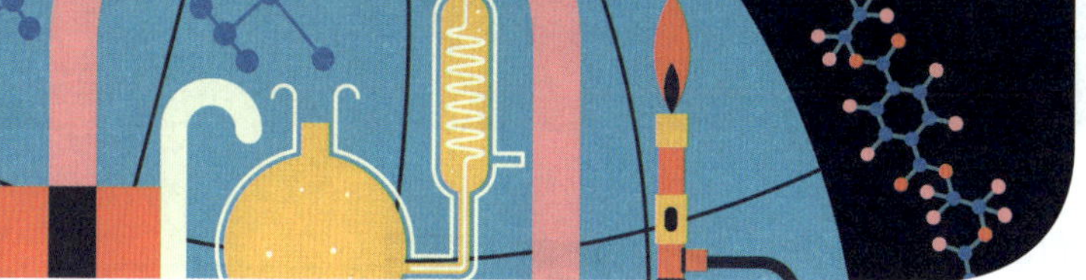



**(b)** Separate unlabelled solid samples of three anhydrous sodium compounds are provided for a student to identify.

These compounds are known to be sodium carbonate, sodium fluoride, and sodium chloride, but it is not known which sample is which.

Outline a logical sequence of test-tube reactions that the student could carry out to identify each of these compounds.

Include the observations the student would expect to make.

Give equations, including state symbols, for any reactions that would take place.

..................................................................................................

..................................................................................................

..................................................................................................

..................................................................................................

..................................................................................................

..................................................................................................

..................................................................................................

..................................................................................................

.......................................................................................... **[6]**

**Exam tip**

For extended response questions, it is vital you include all areas in a succinct way. Performing extra tests will lose you marks.

**3** Group 2 metal compounds are commonly used in medicine and medical procedures.

**(a)** Barium chloride is toxic. If a person ingests barium chloride, the doctor will administer a solution of magnesium sulfate for the person to drink.

Suggest why drinking magnesium sulfate will prevent the toxic effects of barium chloride.

..................................................................................................

.......................................................................................... **[1]**

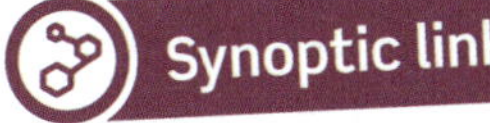

**Synoptic link**

2.1.3

**(b)** Medicines for the treatment of nervous disorders often contain calcium bromide.

Outline a simple practical procedure that could prove the presence of bromide ions in a medicine.

Include any observations you would expect to make.

..................................................................................................

..................................................................................................

..................................................................................................

..................................................................................................

..................................................................................................

.......................................................................................... **[4]**

**Exam tip**

Make sure you include the second step involving ammonia and your observation.

**(c)** A drug manufacturer wants to produce calcium bromide on a large scale. They have a process tested that gives a 93% yield.

Calculate the mass of bromine they will need to use to create 500 kg of calcium bromide.

**Exam tip**

Make sure you pay attention to the yield!

mass of bromine = .................... **[4]**

**4** Group 7 is often known as the Halogens.

**(a)** Define electronegativity and identify the element in Group 7 that has the highest electronegativity.

**Synoptic links**

2.2.2

..........................................................................................

..........................................................................................

..........................................................................................

..........................................................................................

..........................................................................................

.......................................................................................... **[2]**

**(b)** Explain why iodine is a solid at room temperature but bromine is a liquid.

..........................................................................................

..........................................................................................

..........................................................................................

..........................................................................................

..........................................................................................

.......................................................................................... **[2]**

**(c) (i)** The presence of halide ions is determined using nitric acid, silver nitrate, and ammonia.

Explain the purpose of nitric acid in the test.

**Exam tip**

Be specific about what is being removed and why.

..........................................................................................

..........................................................................................

..........................................................................................

.......................................................................................... **[2]**

**(ii)** Explain how ammonia can be used to confirm the identity of a halide ion as an iodide ion.

..........................................................................................

..........................................................................................

..........................................................................................

.......................................................................................... **[2]**

5 Ammonium sulfate is an ionic salt that is used to reduce the pH of soil.

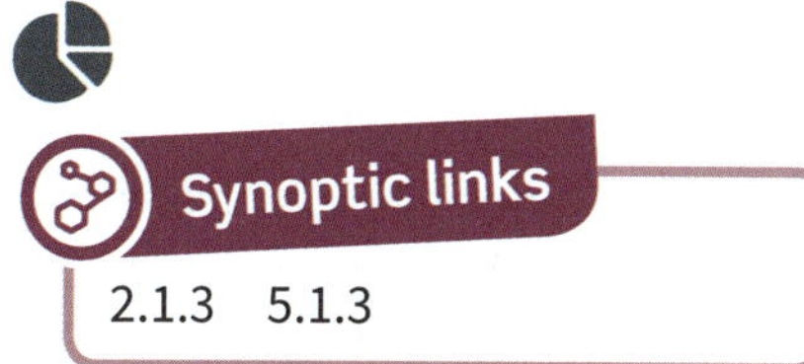


**(a)** Give an equation to show how ammonium ions can behave as a Brønsted–Lowry acid.

**[1]**

**(b) (i)** Describe the simple laboratory test that can be performed to confirm the existence of ammonium ions in a soil sample.

..........................................................................................

..........................................................................................

..........................................................................................

..........................................................................................

..........................................................................................

.......................................................................................... **[3]**

**Exam tip**

Think carefully about the role of an acid and **5 (b)** will become a lot easier.

**(ii)** Give the balanced equation for the reaction that you carried out in **5(b)(i)**.

**[1]**

6 This question is about chlorine, sodium chloride, and sodium bromide.

**(a)** Determine the volume, in $dm^3$, that 57 g of chlorine gas will occupy at 100 °C and 100 kPa.

The gas constant is 8.31 J $mol^{-1}$ $K^{-1}$.

volume = ...................... $dm^3$ **[5]**

**(b)** A colourless solution contains a mixture of sodium chloride and sodium bromide.

Using aqueous silver nitrate and any other reagents of your choice, develop a procedure to prepare a pure sample of silver bromide from this mixture.

Explain each step in the procedure and illustrate your explanations with equations, where appropriate.

.....................................................................................................

.....................................................................................................

.....................................................................................................

.....................................................................................................

.....................................................................................................

.....................................................................................................

.....................................................................................................

.....................................................................................................

.....................................................................................................

.....................................................................................................

.....................................................................................................

.....................................................................................................

.....................................................................................................

.....................................................................................................

.....................................................................................................

.....................................................................................................

............................................................................................... **[6]**

> **Exam tip**
>
> Be systematic! For each step, state the reagent, give the observation, and write the equation.

**(c)** Suggest an ionic equation for the reaction between chlorine and cold dilute sodium hydroxide solution.

Give the oxidation state of chlorine in each of the chlorine-containing ions formed.

.....................................................................................................

.....................................................................................................

.....................................................................................................

.....................................................................................................

.....................................................................................................

.....................................................................................................

.....................................................................................................

.....................................................................................................

............................................................................................... **[2]**

7 The solubility of barium hydroxide was determined experimentally using the following method:

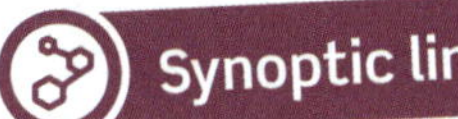

**Synoptic link**

2.1.3

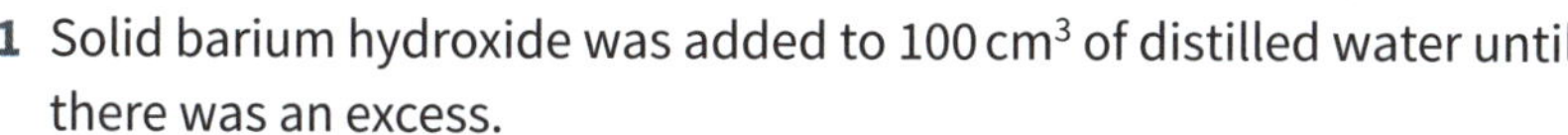

1. Solid barium hydroxide was added to 100 cm$^3$ of distilled water until there was an excess.
2. The mixture was filtered.
3. The filtrate had an excess of sulfuric acid added.
4. The barium sulfate was filtered out, washed with cold water, and dried.
5. The mass of barium sulfate was then recorded.

**(a)** Explain the purpose of filtering the solution before adding the sulfuric acid.

..........................................................................................

..........................................................................................

..........................................................................................

..........................................................................................

.................................................................................... **[1]**

**(b)** Explain why the barium sulfate was washed before being dried.

..........................................................................................

..........................................................................................

..........................................................................................

..........................................................................................

.................................................................................... **[1]**

**(c)** Give an equation to show the reaction between barium hydroxide and sulfuric acid.

**[1]**

**(d)** At the end of the experiment, the mass of barium sulfate produced was 4.31 g.

Calculate the mass of barium hydroxide that would dissolve in 1 dm$^3$ of distilled water.

**Exam tip**

Remember, you only had 100 cm$^3$ in this experiment so do not forget to factor up your answer.

mass of barium hydroxide = .................... g **[5]**

# Knowledge

## 8 Enthalpy changes

### Enthalpy

**Enthalpy**, H, is a measure of the heat energy in a chemical system. As energy must be conserved, the energy needed to break bonds and the energy needed to make the new bonds may result in a change in energy, the **enthalpy change** ($\Delta H$). This is measured in $kJ\,mol^{-1}$.

### Endothermic and exothermic changes

An **endothermic** change is when energy is transferred *from* the surroundings to the system ($\Delta H$ is positive).

An **exothermic** change is when energy is transferred *from* the system to the surroundings ($\Delta H$ is negative).

### Measuring the enthalpy change of combustion for a liquid fuel

The equipment needed to measure enthalpy change of combustion is shown below:

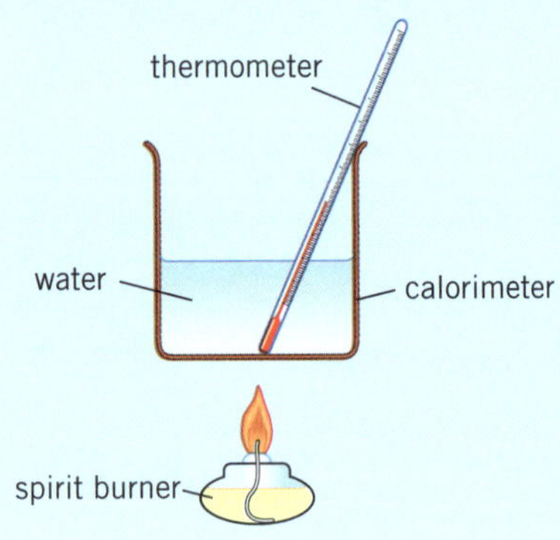


When a reaction takes place the temperature will drop rapidly after the reactants are mixed, and react. Using a graph, a line of best fit can be extrapolated backwards to find an estimate of the temperature immediately after mixing.

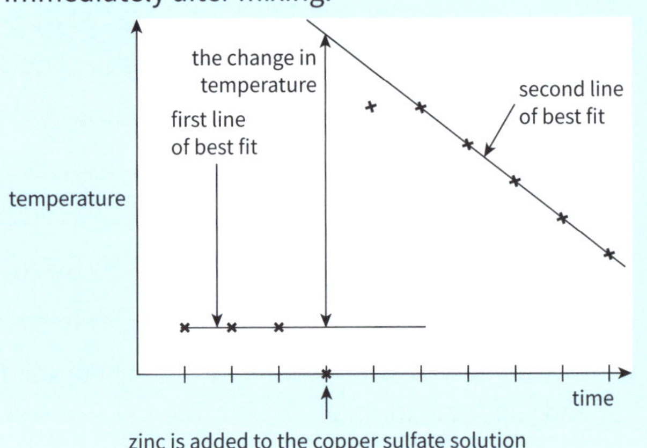


### Average bond enthalpy

**Average bond enthalpy** is the energy required to break one mole of a particular bond in a range of substances. Calculate the energy change using the following method:

**Step 1** Draw all the atoms and bonds in the compounds.

**Step 2** Count the number of bonds of each type on each side of the equation.

**Step 3** Add up the individual bond enthalpy values to determine the total enthalpy needed to break the bonds in the reactants.

**Step 4** Add up the individual bond enthalpy values to determine the total enthalpy released when new bonds are made when forming the product.

**Step 5** Calculate the enthalpy change of reaction.

### Specific heat capacity

**Specific heat capacity** is how much energy is needed to increase the temperature of 1 g of a substance by 1 K:

$$q = mc\Delta T$$

where:

- $q$ = energy change (J)
- $m$ = mass of substance (g)
- $c$ = specific heat capacity ($J\,g^{-1}\,K^{-1}$)
- $\Delta T$ = temperature change (K)

Specific heat capacity can be measured experimentally using a calorimeter.

It is important to reduce energy loss during this experiment, otherwise the data will be inaccurate.

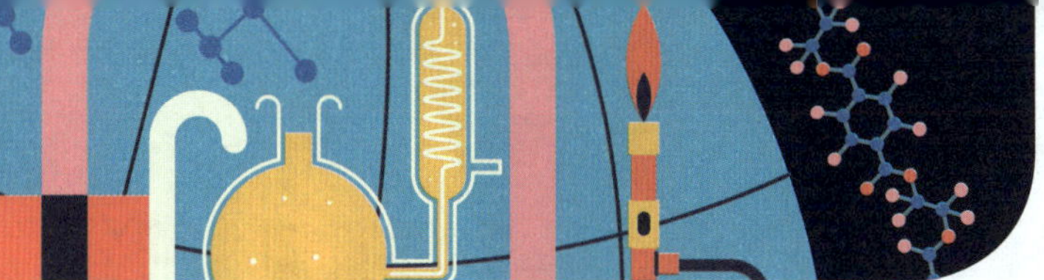

8

## Hess' Law

**Hess' law** tells us the start and end of a reaction are important but the route taken to get there is not important.

$$A \rightarrow B = A \rightarrow C \rightarrow B$$

Hess' law is used to determine the change in enthalpy of reactions that can not be measured directly. Some reactions will use the standard enthalpy of formation ($\Delta_f H^\ominus$) values and some will use the standard enthalpy of combustion ($\Delta_c H^\ominus$) values.

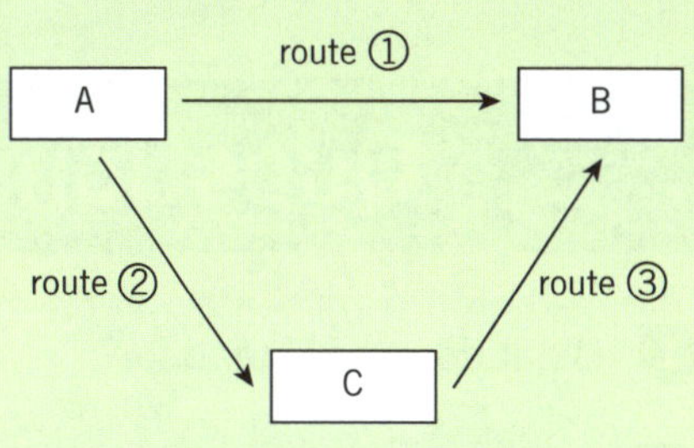


## Example using standard enthalpy of combustion values

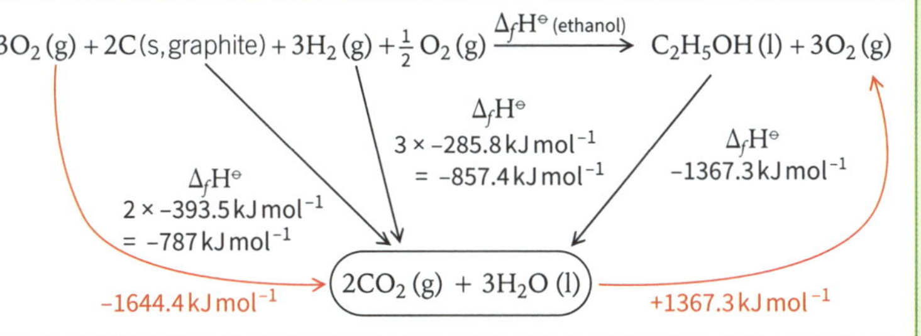


## Example using standard enthalpy of formation values

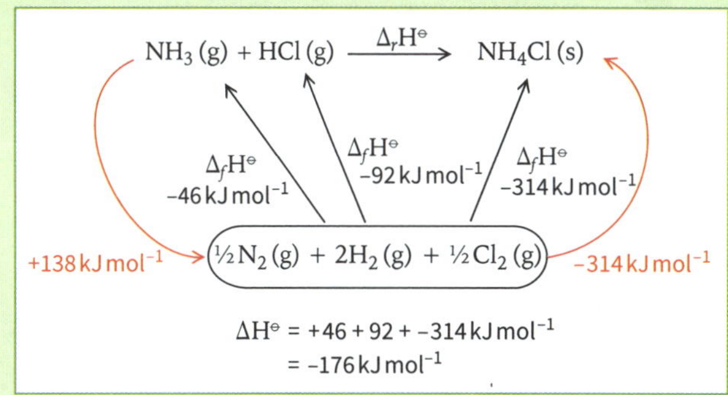


## Activation energy

For a reaction to take place, the particles need to collide with enough energy to over come the **activation energy ($E_a$)**.

This is the minimum value of energy needed to break the bonds and start the reaction.

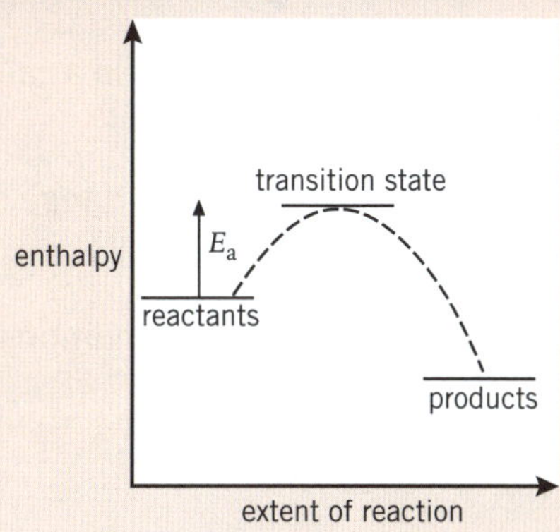


## Definitions

**Standard enthalpy changes** ($\Delta H^\ominus$) are all measured under the standard conditions (100 kPa and a stated temperature, 298 K).

**Enthalpy change of reaction** ($\Delta_r H$) is the enthalpy change for a reaction.

**Enthalpy change of combustion** ($\Delta_c H$) is the enthalpy change when one mole of a substance is burnt completely in oxygen.

**Enthalpy change of formation** ($\Delta_f H$) is the enthalpy change when one mole of a substance is formed from elements.

**Enthalpy change of neutralisation** ($\Delta_{neut} H$) is the enthalpy change when one mole of a water is formed from neutralisation.

For answers and more practice questions visit www.oxfordrevise.com/scienceanswers

Even more practice and interactive revision quizzes are available on kerboodle

# Retrieval

Learn the answers to the questions below, then cover the answers column with a piece of paper and write as many as you can. Check and repeat.

| | Questions | Answers |
|---|---|---|
| 1 | What does $\Delta_r H$ stand for? | enthalpy change of reaction |
| 2 | What is the enthalpy change of formation ($\Delta_f H$)? | the enthalpy change when one mole of a substance is formed from elements |
| 3 | Is a reaction exothermic or endothermic if $\Delta H$ is negative? | exothermic |
| 4 | If energy is released during a reaction is it endothermic or exothermic? | exothermic |
| 5 | When can Hess' law be applied? | when the enthalpy change can not be measured directly |
| 6 | What is the equation for determining the energy release that can be measured by a calorimeter? | $q = mc\Delta T$ |
| 7 | Is a reaction exothermic or endothermic if $\Delta H$ is positive? | endothermic |
| 8 | How can the estimated temperature change immediately after mixing two liquids be found? | extrapolate a graph backwards |
| 9 | What is activation energy? | the minimum value of energy needed to break the required bonds and start the reaction |
| 10 | How can inaccuracies in a calorimeter be reduced? | reduce energy loss/ensure complete combustion |
| 11 | What is the enthalpy change of combustion ($\Delta_c H$)? | the enthalpy change when one mole of a substance is burnt completely in oxygen |
| 12 | What are standard conditions in a reaction? | 100 kPa and a stated temperature |
| 13 | What is the symbol for enthalpy change of neutralisation? | $\Delta_{neut}H$ |
| 14 | What units is enthalpy measured in? | $kJ\ mol^{-1}$ |
| 15 | What is Hess' law? | the start and end of a reaction are important but not the route taken to get there |
| 16 | Is a reaction endothermic or exothermic if energy is taken in during the reaction? | endothermic |
| 17 | What is needed to break bonds in molecules? | energy needs to be taken in |
| 18 | What is specific heat capacity? | the amount of energy needed to increase the temperature of 1 g of a substance by 1 K |

Put paper here

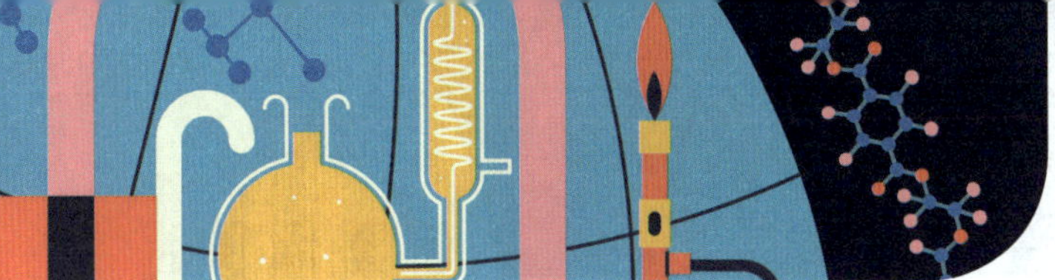



| | Question | Answer |
|---|---|---|
| 19 | How is the enthalpy change in a reaction calculated? | calorimetry/Hess' law/difference between energy released by bond formation and energy taken in by breaking bonds |
| 20 | What is the standard enthalpy of formation ($\Delta_f H^{\ominus}$)? | the enthalpy change when one mole of a substance is formed from elements (in their standard states) under standard conditions |

Put paper here

# Practical skills

Practise your practical skills using the worked example and practice questions below.

## Enthalpy determination

Measuring enthalpy changes is an important skill that can give you information about the stability of compounds. Hess' law can be used to work out enthalpy changes for chemical reactions that do not or cannot actually happen.

When measuring temperature changes, it is important that as little heat as possible is lost. This can be done in a number of ways: for example, using a polystyrene cup with a lid. Temperature readings also need to be taken quickly so the heat has not had time to escape.

When calculating percentage error from for a temperature change, you take two measurements (initial and final), so you need to double the percentage error.

## Worked example

A student reacts powdered zinc with 1.50 mol dm$^{-3}$ sulfuric acid.

**Questions**

1 The temperatures changes from 20.0 °C to 35.5 °C. The error in each temperature reading is ± 0.5 °C. Calculate the percentage error in the temperature change.

2 The student intends to calculate the enthalpy change of reaction. What other measurements in the laboratory does the student need to take?

3 The student carries out the reaction in a glass beaker and not in a polystyrene cup. Will the student's calculated value of $\Delta_r H$ be more exothermic or less exothermic than the true value?

**Answers**

1 $\Delta T = 35.5 - 20.0 = 15.5$ °C

$\%\ \text{error} = \frac{2 \times 0.5}{15.5} = 6.5\%$

2 volume of $H_2SO_4$ and mass of zinc

3 less exothermic as the measured temperature change will be less as heat will be lost

## Practice

A student is trying to measure $\Delta_c H$ for methanol, using the following method:

- 150 cm$^3$ water in a beaker is heated for three minutes using a spirit burner.
- The flame is extinguished.
- The student weighs the spirit burner before and after heating.
- The temperature changes from 20.5 °C to 69 °C.

1 The error in each temperature reading is ± 0.5 °C. Calculate the temperature change and its percentage error.

2 Would it be better to measure the volume of water used with a 15 cm$^3$ measuring cylinder, with a maximum error of 1 cm$^3$, or with the repeated use of a 25 cm$^3$ measuring cylinder with a maximum error of 0.2 cm$^3$? Use calculations to support your answer.

3 How would putting a lid on the beaker change the calculated value of $\Delta_c H$? Explain your answer.

4 How would starting with water at 60 °C change the calculated value of $\Delta_c H$? Explain your answer.

# Practice

## Exam-style questions

1 A student investigated the amount of heat energy released when propan-1-ol was burnt.

Synoptic link

2.1.3

They set up a simple calorimetry experiment as shown in **Figure 1.1**.

Their results are shown in **Table 1.1**.

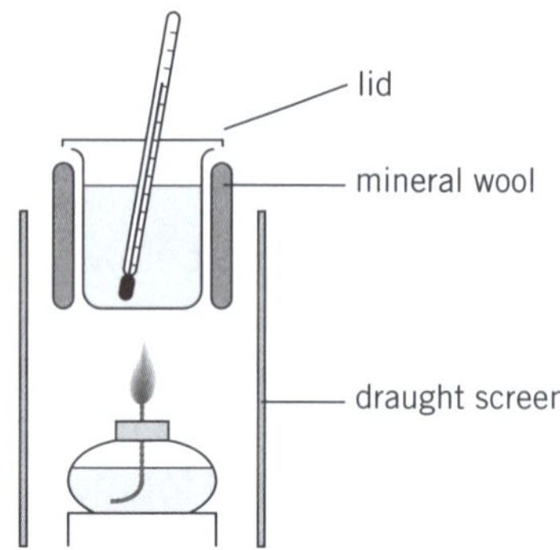


**Figure 1.1**

| | |
|---|---|
| **Volume of water ($cm^3$)** | 200 |
| **Initial temperature (°C)** | 21.2 |
| **Maximum temperature (°C)** | 37.2 |
| **Temperature change (°C)** | |
| **Initial mass of spirit burner (g)** | 12.45 |
| **Final mass of spirit burner (g)** | 11.28 |
| **Mass of propan-1-ol burnt (g)** | 1.17 |

**Table 1.1**

**(a)** Calculate the temperature change.

temperature change = .....................°C **[1]**

**(b)** Define enthalpy of combustion.

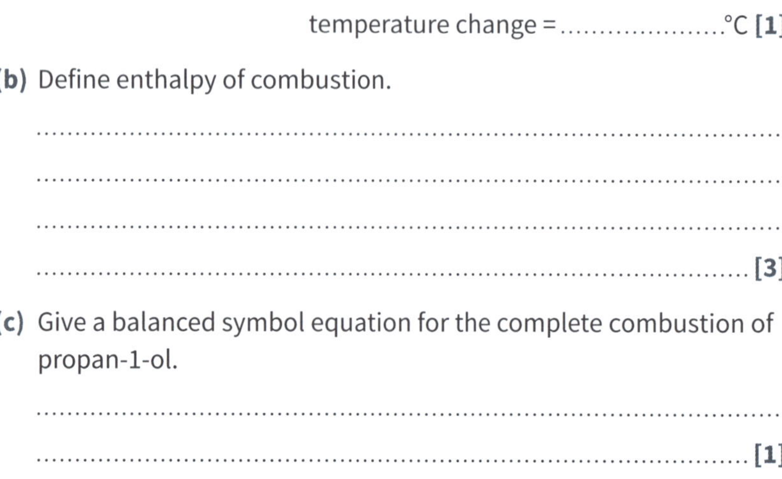

.......................................................................................

.......................................................................................

.......................................................................................

....................................................................................... **[3]**

**(c)** Give a balanced symbol equation for the complete combustion of propan-1-ol.

.......................................................................................

....................................................................................... **[1]**

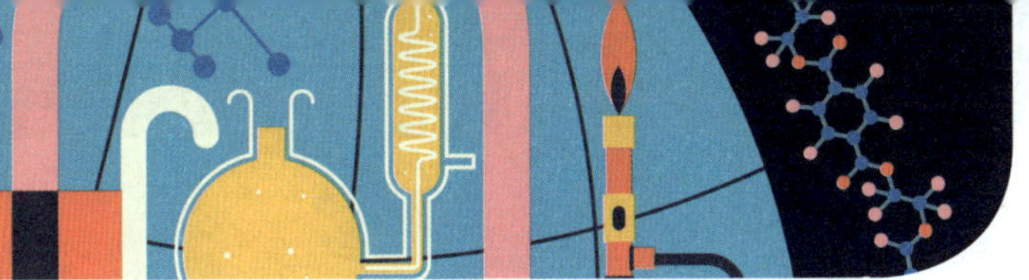



**(d)** Calculate the enthalpy of combustion of propan-1-ol from the student's results (assume the specific heat capacity of the calorimeter is negligible).

Give your answer to an appropriate number of significant figures.

The density of water is 1.00 g $cm^{-3}$

The specific heat capacity of water is 4.18 J $g^{-1}$ °$C^{-1}$

enthalpy of combustion = .................... kJ $mol^{-1}$ **[5]**

**(e)** The accepted value for the enthalpy of combustion of propan-1-ol is −2021 kJ $mol^{-1}$.

Give **three** reasons why the student did not achieve this result in their experiment.

..........................................................................................

..........................................................................................

..........................................................................................**[3]**

**Exam tip**

Be specific, human errors will not be credited.
Think carefully about the limitations of the experiment.

**2** A student carried out an investigation into the enthalpy change when an acid is neutralised by an alkali.

The student poured 25 $cm^3$ of 2.00 mol $dm^{-3}$ sodium hydroxide, NaOH, into a polystyrene cup. They recorded the temperature every minute for 4 minutes. At the fifth minute, they added 25 $cm^3$ of 2.00 mol $dm^{-3}$ hydrochloric acid, stirred the mixture, and did not record the temperature. From the sixth to the tenth minute they continued to record the temperature.

Their results are shown in **Table 2.1**.

| Time (min) | Temperature (°C) |
|---|---|
| 0 | 21.2 |
| 1 | 21.2 |
| 2 | 21.0 |
| 3 | 21.2 |
| 4 | 21.2 |
| 5 | |
| 6 | 34.6 |
| 7 | 34.4 |
| 8 | 34.0 |
| 9 | 33.8 |
| 10 | 33.4 |

**Table 2.1**

(a) Plot a graph of the data in **Table 2.1**.

Calculate the instantaneous temperature rise at the 5th minute.

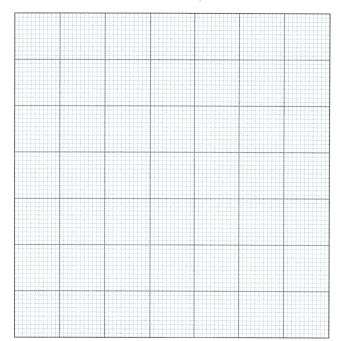

**[6]**

(b) Use your answer from **2(a)** to calculate the enthalpy of neutralisation for the reaction.

The specific heat capacity of water is 4.18 J $g^{-1}$ °$C^{-1}$

Density of water is 1.00 g $cm^{-3}$

(If you did not determine a temperature rise in **2(a)**, use 11.0 °C. This is not the correct answer.)

**Exam tip**

Take care to use the final mass of solution after both reactants have been added.

enthalpy of neutralisation = .................... kJ $mol^{-1}$ **[5]**

(c) Sodium hydroxide can also be neutralised by sulfuric acid in the following reaction:

$$2NaOH + H_2SO_4 \rightarrow Na_2SO_4 + 2H_2O$$

Using your result for **2(b)** and the equation, suggest a value for the enthalpy change of neutralisation.

Give a reason for your answer.

..........................................................................................

..........................................................................................

..........................................................................................

.......................................................................................... **[2]**

**Synoptic links**

4.1.1 4.1.2

**3** Isobutane (2 methyl propane) is commonly used as a gas source for barbecues and camping stoves.

**(a)** Give the skeletal formula for isobutane.

**[1]**

**(b)** Isobutane is an isomer of butane.

Identify the type of structural isomerism illustrated by isobutane.

..........................................................................................

..........................................................................................**[1]**

**(c)** Isobutane burns in excess oxygen to form carbon dioxide and water.

Give the balanced equation **[1]**

............ $C_4H_{10}$ + ............ $O_2 \rightarrow$ ............ $CO_2$ + ............ $H_2O$

**(d)** Define Hess' law.

..........................................................................................

..........................................................................................

..........................................................................................

..........................................................................................**[1]**

**(e)** **Figure 3.1** shows the reaction of butane being converted into isobutane.

```
     H   H   H   H                    H
     |   |   |   |                    |
  H—C—C—C—C—H      ⇌           H—C—H
     |   |   |   |              H     |    H
     H   H   H   H              |     |    |
                             H—C——C——C—H
                                |     |    |
                                H     H    H
```

butane isobutane

**Figure 3.1**

Use the enthalpy of combustion data in **Table 3.1** to calculate the enthalpy for the reaction of butane into isobutane.

| Molecule | $\Delta_c H^\circ$ (kJ mol$^{-1}$) |
|---|---|
| butane | −2878 |
| isobutane | −2869 |

**Table 3.1**

enthalpy = ...................... kJ mol$^{-1}$ **[2]**

**(f)** The conversion of butane into isobutane is a reversible reaction.

Give the enthalpy change for the conversion of isobutane into butane.

..............................................................................................

..............................................................................................

..............................................................................................**[1]**

> **Exam tip**
> Remember, Le Châtelier's principle can be applied here to get to the answer quickly.

**(g)** Suggest why butane does not spontaneously convert into isobutane despite its favourable enthalpy.

..............................................................................................

..............................................................................................

..............................................................................................**[1]**

**4** Methanol is an important chemical that has a global production of over 70 million tonnes annually. It is produced by the reaction of synthesis gas – a mixture of carbon monoxide and hydrogen – according to the following equation:

$$CO + 2H_2 \rightarrow CH_3OH$$

**(a)** Use the relevant bond enthalpy data in **Table 4.1** to calculate the enthalpy change for the reaction.

| Bond | Enthalpy (kJ mol$^{-1}$) |
|---|---|
| C≡O | 1072 |
| C—H | 413 |
| H—H | 432 |
| C=O | 799 |
| O—H | 467 |
| C—O | 358 |

**Table 4.1**

$\Delta_r H$ = .......................**[3]**

**(b)** The accepted value for the reaction is, in fact, −91 kJ mol$^{-1}$. Explain why the value you calculated in **4(a)** is not precisely the accepted value.

..............................................................................................

..............................................................................................

..............................................................................................

..............................................................................................**[2]**

**(c)** Methanol can be combusted in spirit burners. In the presence of excess oxygen, carbon dioxide, and water are produced:

$2CH_3OH + 3O_2 \rightarrow 2CO_2 + 4H_2O \qquad \Delta_cH^{\ominus} = -715\,kJ\,mol^{-1}$

Use the enthalpy of combustion and the relevant data in **Table 4.1** to calculate the bond enthalpy of the oxygen double bond in this reaction.

...................... **[3]**

**(d)** Carbon monoxide is used in the blast furnace in the extraction of iron. **Table 4.2** contains some standard enthalpy of formation data.

| **Compound** | $CO(g)$ | $Fe_2O_3(s)$ |
|---|---|---|
| **Enthalpy of formation ($kJ\,mol^{-1}$)** | −111 | −822 |

**Table 4.2**

$Fe_2O_3(s) + 3CO(g) \rightarrow 2Fe(s) + 3CO_2(g) \qquad \Delta H = -19\,kJ\,mol^{-1}$

Use these data and the equation for the reaction of iron(III) oxide with carbon monoxide to calculate a value for the standard enthalpy of formation for carbon dioxide.

$\Delta_fH^{\ominus}$ = ...................... **[3]**

**5** Ethane can undergo fluorination in the following reaction:

$CH_3CH_3 + 2F_2 \rightarrow CH_2FCH_2F + 2HF \quad \Delta_rH^{\ominus} = -1134\,kJ\,mol^{-1}$

**(a)** Give the equation that would represent the enthalpy of formation of ethane. Include all state symbols.

**[1]**

4.1.1

**(b)** Name the compound $C_2H_4F_2$.

........................................................................................**[1]**

**(c)** Suggest why the enthalpy of formation of fluorine is quoted as 0.

........................................................................................

........................................................................................

........................................................................................

........................................................................................**[1]**

**Exam tip**

Make sure you apply all the IUPAC rules in order.

**(d)** Use the value of the enthalpy change for the reaction above and the data in **Table 5.1** to calculate the enthalpy of formation of $C_2H_4F_2$. **[3]**

| **Compound** | $C_2H_6(g)$ | $HF(g)$ |
|---|---|---|
| **Enthalpy of formation ($kJ\,mol^{-1}$)** | −84 | −273 |

**Table 5.1**

6 Ethene can be hydrogenated to form ethane. **Table 6.1** shows some bond enthalpy data.

**Synoptic link**

2.2.2

| **Bond** | H—H | C—C | C=C | N≡N | N—H |
|---|---|---|---|---|---|
| **Mean bond enthalpy ($kJ\,mol^{-1}$)** | 436 | 348 | 612 | 944 | 388 |

**Table 6.1**

**(a)** Use the following equation and data from the table to calculate a value for the C—H bond enthalpy in ethane. **[3]**

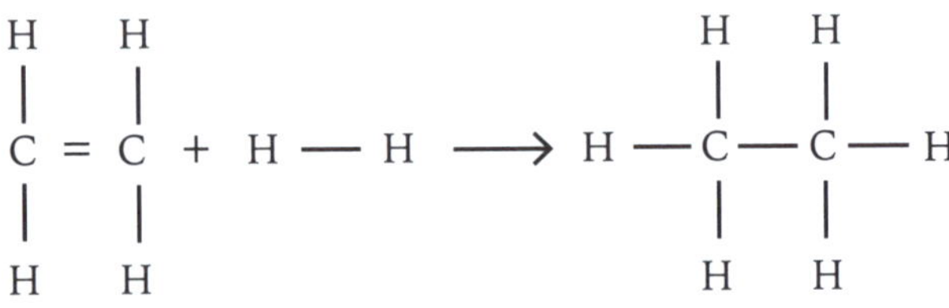

$\Delta H = -136\ kJ\ mol^{-1}$

**(b)** State why the enthalpy of formation of hydrogen is $0\ kJ\ mol^{-1}$. **[1]**

**Exam tip**

Do not just write 'by definition', construct a full sentence.

**(c)** The enthalpy changes for the formation of isolated atoms of hydrogen and atomic carbon from their respective elements in their standard states are as follows:

$\frac{1}{2}H_2(g) \rightarrow H(g)$ $\quad \Delta H^{\ominus} = +218\ kJ\ mol^{-1}$

$C(s) \rightarrow C(g)$ $\quad \Delta H^{\ominus} = +715\ kJ\ mol^{-1}$

By reference to its structure, suggest why a large amount of heat energy is required to produce free carbon atoms from solid carbon. **[3]**

**(d)** Suggest which compound, ethane or ethene, will have the larger enthalpy of combustion.

Give a reason for your choice. **[2]**

7 Hydrazine, $N_2H_4$, is a derivative of ammonia. It decomposes in an exothermic reaction. Hydrazine is also used in rocket fuel, where it is reacted with hydrogen peroxide.

**Synoptic link**

2.1.3

**(a)** Give a balanced symbol equation to show the decomposition of hydrazine to form ammonia and nitrogen only. **[1]**

**(b)** Calculate the enthalpy change of this decomposition using the data in **Table 7.1**. **[4]**

| Compound | N—H | N—N | N≡N |
|---|---|---|---|
| Mean bond enthalpy ($kJ\,mol^{-1}$) | 388 | 163 | 944 |

**Table 7.1**

**(c)** When a rocket burns, the hydrazine reacts as follows:

$$N_2H_4(l) + 2H_2O_2(l) \rightarrow N_2(g) + 4H_2O(g)$$

Use the data in **Table 7.2** to calculate the enthalpy change for this reaction. **[3]**

| Compound | $N_2H_4$ | $H_2O_2$ | $H_2O$ |
|---|---|---|---|
| Enthalpy of formation ($kJ\,mol^{-1}$) | +96 | −187 | −286 |

**Table 7.2**

**(d)** A rocket engineer is trying to build a rocket engine. To calculate how much hydrazine they need to place in the rocket, the engineer completes a calorimetry experiment. The engineer places 1.45 g of hydrazine with excess hydrogen peroxide inside a bomb calorimeter containing 500 $cm^3$ of water at 298 K.

Use your answer to **7(c)** to calculate the maximum temperature recorded in the water.

Give your answer to an appropriate number of significant figures. **[4]**

**Exam tip**

Set up your working clearly. Make sure the examiner knows what each value corresponds to.

**(e)** The rocket needs to produce 4.6 $m^3$ of gas when it is fired. Calculate the mass of hydrazine needed to produce 4.6 $m^3$ of gas under standard conditions.

Give your answer to an appropriate number of significant figures.

The gas constant = 8.31 $J\,mol^{-1}\,K^{-1}$ **[4]**

**8** Propanone is a structural isomer of propanal.

**(a)** Define the term standard molar enthalpy of formation, $\Delta_f H^{\ominus}$. **[3]**

**(b)** State Hess's law. **[1]**

**(c)** Propanone, $CH_3COCH_3$, burns in oxygen as shown by the equation:

$$CH_3COCH_3(l) + 4O_2(g) \rightarrow 3H_2O(l) + 3CO_2(g)$$

Use the data given in **Table 8.1** to calculate the standard enthalpy of combustion of propanone.

**Exam tip**

It is perfectly acceptable to draw a cycle and use it to calculate the answer.

| Compound | $CO_2(g)$ | $H_2O(l)$ | $CH_3COCH_3(l)$ |
|---|---|---|---|
| Enthalpy of formation ($kJ\,mol^{-1}$) | −394 | −286 | −248 |

**Table 8.1**

# Knowledge

## 9 Reaction rates and equilibrium (qualitative)

### Simple collision theory

When particles collide with sufficient energy, a reaction can happen.

For a collision to be *effective*:

- it needs to have enough energy to break the bonds (overcome the activation energy $E_a$)
- the particles need to collide in an orientation that will allow a reaction.

The rate of reaction can be affected by a number of factors:

- concentration
- temperature
- using a catalyst
- pressure
- surface area of reactants.

### The Boltzmann distribution

The range of energies that particles within a gas or liquid have can be displayed on a Maxwell–Boltzmann distribution curve.

The distribution shows the following characteristics:

- the area under the graph is the sum of all the energies
- there are no particles with zero energy
- the majority of particles have an intermediate energy
- only a few particles have enough energy to overcome the activation energy.

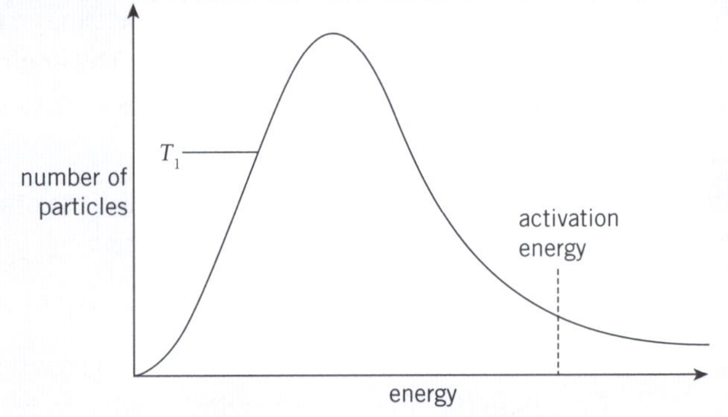


### Catalysts

**Catalysts** can be either **homogeneous** (in the same state as the reactants) or **heterogeneous** (in a different state to the reactants). Catalysts can lower the activation energy ($E_a$) that is needed for a reaction to start, by providing an alternative pathway.

### Economic importance and sustainability benefits of catalysts

**Economic:** Reactions are faster and occur at lower temperatures. Energy costs are reduced, so catalysts are often used in industry.

**Sustainability:** The alternative pathway provided by a catalyst has lower energy requirements and more particles on a Maxwell–Boltzmann distribution curve will pass this threshold. This will reduce the demand for fossil fuels and reduce $CO_2$ emissions.

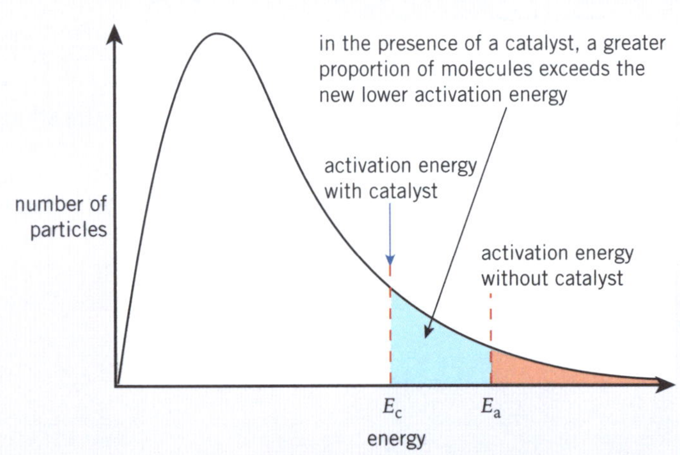

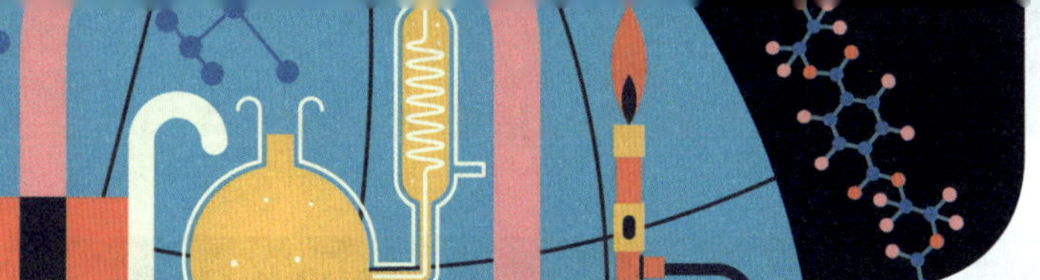



## The effect of temperature

Increasing the temperature of a reaction, from $T_1$ to $T_2$:

- Will increase the average energy of the particles. This will shift the Maxwell–Boltzmann distribution curve to the right, increasing the number of particles that have enough energy to overcome the activation energy.
- The particles will move faster and so the frequency of collisions increases.

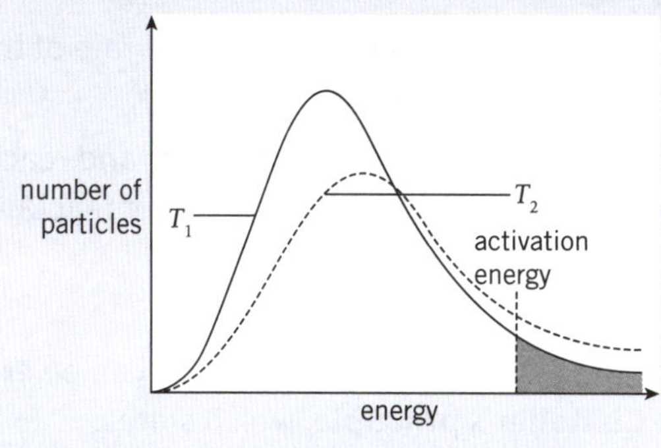


## Rate of reaction

The **rate of reaction** is how fast a reactant is being used up or how fast a product is being made.

$$\text{rate (mol dm}^{-3}\text{ s}^{-1}) = \frac{\text{change in concentration (mol dm}^{-3})}{\text{time } (s)}$$

The rate of reaction at a particular time can be calculated from the gradient of the tangent to the curve at that particular point on a concentration–time graph.

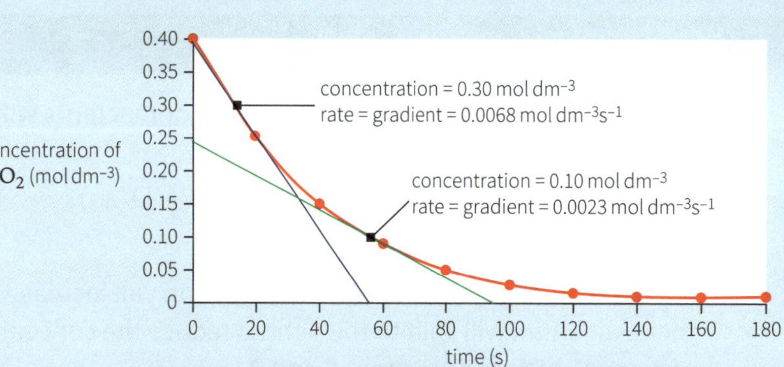


## Increasing the rate of reaction

To increase the rate of reaction there needs to be an increase in the number of *effective* collisions taking place in a set time. As well as increasing the temperature, this can be done by:

- **Increasing the concentration of a solution:** the number of particles will increase in a set volume so the particles are closer together, making collisions more likely.
- **Increasing the pressure of a gas:** the same number of particles will be compressed into a smaller volume so the particles are closer together, increasing the chances of a collision.
- **Increasing the surface area of a reactant:** the more particles that are on the surface, the more particles that are available to react. This can be achieved by turning a solid lump into a powder.

## Investigating the rates

The rate of a reaction can be followed over time in a number of ways, such as by loss of mass; by change in colour or turbidity; by change in pH; or by production of gas.

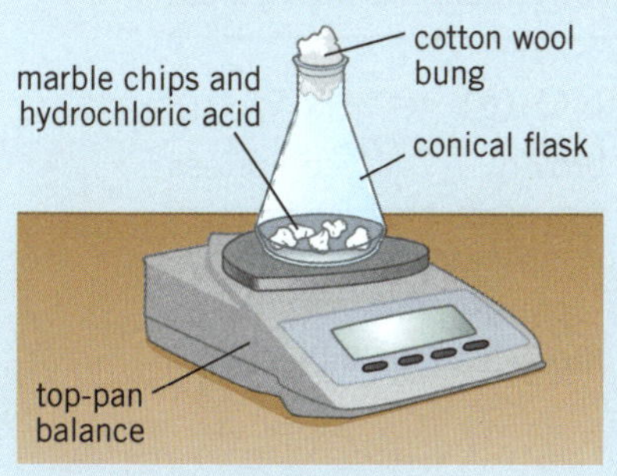


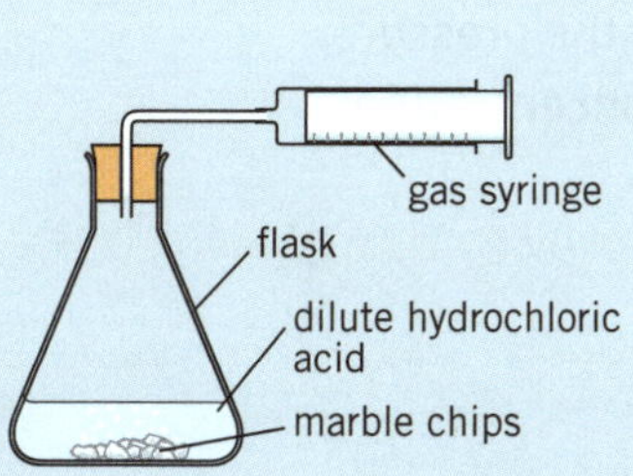


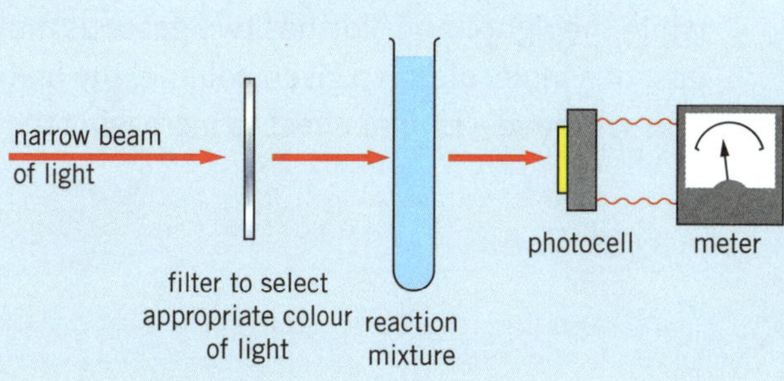

## Dynamic equilibrium

When a reversible reaction takes place within a closed system, an **equilibrium** is established.

An equilibrium is reached when the rate of the forward reaction is equal to rate of the reverse reaction.

The concentrations of the products and reactants stay the same. An equilibrium is *not* a 50 : 50 ratio of reactant and products.

The process of shifting equilibrium can be explained by **Le Châtelier's principle**, which states:

*If a system at equilibrium is changed by an external factor, the system adjusts to reduce the effect of that change.*

## Changing concentration

Any change to the concentration of products or reactants will shift the equilibrium:

$$A(aq) + B(aq) \rightleftharpoons C(aq) + D(aq)$$

The reaction will work to counteract that change.

1 If more **A** is added, then the forward reaction will increase and the equilibrium will shift to the right to reduce the concentration of **A** and **B** and produce more **C** and **D**.

2 Removal of **C** (or **D**) will have the same effect, the reaction works to increase their concentration shifting the equilibrium to the right.

3 Increasing the concentration of **C** or **D** will shift the equilibrium to the left and produce more **A** and **B**.

## Changing pressure

Changing the pressure of a gaseous reaction in equilibrium will affect the position of the equilibrium, only if the reaction has a different number of molecules on each side.

$$N_2O_4(g) \rightleftharpoons 2NO_2(g)$$

Dinitrogen tetraoxide ($N_2O_4(g)$) is a colourless gas. In a closed system it sets up an equilibrium with nitrogen dioxide ($NO_2(g)$), which is a brown gas.

The left-hand side of the reaction has one gaseous molecule, while the right-hand side has two gaseous molecules. The more gaseous molecules in a given volume, the higher the pressure. This will have a similar effect to increasing the concentration.

Increasing the pressure will make the position of equilibrium shift to the side with the fewer gaseous molecules to counteract the change.

In this example, more $N_2O_4(g)$ will be formed. This can be seen by the colour change.

increased pressure causes shift towards *fewer* gaseous molecules →

$$2NO_2(g) \rightleftharpoons N_2O_4(g)$$

brown | colourless

← decreased pressure causes shift towards *more* gaseous molecules

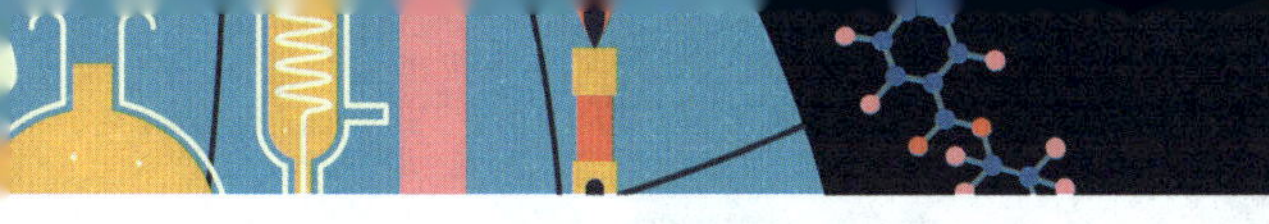



## Changing temperature

Changing the temperature at which a reaction takes place will have an effect on the equilibrium depending on if the reaction is endothermic ($\Delta H$ is positive) or exothermic ($\Delta H$ is negative).

- If a reaction is exothermic in the forward direction, the reverse reaction will be endothermic. The value of the enthalpy change ($\Delta H$) will be the same in either direction, it is just the sign that changes.
- An *increase* in temperature shifts the equilibrium in the endothermic direction, aiming to lower the temperature.
- A decrease in temperature shifts the equilibrium to the exothermic direction.

Increasing the temperature a reaction takes place at will also increase the rate of reaction; this is an important consideration for industry.

## Using a catalyst

A catalyst will have *no effect* on the position of equilibrium. A catalyst will speed up the rate of reaction, by providing an alternative path with a lower activation energy.

A catalyst may be used in industry to reduce the time taken to get to equilibrium.

## Equilibrium and industry

Industry needs to take different parts of chemistry into consideration to achieve the best economic value.

The conversion of nitrogen gas and hydrogen gas into ammonia is an exothermic reaction, so lower temperatures favour the production of ammonia, but would have a slower rate of reaction.

Higher pressures favour the production of ammonia, but this would make the reaction more expensive.

Different factors need to be weighed up when determining the optimal conditions for an industrial reaction.

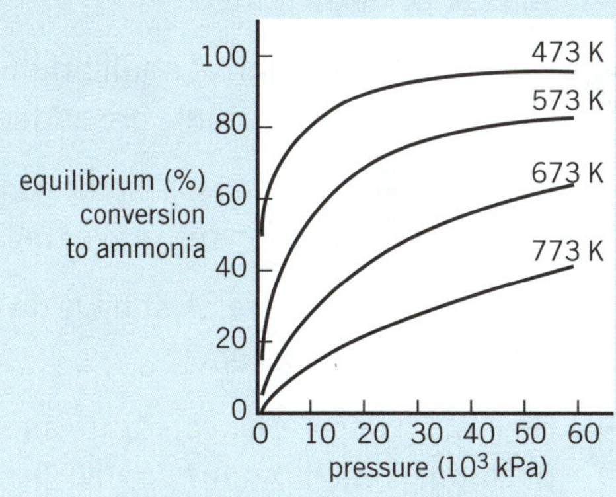


## The equilibrium constant, $K_c$

$K_c$ is the **equilibrium constant**, for a reversible reaction. $K_c$ is used to show the ratio between products and reactants at equilibrium. $K_c$ will change with temperature but not with a change in concentration or pressure. A catalyst will have no effect on $K_c$.

At equilibrium:

- if $K_c$ is greater than 1 there will be more products
- if $K_c$ is below 1 there will be more reactants.

For the reaction: $aA + bB \rightleftharpoons cC + dD$

where [A], [B], [C], [D] are the equilibrium concentrations.

$$K_c = \frac{[C]^c[D]^d}{[A]^a[B]^b}$$

For answers and more practice questions visit www.oxfordrevise.com/scienceanswers

Even more practice and interactive revision quizzes are available on kerboodle

# Retrieval

Learn the answers to the questions below, then cover the answers column with a piece of paper and write as many as you can. Check and repeat.

| | Questions | Answers |
|---|---|---|
| 1 | How can a reaction can be followed? Give three ways. | any 3 from: by loss of mass; by change in colour or turbidity; pH change; or by production of gas |
| 2 | What effect do catalysts have on activation energy? | they lower it |
| 3 | What effect would increasing the temperature that a reaction takes place at have on a Maxwell–Boltzmann distribution curve? | it will shift to the right and the peak of the curve will lower |
| 4 | What is the expression for the equilibrium constant for the following reaction: $aA + bB \rightleftharpoons cC + dD$? | $K_c = \frac{[C]^c[D]^d}{[A]^a[B]^b}$ |
| 5 | How can the total surface area of a lump of a substance be increased? | by turning it into a powder |
| 6 | How will the position of equilibrium be affected if more reactants are added? | it will shift to the products side to reduce the concentration of reactants |
| 7 | What effect would decreasing the concentration of a reactant have on the rate of reaction? | it will decrease |
| 8 | What effect does a catalyst have on the position of equilibrium? | none |
| 9 | How will a decrease in pressure affect the position of equilibrium? | it will shift to the side with more gaseous molecules |
| 10 | What is Le Châtelier's principle? | if a system at equilibrium is disturbed, the equilibrium moves in the direction that tends to reduce the change |
| 11 | How is the total sum of all the energies of particles in a reaction calculated? | by calculating the area under a Maxwell–Boltzmann distribution curve |
| 12 | How do catalysts speed up reactions? | they provide alternative reaction pathways |
| 13 | What effect does a catalyst have on $K_c$? | none |
| 14 | If there is an increase in temperature in a reversible reaction, will the position of equilibrium shift to the exothermic or the endothermic direction? | the endothermic reaction |
| 15 | How does concentration effect $K_c$? | it does not |
| 16 | Why is a high temperature used for ammonia production? | this is a compromise between fast rate of reaction and a reasonable yield |
| 17 | How is the rate of reaction found from a graph? | from the gradient of the curve of a concentration–time graph |

Put paper here

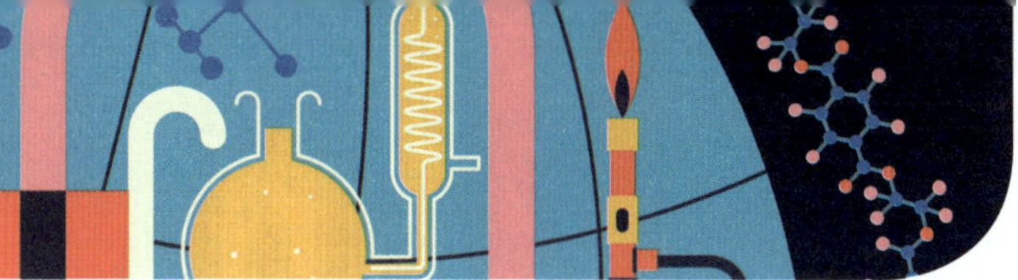



| | | |
|---|---|---|
| 18 | What does a Maxwell–Boltzmann distribution curve show? | the distribution of energies that particles have |
| 19 | What is a homogeneous catalyst? | a catalyst in the same state as the reactant |
| 20 | What does $\Delta H$ represent? | enthalpy change |

Put paper here

## Practical skills

Practise your practical skills using the worked example and practice questions below.

### Continuous monitoring method

You can measure the rate of a reaction continuously by measuring at regularly time intervals:

- the pH, to determine $H^+$ ion concentration
- mass, to determine if a gas is produced or required
- volume of gas produced, which would normally be collected in a gas syringe
- and using a colourimeter and a calibration curve to measure the concentration of a coloured reagent or product.

In each case, calculations are needed after the experiment to determine concentrations.

### Worked example

A student wishes to determine the rate equation for the decomposition of hydrogen peroxide to oxygen and water.

**Questions**

1 Write an equation for the decomposition of $H_2O_2$.

2 The student decides to collect the $O_2$ over water. Should they collect $O_2$ in a measuring cylinder or burette? Explain why.

3 How could the student calculate $[H_2O_2]$ from volume of $O_2$?

**Answers**

1 $2H_2O_2 \rightarrow O_2 + 2H_2O$

2 burette, as it is more more precise

3 for each reading:

$$\text{mol}(O_2) = \frac{\text{vol}(O_2)}{24\,000}$$

$$\text{mol}(H_2O_2) = 2 \times \frac{\text{vol}(O_2)}{24\,000}$$

$$[H_2O_2] = \frac{\text{mol}(H_2O_2)}{\text{vol}(H_2O_2)}$$

$$= \frac{2 \times \text{vol}(O_2)}{2400 \times \text{vol}(H_2O_2)}$$

$$= \frac{\text{vol}(O_2)}{1200 \times \text{vol}(H_2O_2)}$$

### Practice

A student carries out an experiment to investigate the reaction between calcium carbonate and HCl(aq). The student uses 8 g of $CaCO_3$ and 16 $cm^3$ of 2.00 mol $dm^{-3}$ HCl(aq). They are mixed in a conical flask on a top pan balance and the mass is recorded.

1 The student adds some cotton wool loosely in the mouth of the conical flask. Explain why.

2 Explain how the student could:

a Calculate [HCl] for each data point.

b Use a graph to show that rate = $k$[HCl].

c Determine $k$.

For answers and more practice questions visit www.oxfordrevise.com/scienceanswers

Even more practice and interactive revision quizzes are available on kerboodle

# Practice

## Exam-style questions

**1** A student investigated the reaction between hydrochloric acid and magnesium and obtained the data in **Table 1.1**. This is an exothermic reaction.

| Concentration of hydrochloric acid ($mol\,dm^{-3}$) | Rate of reaction ($mol\,dm^{-3}\,s^{-1}$) |
|---|---|
| 0.1 | $8.00 \times 10^{-4}$ |
| 0.2 | $1.50 \times 10^{-3}$ |
| 0.3 | $2.40 \times 10^{-3}$ |
| 0.4 | $4.00 \times 10^{-3}$ |
| 0.5 | $6.40 \times 10^{-3}$ |
| 0.6 | $9.50 \times 10^{-3}$ |
| 0.7 | $1.30 \times 10^{-2}$ |

**Table 1.1**

**(a)** Plot a graph of concentration of hydrochloric acid against the rate of reaction.

**[3]**

**(b)** Use your understanding of collision theory to explain the shape of the line on your graph.

..........................................................................................

..........................................................................................

..........................................................................................

.......................................................................................... **[3]**

**Exam tip**

First explain how increasing concentration affects the rate and then think carefully about the energetics of the reaction to explain why it curves.

**(c)** Explain why the magnesium ribbon used needed to be cleaned before investigating the rate of reaction with hydrochloric acid.

..........................................................................................

..........................................................................................

.......................................................................................... **[2]**

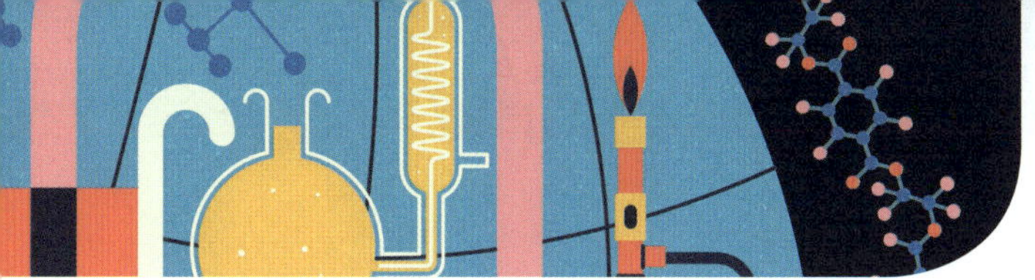



**2** This question is about equilibrium.

**(a)** Explain what happens to a system at equilibrium when the concentration of the reactants is increased.

........................................................................................

........................................................................................**[1]**

**(b)** State the meaning of the term catalyst.

Explain how a catalyst works.

........................................................................................

........................................................................................

........................................................................................**[2]**

**(c)** State the effect, if any, a catalyst has on the time taken to reach equilibrium.

........................................................................................

........................................................................................**[1]**

**Exam tip**

It is vital you know how a catalyst affects an equilibrium as this is often assessed.

**(d)** State the effect, if any, a catalyst has on the position of equilibrium.

........................................................................................

........................................................................................**[1]**

**(e)** The reaction between ethene and steam to form ethanol occurs according to the following equation:

$C_2H_4(g) + H_2O(g) \rightleftharpoons CH_3CH_2OH(g) \qquad \Delta_r H^{\ominus} = -42\ kJ\ mol^{-1}$

State the conditions that will result in the highest percentage yield of ethanol.

........................................................................................

........................................................................................

........................................................................................

........................................................................................

........................................................................................

........................................................................................**[6]**

**3** **Figure 3.1** shows the Boltzmann distribution for a sample of gas at a fixed temperature.

$E_a$ is the activation energy for the decomposition of this gas.

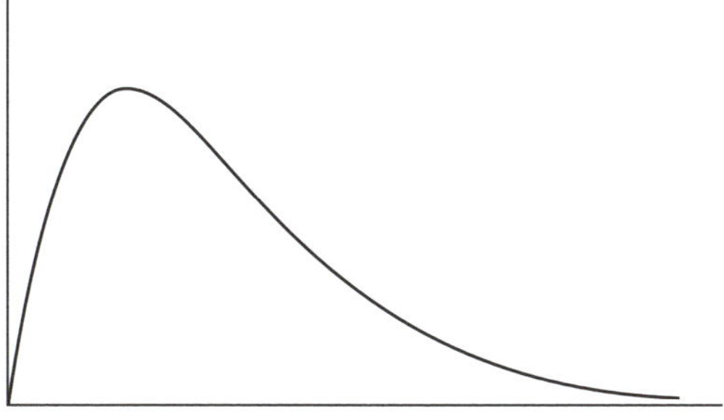

**Figure 3.1**

(a) (i) On the diagram, give labels to both axes. **[2]**

(ii) $E_{mp}$ is the most probable value for the energy of the molecules.

On the appropriate axis of the diagram, mark the value of $E_{mp}$ for this distribution. **[1]**

(iii) Suggest a new distribution for the same sample of gas at a lower temperature by adding it to the diagram. **[1]**

**Exam tip**

Take care to make sure the line only crosses the existing line once.

(b) With reference to the Maxwell–Boltzmann distribution, explain why a decrease in temperature decreases the rate of decomposition of this gas.

..............................................................................................

..............................................................................................

.............................................................................................. **[2]**

(c) State the effect that doubling the pressure has on the shape of the Maxwell–Boltzmann distribution at constant temperature. Explain your answer.

..............................................................................................

..............................................................................................

.............................................................................................. **[2]**

**4** Esters are organic compounds known for their smells. They are often found in essential oils.

A student wanted to investigate the equilibrium position of the synthesis of propyl methanoate. They placed propan-1-ol and reagent **X** into a 2 $dm^3$ vessel and heated it.

**Synoptic links**

6.1.3 5.1.2

(a) (i) Name reagent **X**.

..............................................................................................

.............................................................................................. **[1]**

(ii) Give a balanced symbol equation for the reaction.

**[1]**

(iii) State the practical set up that is required to ensure the reaction can complete.

Explain why this is needed.

..............................................................................................

..............................................................................................

.............................................................................................. **[2]**

**Exam tip**

There are two possible set ups. One is for the production of carboxylic acids, the other for esters.

**(b)** The student placed 2.00 moles of propan-1-ol and 1.50 moles of **X** into the vessel and left the mixture to reach equilibrium. When they returned, they determined the amount of propyl methanoate was 1.05 moles.

Calculate the value of the equilibrium constant, $K_c$, and state its units.

> **Exam tip**
> In this particular case, the volume of the container is not needed to calculate the answer.

$K_c$ = ....................

units = .................... **[5]**

**5** Compounds **A** and **B** react together to form an equilibrium mixture containing compounds **C** and **D** in the following equation:

$$\mathrm{A(aq)} + 2\mathrm{B(aq)} \rightleftharpoons \mathrm{C(aq)} + 3\mathrm{D(aq)} \qquad \Delta H = +14\,\mathrm{kJ\,mol^{-1}}$$

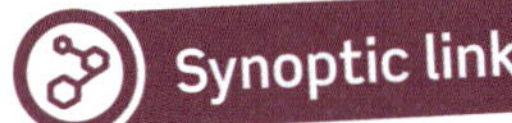


5.1.2

**(a)** State the effect on the position of equilibrium caused by the addition of a catalyst.

..........................................................................................

.......................................................................................... **[1]**

**(b)** Explain the effect on the concentration of **B** if the solution was cooled.

..........................................................................................

..........................................................................................

..........................................................................................

.......................................................................................... **[3]**

**(c)** A beaker contained 100 cm$^3$ of a 0.66 mol dm$^{-3}$ aqueous solution of **B**. $1.8\times10^{-2}$ mol of **A** was added and the mixture was allowed to reach equilibrium.

The mixture was found to contain $4.8\times10^{-2}$ mol of **B**.

**(i)** Calculate the number of moles of **A**, **C**, and **D** in the equilibrium mixture.

> **Exam tip**
> Use the balanced equation in the question stem to find the unknown amounts.

number of moles of **A** = ....................

number of moles of **B** = ....................

number of moles of **C** = .................... **[5]**

**(ii)** Give the expression for $K_c$.

Calculate its value and give its units.

$K_c$ = ....................

units = .................... **[5]**

**6** The following reaction happens between three compounds **X**, **Y**, and **Z**.

$$\mathbf{X}(aq) + 2\mathbf{Y}(aq) + \mathbf{Z}(aq) \rightarrow \mathbf{D}(aq) + \mathbf{E}(aq)$$

**(a)** Explain, using collision theory, how increasing the temperature will affect the rate of reaction. **[3]**

**(b)** In reality, this reaction can form an equilibrium if allowed to occur within a closed system.

$\mathbf{X}(aq) + 2\mathbf{Y}(aq) + \mathbf{Z}(aq) \rightleftharpoons \mathbf{D}(aq) + \mathbf{E}(aq)$ $\Delta H = -15\,kJ\,mol^{-1}$

1.50 moles of **X**, 2.00 moles of **Y**, and 0.80 moles of **Z** were placed in 500 $cm^3$ of distilled water and allowed to reach equilibrium. At equilibrium 0.60 moles of **E** were present.

Give an expression for the equilibrium constant $K_c$.
Calculate the value of $K_c$ value and give its units. **[5]**

**(c)** State the effect on the magnitude of $K_c$ if the solution at equilibrium is heated.

Explain your answer. **[3]**

**Synoptic link**

5.1.2

**Exam tip**

All explanations for changes to equilibria must be linked back to Le Châtelier's principle.

**7** Sulfur is an element in Group 6 of the Periodic Table. It is a yellow crystalline solid that does not conduct electricity.

**(a)** Give the full electron ground state configuration of sulfur. **[1]**

**(b)** Sulfur exists as molecules consisting of rings of eight sulfur atoms bonded together. Oxygen is also in Group 6 but forms diatomic molecules. Oxygen is a gas at room temperature.

Explain the differences in state at room temperature. **[4]**

**(c)** Sulfur trioxide is an important chemical feedstock. It is produced in the Contact process, where sulfur dioxide reacts with oxygen in a reversible reaction in the presence of a catalyst. The reaction is:

$$2SO_2 + O_2 \rightleftharpoons 2SO_3$$

Define dynamic equilibrium. **[2]**

**Synoptic links**

2.2.1 2.2.2 5.1.2

**Exam tip**

Any explanation of melting point or boiling point must include explanation of intermolecular forces.

**(d)** State the effect the addition of the catalyst has on the position of equilibrium. **[1]**

**(e)** Sulfur dioxide and oxygen were mixed in a 2 : 1 mol ratio and sealed in a flask with a catalyst.

The following equilibrium was established at temperature $T_1$.

$$2SO_2(g) + O_2(g) \rightleftharpoons 2SO_3(g) \qquad \Delta H^{\ominus} = -196\,kJ\,mol^{-1}$$

The partial pressure of sulfur dioxide in the equilibrium mixture was 24 kPa and the total pressure in the flask was 104 kPa.

Give the partial pressure and mole fraction of oxygen. **[2]**

**(f)** Calculate the partial pressure of sulfur trioxide in the equilibrium mixture. **[1]**

**(g)** Calculate $K_p$ for this reaction.

Give the units for $K_p$. **[4]**

**(h)** State how decreasing the volume of the container would affect the position of equilibrium.

Give a reason for your answer. **[2]**

**8** **Figure 8.1** shows the Boltzmann distribution for the production of methanol.

$$CO(g) + 2H_2(g) \rightleftharpoons CH_3OH(g) \qquad \Delta H = -90\,kJ\,mol^{-1}$$

$E_a$ labels the position of the activation energy for this reaction.

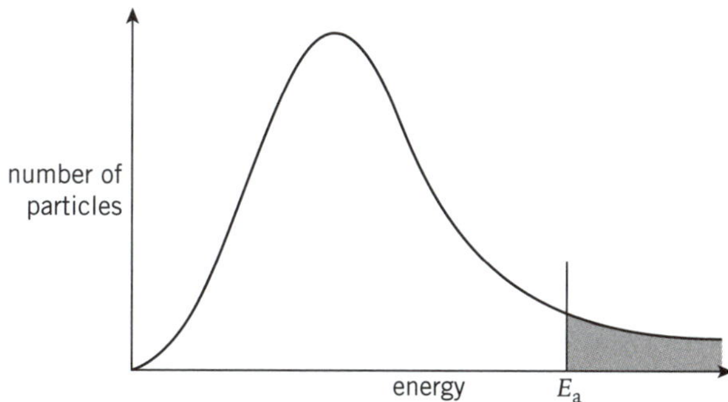


**Figure 8.1**

**(a)** Show how the activation energy would change if a catalyst was added by adding a line to the diagram.

Give your line the label $E_{cat}$. **[1]**

**(b)** Show how the distribution of molecules will change if the temperature of the reaction was increased, by adding a line to the diagram.

Explain why this will increase the rate of reaction. **[4]**

**(c)** Explain why in a manufacturing plant this reaction is carried out at a temperature of 300 °C rather than a higher temperature. **[2]**

**Exam tip**

In this kind of question, you need to explain the compromise in terms of the yield and the rate.

# Knowledge

## 10 Basic concepts

### Terms used in organic chemistry

A **homologous series** is a group of organic compounds that have the same functional group(s), but differ by $CH_2$ every time.

A **functional group** is the set of atoms that are responsible for the characteristics of a compound.

An **alkyl group** is a branch of a compound that has the general formula $C_nH_{2n+1}$.

An **aliphatic** compound has all of its carbon atoms joined together in a straight chain, with or without branches and non-aromatic rings.

A **alicyclic compound** is arranged in non-aromatic rings with or without branching, or side chains.

An **aromatic compound** contains a benzene ring.

A **saturated compound** has only carbon–carbon single bonds.

An **unsaturated compound** has one or more carbon–carbon double bonds, carbon–carbon triple bonds, or an aromatic ring.

The **general formula** of a compound can be used to predict the formula of any member of a homologous series.

### Naming of alkanes and alkyl groups

You need to know how to name the first ten members of the alkanes and their **alkyl groups** (which can be represented with the letter R).

Alkyl groups are the side chains attached to the parent chain. They have one less hydrogen than their parent chain. For example, $CH_3$ is the methyl group:

```
     H   CH3  H    H    H
     |    |   |    |    |
H —  C —  C — C —  C —  C — H
     |    |   |    |    |
     H    H   H    H    H
```

| Number of carbons | Main stem of name | Alkyl group |
|---|---|---|
| 1 | meth- | methyl- |
| 2 | eth- | ethyl- |
| 3 | prop- | propyl- |
| 4 | but- | butyl- |
| 5 | pent- | pentyl- |
| 6 | hex- | hexyl- |
| 7 | hept- | heptyl- |
| 8 | oct- | octyl- |
| 9 | non- | nonyl- |
| 10 | dec- | decyl- |

### Naming organic compounds

The IUPAC (International Union of Pure and Applied Chemistry) determine the rules for how organic compounds are named. This allows the structure of a compound to be determined exactly from the name.

The name of an organic compound is made up of several parts and rules:

- A hyphen/dash is used in between words and numbers and a comma is used in between numbers (e.g. pentan-1,2-diol)
- The prefix or suffix specifies some of the branched chains and functional groups.
- The main stem of the name specifies the number of carbons in *the longest continuous chain*.
- If more than one functional group is present, the compound will be named for the highest priority functional group.

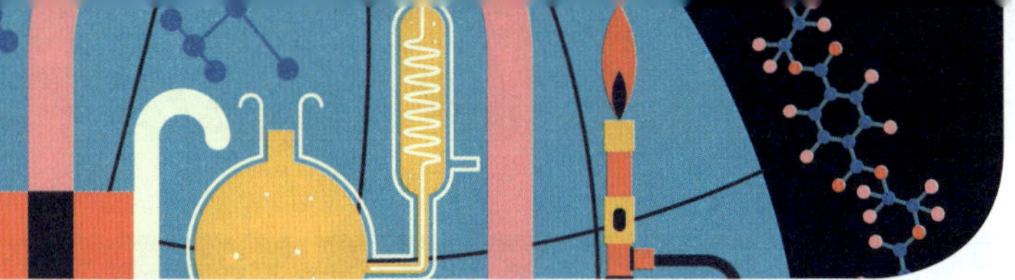



## Deciding on the stem of the name

The stem of the name comes from the *longest continuous carbon chain*. This is not always straight; they can go around corners.

This is 2,3-dimethyl-pentane, *not* 2-ethyl-3-methylbutane.

```
    H     H     H   H
    |     |     |   |
H — C ——— C ——— C — C — H
    |     |     |   |
    H H — C — H |   H
          |     |
          H H — C — H
                |
            H — C — H
                |
                H
```

## Cyclo- prefix

Organic compounds do not have to be in straight lines; they can go around in a loop; this is indicated by the *cyclo-* prefix. You should also be able to determine this from the number of hydrogens that pentadiene is $C_5H_8$ whereas cyclopentadiene would be $C_5H_6$. In $C_5H_6$ there are not enough carbons to be an alkene with two double bonds in a straight chain, so the 'missing' two hydrogens must come from the joining of a carbon chain in a ring.

This is benzene:

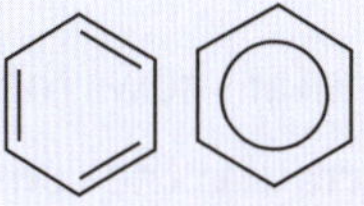

This is cyclohexane:

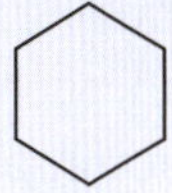

## Homologous series and their suffixes

| Homologous series | Suffix or prefix |
|---|---|
| alkanes | -ane |
| alkenes | -ene |
| haloalkanes | chloro-<br>bromo-<br>iodo-<br>fluoro- |
| alcohols | -ol |
| aldehydes | -al |
| ketones | -one |
| carboxylic acids | -oic acid |

## Indicating the location of functional groups

The position and type of branched chain on a compound can be identified by the name. The name of the compound will have the functional groups on the lowest carbon number on the longest carbon chain. So, occasionally you must number compounds from the right instead of from the left.

For example, the $-OH$ group is positioned on the second carbon from the left on butan-2-ol, not butan-3-ol.

```
    H   H   OH  H
    |   |   |   |
H —4C —3C —2C —1C — H
    |   |   |   |
    H   H   H   H
```

butan-2-ol

The numbers in a name indicate the location of the functional groups. For example, 'dichloroethane' does not give enough information to determine the structure:

```
     Cl H
     |  |
Cl —1C—2C— H   is called 1,1-dichloroethane
     |  |
     H  H

     Cl Cl
     |  |
and H—1C—2C— H   is called 1,2-dichloroethane.
     |  |
     H  H
```

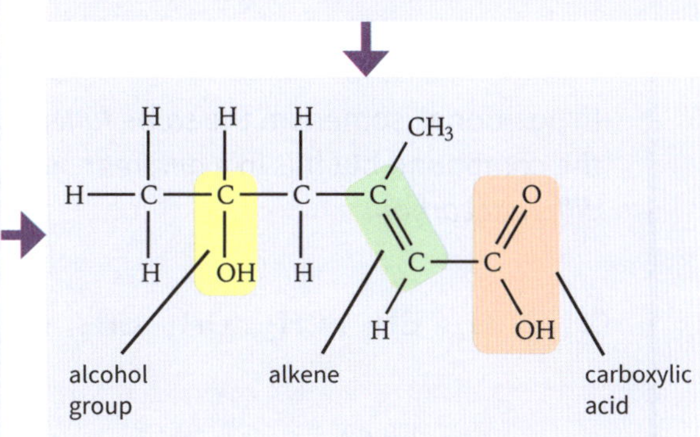


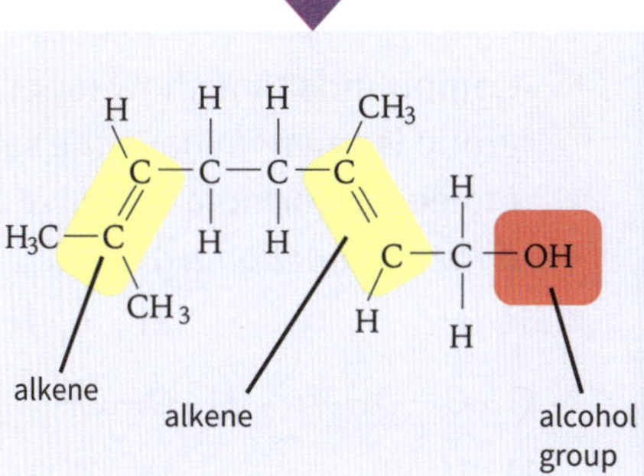


A compound that has multiple different functional groups needs to have each of the groups listed alphabetically. For example, 2-bromo-1-iodopropane is not listed numerically.

```
     I  Br  H
     |  |   |
H —1C —2C —3C — H    2-bromo-1-iodopropane
     |  |   |
     H  H   H
```

## Types of formula for organic compounds

Organic compounds can be represented by different types of formula.

**General formula:** shows the simplest algebraic formula.

**Displayed formula:** the compound fully drawn out, showing *every* bond and atom.

**Structural formula:** moving from left to right of the displayed formula of a compound, each carbon is written separately with the other atoms that accompany it, no bonds are shown.

**Skeletal formula:** just shows the bonds and non-carbon atoms. Hydrogen atoms are only shown if not attached to a carbon atom.

**Empirical formula:** the simplest whole number ratio of atoms of each element within a compound.

**Molecular formula:** shows the actual number of each element that make up a compound.

## Butan-1,3-diol

**Empirical formula:** $C_2H_5O$

**Molecular formula:** $C_4H_{10}O_2$

**General formula:** $C_4H_9OH$

**Structural formula:** $CH_2OHCH_2CHOHCH_3$

**Displayed formula:**

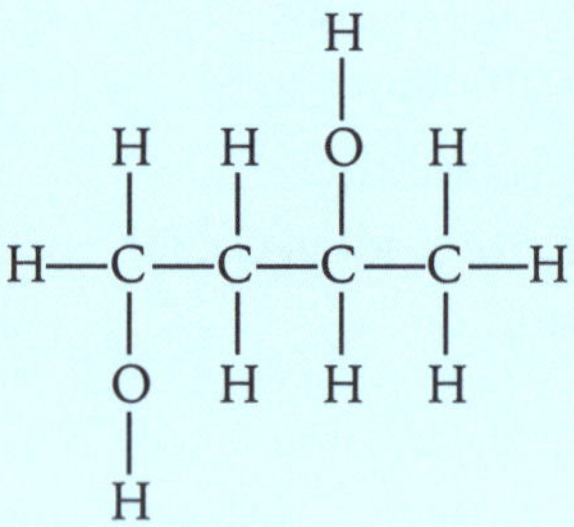

**Skeletal formula:**

OH OH

## Structural isomerism

**Structural isomers** are compounds with the same molecular formula, but they are arranged differently (have different structural formulae). These could be:

- **Functional group isomerism** where the compounds have different functional groups.
- **Positional isomerism** where the compounds have the same functional groups, but they are attached in different places.
- **Chain isomerism** when a compound has undergone branching.

Compounds that show functional group isomerism have the same molecular formula but have a *different functional group*.

```
   H  H             H       H
   |  |             |       |
H—C—C—OH    or   H—C—O—C—H
   |  |             |       |
   H  H             H       H
```

ethanol (an alcohol) methoxymethane (an ether)

In positional isomerism the *same functional group* is in the compound but it is in a *different place, attached to a different carbon*.

$CH_3—CH_2—CH_2—CH_2—OH$

butan-1-ol

```
CH3—CH2—CH—CH3
         |
         OH
```

butan-2-ol

```
       CH3
        |
CH3 —C —CH3
        |
        OH
```

1,1-dimethylethanol

Compounds that show chain isomerism have *different branches off the hydrocarbon chain*.

```
           H
           |
         H—C—H
   H  H    |    H  H
   |  |    |    |  |
H—C—C—C—C—C—H
   |  |    |    |  |
   H  H    H    H  H
```

3-methylpentane

```
          H
          |
        H—C—H
   H  H   |   H
   |  |   |   |
H—C—C—C—C—H
   |  |   |   |
   H  H   |   H
        H—C—H
          |
          H
```

2,2-dimethylbutane

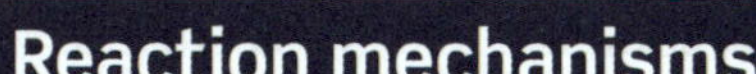

## Reaction mechanisms

Covalent bonds can break either by **homolytic fission** or **heterolytic fission**.

- In homolytic fission, both atoms get one electron from the bond.
- In heterolytic fission, one of the atoms in the bond gets both electrons.

A **reaction mechanism** is a step-by-step sequence of reactions. It shows *how the reaction takes place*. When drawing reaction mechanisms, the following rules apply:

- An atom or group of atoms with an unpaired electron is called a **radical** and is represented by a single dot.
- The formation of a covalent bond is shown by a curly arrow.
- Curly arrows start where the electrons started.
- Curly arrows need to end where the electrons end up.
- Half curly arrows represent one electron.
- Breaking of covalent bonds can be shown by a curly arrow starting from the bond.
- Curly arrows always need to start from a bond, or a lone pair of electrons, or a negative charge.

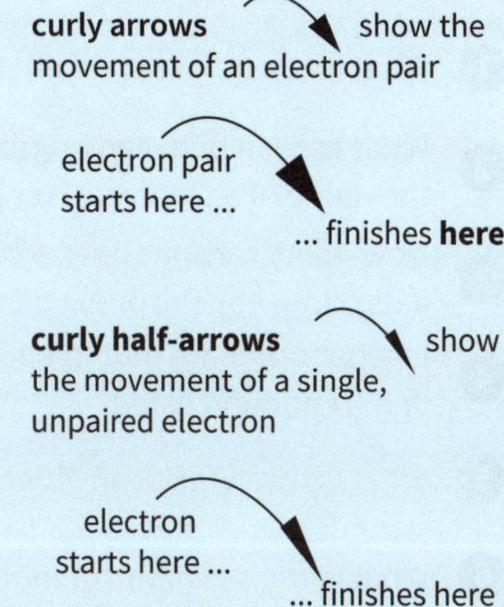
**curly arrows** show the movement of an electron pair
electron pair starts here ...
... finishes **here**
**curly half-arrows** show the movement of a single, unpaired electron
electron starts here ...
... finishes here

## Heterolytic fission

In heterolytic fission a covalent bond breaks (shown by the start of a curly arrow) and one of the bonded atoms ends up with both electrons.

The use of a full arrow shows the movement of two electrons. This results in two ions being formed: the atom that receives both electrons becomes a negative ion and the atom that loses electrons subsequently has a positive charge.

$$H_3C^{\delta+}\text{—}Cl^{\delta-} \xrightarrow{\text{heterolytic fission}} CH_3^{\oplus} + Cl^{\ominus}$$

## Homolytic fission

In homolytic fission the covalent bond breaks evenly. The bond breaking is represented by the curly single headed arrow, but, in this case, the arrow is going in two directions, with one electron going to each atom.

Each atom then becomes a radical, each having an unpaired electron. Unpaired electrons can be represented by a dot.

$$Cl\text{—}Cl \xrightarrow{uv} 2Cl^{\bullet}$$

For answers and more practice questions visit www.oxfordrevise.com/scienceanswers

Even more practice and interactive revision quizzes are available on kerboodle

# Retrieval

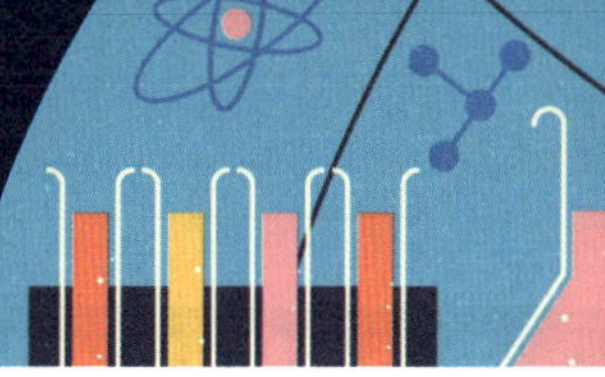

Learn the answers to the questions below, then cover the answers column with a piece of paper and write as many as you can. Check and repeat.

| | Questions | Answers |
|---|---|---|
| 1 | What is the definition of empirical formula? | the simplest whole number ratio of atoms of each element within a compound |
| 2 | What goes in between numbers and letters in the name of a chemical compound? | – / a dash |
| 3 | How many carbons does a compound have if its name has the prefix pent- ? | 5 |
| 4 | In what order are multiple functional groups in a compound listed? | alphabetically |
| 5 | What is the suffix of an aldehyde? | -al |
| 6 | What is the definition of molecular formula? | the actual numbers of atoms of each element that make up a compound |
| 7 | What does a single dot represent in a reaction mechanism? | an unpaired electron |
| 8 | What is functional group isomerism? | same molecular formula but different functional groups |
| 9 | What type of isomerism is shown if two compounds have the same molecular formula but have different branches? | chain |
| 10 | What type of fission occurs when a bond breaks and each bonded atom gets one electron from the bond? | homolytic |
| 11 | How are functional groups within a compound numbered? | from the lowest possible number carbon on the longest carbon chain |
| 12 | What does a curly arrow represent in a reaction mechanism? | movement of electrons |
| 13 | What is the definition of heterolytic fission? | breaking of a covalent bond, when one of the bonded atoms receives both the electrons from the bond |
| 14 | What is an aromatic compound? | one that contains a benzene ring |
| 15 | What is the definition of structural formula? | moving from left to right of a compound, each carbon is written separately with the other atoms that accompany it, no bonds are shown |
| 16 | What goes in between numbers in the name of a chemical compound? | a comma |
| 17 | What prefix is used if a branch off a carbon chain has ten carbons in it? | decyl- |
| 18 | How is the breaking of a covalent bond shown in a reaction mechanism? | a curly arrow starting from the bond |

Put paper here

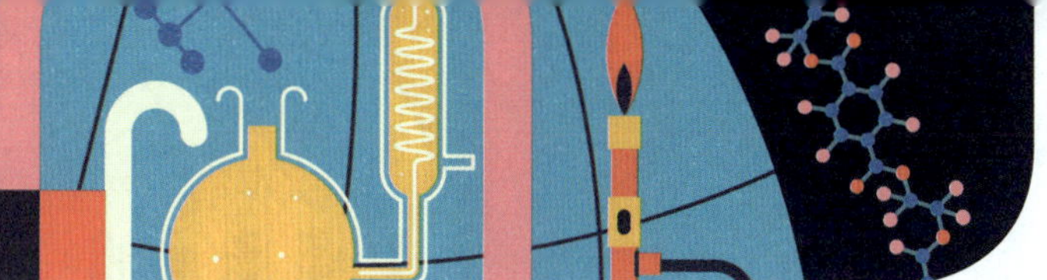



| | Question | | Answer |
|---|---|---|---|
| 19 | What types of bond are in an unsaturated compound? | Put paper here | carbon–carbon double bonds, carbon–carbon triple bonds, or an aromatic ring |
| 20 | What is a cyclic compound? | | contains a non-aromatic ring |

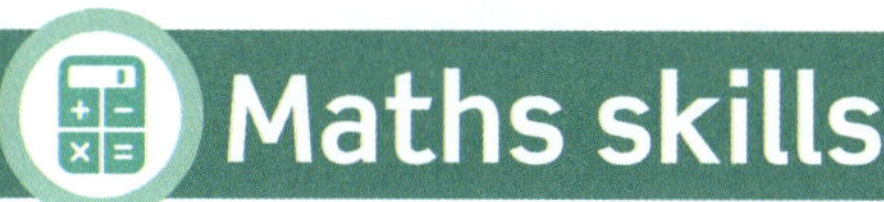

# Maths skills

Practise your maths skills using the worked example and practice questions below.

### Mathematical symbols

You are likely to have come across many more mathematical symbols than before.

$<<$ and $>>$ mean much greater/less than.

For example:

$6 \times 10^{20} >> -0.4$ or

$[H^+(aq)] >> [OH^-(aq)]$ in $HCl(aq)$

The symbol $\propto$ means proportional to and is often used in expressions for the rate of reaction.

- $x \propto y$

  This means that $x$ is proportional to $y$. If one of them increases by a factor then the other increases by the same factor. For example, if $x$ doubles then $y$ doubles.

- $x \propto \frac{1}{y}$

  This means that $x$ is inversely proportional to $y$. So if $x$ doubles, $y$ halves.

- $x^2 \propto y$

  If $x$ doubles, $y$ quadruples.

  The symbols $\approx$ and $\sim$ mean approximately equal to and similar to. This means that values are of similar orders of magnitude.

### Worked example

Which of the following symbols ($=, <, <<, >>, >, \propto, \sim/\approx$) could replace the box in the following examples?

**Questions**

1 $V \square n$ ($V$ is volume of gas, $n$ is moles of gas)

2 $[HA(aq)] \square [H^+(aq)]$ where HA is a weak acid.

**Answers**

1 As $V$ and $n$ are different quantities you cannot compare them, so $=, <, <<, >, >>, \sim$ are left out. The only symbol left is $\propto$. Does this make sense? Is the volume of a gas proportional to the number of moles of gas? The volume of one mole of gas is 24 $dm^3$, the volume of two moles of gas is 48 $dm^3$. If one doubles, so does the other and the constant of proportionality is 24 $dm^3\,mol^{-1}$. So, $V \propto n$.

2 In a weak acid, not many of the acid molecules dissociate. Therefore $[HA(aq)] >> [H^+(aq)]$. It has to be $>>$ and not $>$ because otherwise the assumption in calculating $K_a$ would not hold.

### Practice

1 Which of the following statements are true?

  **a** $\pi^2 \propto 10$

  **b** $\pi^2 = 10$

  **c** $\pi^2 \sim 10$

  **d** $\log_{10} 1\,000\,000 = 6$

  **e** $e^{-100} >> 100$

  **f** For a reaction that is zero order with respect to X, rate $\propto [X]$

  **g** For any chemical reaction, rate $\propto T$ ($T$ is temperature in kelvin)

  **h** $\Delta_c H^\ominus(C_3H_8) >> \Delta_c H^\ominus(C_2H_6)$ (ignore minus signs)

2 Which symbol best replaces the box?

  **a** $A_r$ (Se) $\square$ $A_r$ (S)

  **b** Number of carbon atoms in a diamond ring $\square$ number of polymers chains in a plastic supermarket bag.

  **c** Number of girls taking A level chemistry in England $\square$ number of boys taking A level chemistry in England.

# Practice

## Exam-style questions

1 A straight-chain saturated hydrocarbon, **P**, contains five carbon atoms per molecule.

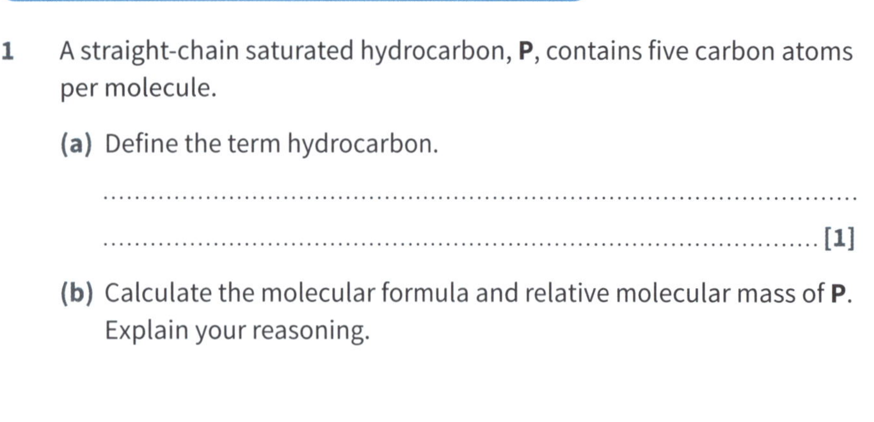

**Synoptic link**

2.1.1e

**(a)** Define the term hydrocarbon.

..........................................................................................

.................................................................................. **[1]**

**(b)** Calculate the molecular formula and relative molecular mass of **P**. Explain your reasoning.

**[3]**

**(c)** There are three structural isomers of **P**.

Define the term structural isomerism.

..........................................................................................

.................................................................................. **[1]**

**(d)** Give the structural formulae and give the systematic name for each of the three structural isomers of **P**.

**[1]**

**(e)** Calculate the number of atoms in 0.36 g of **P**.

Avogadro's constant = $6.022 \times 10^{23}\ mol^{-1}$

**Exam tip**

The question asks about the number of atoms, not molecules.

number of atoms = .................... **[3]**

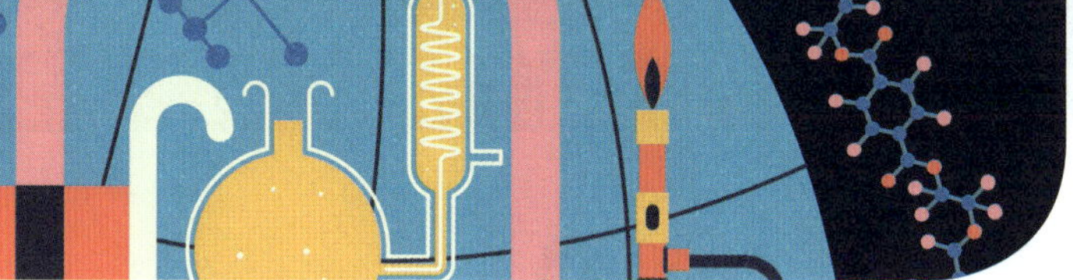



2 **Figure 2.1** shows the skeletal formulae for three isomeric unsaturated hydrocarbons.

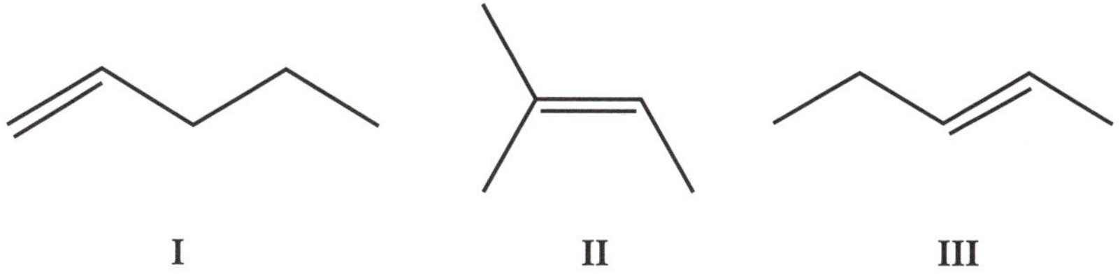


**Figure 2.1**

**(a)** Give the systematic name for each isomer.

**I** ..........................................................................................

**II** ..........................................................................................

**III** .......................................................................................... **[3]**

**(b) (i)** Explain the term stereoisomer.

..........................................................................................

.......................................................................................... **[2]**

**(ii)** One of the isomers from **I** to III has a stereoisomer.

Identify which isomer (**I** or **II** or **III**) and give the skeletal formula of its stereoisomer.

**[2]**

**(iii)** Describe the requirements of this type of stereoisomerism.

..........................................................................................

..........................................................................................

.......................................................................................... **[2]**

**(c)** **Figure 2.2** shows the skeletal formula of a saturated hydrocarbon (**IV**), which is an isomer of these three hydrocarbons. Give the systematic name of isomer **IV**.

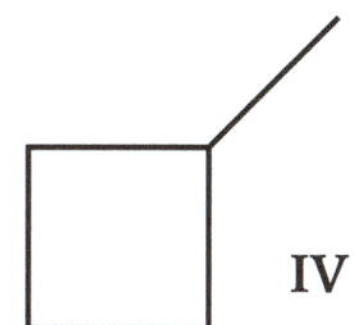


**Figure 2.2**

.......................................................................................... **[1]**

Identify the longest carbon chain and then your functional group is on the lowest numbered carbon.

3 In an experiment to find the relative molecular mass of an organic compound **Q**, 0.20 g of liquid **Q** was injected into a gas syringe to form 85.0 $cm^3$ of gaseous **Q**. The gas syringe was encased in a syringe oven at a temperature of 450 K. The pressure was measured as 1.01 kPa.

Gas constant, $R = 8.31\,J\,K^{-1}\,mol^{-1}$

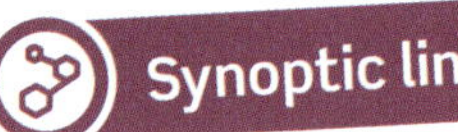

2.1.3f

Use the ideal gas equation and rearrange to find $n$, the number of moles.

**(a)** Using this data, calculate the relative molecular mass of **Q**. Show your working.

molecular mass of **Q** = .................... **[4]**

**(b)** The actual relative molecular mass of **Q** was found to be 88 from its molecular ion peak on its mass spectrum.

Describe what is meant by the molecular ion peak.

..........................................................................................

..........................................................................................

..........................................................................................

.......................................................................................... **[1]**

**(c)** Calculate the percentage error of the method described above. Give your answer to 3 significant figures.

percentage error = ....................% **[2]**

**(d)** The incomplete diagrams in **Figure 3.1** show the displayed formulae of two compounds having a relative molecular mass of 88.

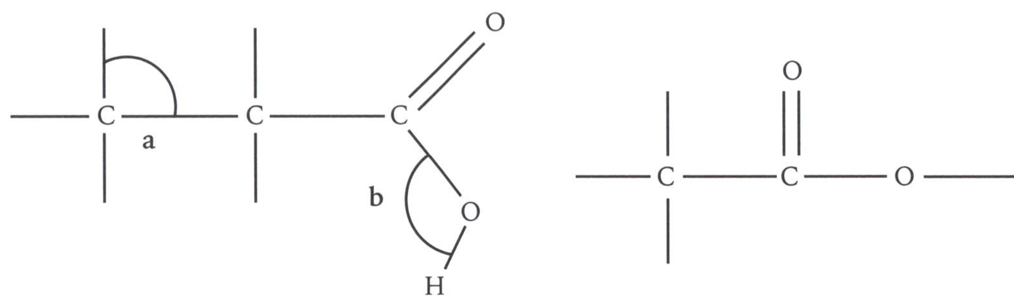


**Figure 3.1**

(i) Give both displayed formulae by inserting the missing carbon and hydrogen atoms into **Figure 3.1**. [3]

(ii) Suggest values for the angles **a** and **b** in the diagram.

........................................................................................

........................................................................................

........................................................................................

........................................................................................

........................................................................................ [2]

4 In carbon compounds there are polar and non-polar bonds.

(a) Explain why the C—Cl bond is a polar bond.

........................................................................................

........................................................................................

........................................................................................

........................................................................................ [3]

**Synoptic link**

2.2.2i

(b) (i) Explain why the compound $CCl_4$ (tetrachloromethane) contains four polar C—Cl bonds but is not a polar molecule.

........................................................................................

........................................................................................

........................................................................................

........................................................................................

........................................................................................ [2]

(ii) Draw the $CHCl_3$ (trichloromethane) molecule with the symbols δ+ and δ− to indicate the dipole present in the molecule.

**Exam tip**

Draw all the bonds and place symbols where there are the highest and lowest concentrations of electrons.

[2]

**(c)** **Figure 4.1** shows the structure of the compound propanone. The diagram includes the dipoles present in the molecule.

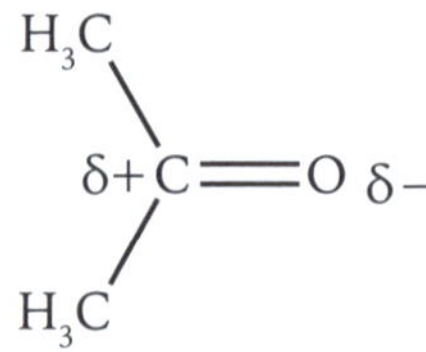


**Figure 4.1**

When propanone is added to trichloromethane there is a small temperature rise.

Suggest a reason for this observation and draw a diagram to support your answer.

..........................................................................................

.......................................................................................... **[3]**

**5** The compounds butan-1-ol (boiling point 118 °C) and pentane (boiling point 36 °C) have similar relative molecular masses but different physical properties.

2.2.2k 2.2.2l 2.2.2j 1.2.1a

**(a)** Give the structural formulae of both compounds.

**[2]**

**(b)** Explain the difference in boiling points of the two compounds.

..........................................................................................

..........................................................................................

..........................................................................................

.......................................................................................... **[3]**

**(c)** A group of students was asked to investigate the polarity of both compounds. They were given a silk rag, a nylon rod, and two burettes.

Explain how they could show the difference in polarity using this apparatus and predict the results of their investigation.

..........................................................................................

..........................................................................................

..........................................................................................

..........................................................................................

..........................................................................................

.......................................................................................... **[5]**

6 This question concerns the two compounds with molecular formulae $C_2H_4Br_2$ and $C_2H_2Br_2$.

**(a)** Give the structural formulae and systematic names of the isomers of $C_2H_4Br_2$

**[2]**

**(b) (i)** Give the displayed formulae and systematic names of the isomers of $C_2H_4Br_2$.

**Exam tip**

Consider the mark allocation. Three marks suggests three isomers.

**[3]**

**(ii)** For each of the three isomers of $C_2H_2Br_2$, explain whether or not they are polar molecules.

........................................................................................

........................................................................................

........................................................................................

........................................................................................ **[3]**

**(iii)** Suggest the values of the bond angles in both $C_2H_4Br_2$ and $C_2H_2Br_2$.

Make it clear which compound you are referring to.

........................................................................................

........................................................................................

........................................................................................ **[2]**

# Knowledge

## 11 Hydrocarbons

### Properties of alkanes

#### Bonding in alkanes

Alkanes are **saturated hydrocarbons**, meaning that all of the carbon–carbon bonds are single bonds.

Their general formula is $C_nH_{2n+2}$. Bonding in alkanes is the overlapping of orbitals in bonding atoms, forming **sigma bonds (σ-bonds)**. These are single bonds, either C—H (where the s-orbital from hydrogen overlaps with a p-orbital from carbon) or C—C (where two p-orbitals overlap). These bonds have free rotation.

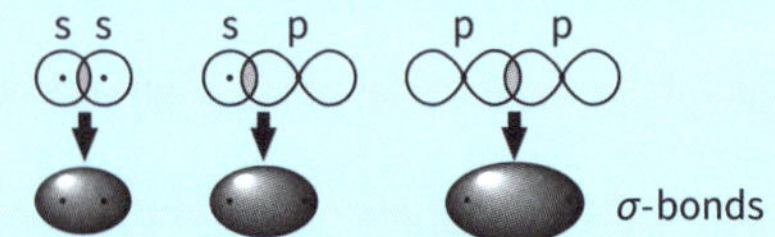


### Shape of alkane molecules

Around each carbon atom the bonded atoms are arranged in a *tetrahedral shape*. This can lead to several tetrahedral shapes within one compound, giving a chevron-shaped appearance. Due to the free rotation around the single bonds, the atoms are not always 'neatly' arranged like this. The shape is due to the **electron repulsion** of the four **electron pairs** around each carbon atom.

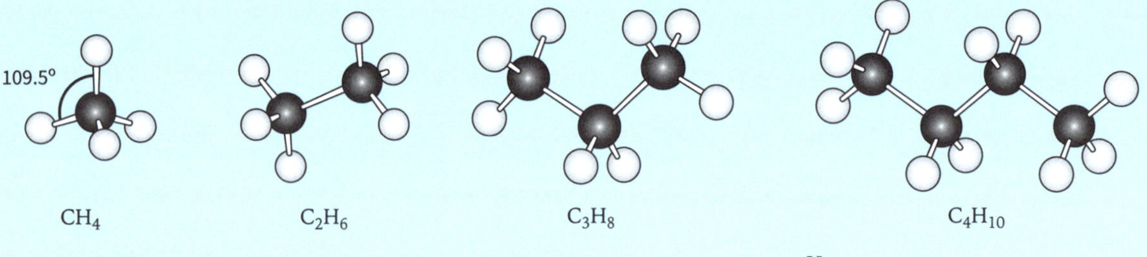


Compounds are drawn in 3D, using lines to show bonds in line with the page. Solid wedges are used to show bonds out in front of the page and dashed wedges are used to show bonds that are going behind the page.

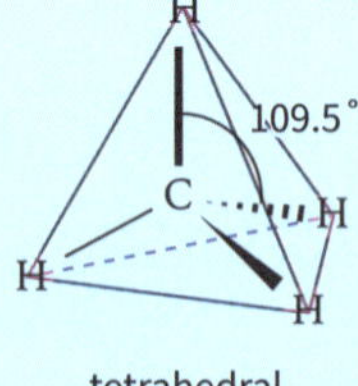

tetrahedral

### Boiling points of alkanes

The boiling points of alkanes (and other organic compounds) *increase as the length of the carbon chain increases*.

- The boiling points depend on the induced dipole–dipole (**London forces**) interaction between the molecules. Longer chains have more points of contact, so will have more London forces between compounds, so more energy is needed to overcome these forces. *The greater the intermolecular forces, the higher the boiling point.*
- **Branched alkanes** have lower boiling points compared to alkanes with similar length straight carbon chains. Branched compounds have fewer points of contact than their straight chain isomers. The more branched a compound is, the fewer London forces, decreasing the overall intermolecular forces. *Less energy is needed to overcome these weaker forces so this lowers the boiling point.*

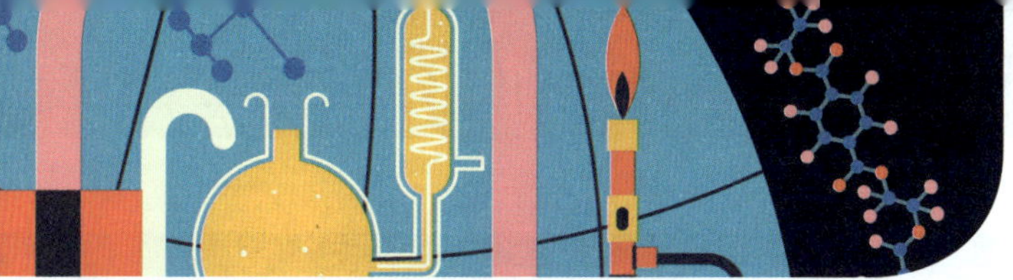



## Reactivity of alkanes

Alkanes have a *low reactivity* due to:

- the high bond enthalpy of the bonds (C—C and C—H sigma bonds are strong)
- the lack of polarity of the σ-bonds (C—C bonds are non-polar)
- the C—H bonds having very similar electronegativity, so they act as if they were non-polar.

## Combustion of alkanes

Alkanes can be used as fuels and burn readily in oxygen. Alkanes, and other organic compounds, can undergo complete or incomplete combustion.

**Complete combustion** is in *excess oxygen* and produces carbon dioxide and water as the only products.

**Incomplete combustion** is in a *limited supply* of oxygen, and a range of products are formed, including toxic carbon monoxide, carbon particles (soot), as well as carbon dioxide and water.

## Reactions of alkanes with the Halogens

The reaction between methane and chlorine (and also for bromine) is a chain reaction with a series of steps. Chloromethane and hydrogen chlorine (hydrochloric acid) are produced:

$$CH_4(g) + Cl_2(g) \rightarrow CH_3Cl(g) + HCl(g)$$

The process is called **radical substitution**. There are three stages: **initiation, propagation,** and **termination**.

## Initiation

The initiation stage uses UV radiation to break the Cl—Cl bond by homolytic fission. The bond breaks evenly, forming two chlorine free radicals, each with one unpaired electron. These are highly reactive.

$$:\ddot{Cl}:\ddot{Cl}: \xrightarrow{uv} 2\ :\ddot{Cl}\cdot$$

## Propagation

**Step 1** The chlorine radical removes hydrogen from methane to form hydrogen chloride and a methyl radical.

$$:\ddot{Cl}\cdot + H:CH_3 \longrightarrow H:\ddot{Cl}: + \cdot CH_3$$

**Step 2** The highly reactive methyl radical reacts with another chlorine molecule and breaks the Cl—Cl bond. $CH_3Cl$ and another chlorine radical are formed.

$$\cdot CH_3 + :\ddot{Cl}:\ddot{Cl}: \longrightarrow :\ddot{Cl}:CH_3 + :\ddot{Cl}\cdot$$

## Termination

The end of the reaction is called termination. This stage removes all of the radicals from the reactions and stops the reaction. A range of radicals are formed, which can react with each other in a number of ways. In each case, two radicals will react to give a compound with fully paired electrons. This could be chlorine gas again, more chloromethane, or ethane.

$$:\ddot{Cl}\cdot + \cdot\ddot{Cl}: \longrightarrow :\ddot{Cl}:\ddot{Cl}:$$

$$:\ddot{Cl}\cdot + \cdot CH_3 \longrightarrow :\ddot{Cl}:CH_3$$

$$H_3C\cdot + \cdot CH_3 \longrightarrow H_3C:CH_3$$

## Properties of alkenes

### Bonding in alkenes

Alkenes are **unsaturated hydrocarbons** with a carbon–carbon double bond. Alkenes can have more then one double bond, but for compounds with *one carbon–carbon double bond* they have the general formula $C_nH_{2n}$.

A double bond is made up of a $\sigma$-bond and a $\pi$-bond.

- The $\sigma$-bond is the overlapping of the orbitals directly between the atoms.
- The $\pi$-bond is the overlapping of the p-orbitals above and below the bonding carbon atoms.

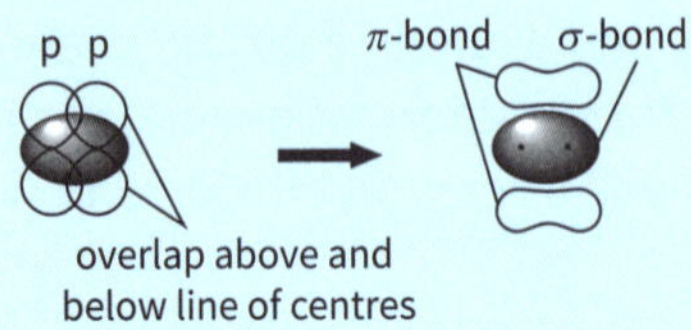


## Shape of alkene molecules

Each carbon atom has three electron-dense regions, which repel each other as far apart as possible, with a 120° bond angle between each carbon atom. This leads to the shape around each carbon in a double bond being *trigonal planar*, with all bonds being in the same plane.

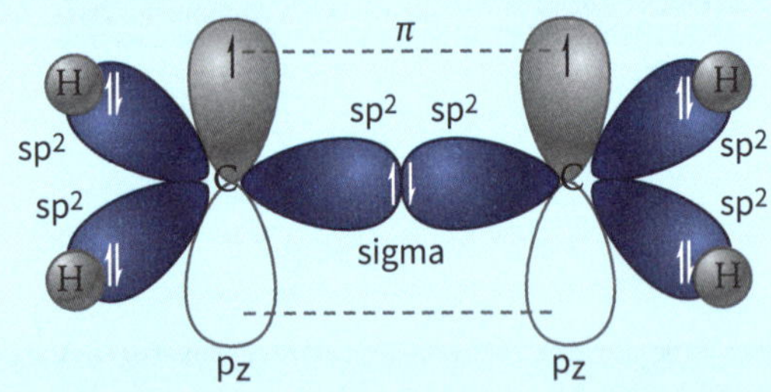


## Stereoisomerism

Stereoisomers are compounds with the same structural formula but the atoms have a different arrangement in space. There can be ***E-Z* isomerism** and **optical isomerism**.

$CH_3$ $CH_3$ / C=C / H H

*Z*-but-2-ene

$CH_3$ H / C=C / H $CH_3$

*E*-but-2-ene

## *E-Z* isomerism

This occurs due to the restricted rotation around double bonds and when there are two different groups attached to each carbon around a double bond.

- *E-isomers* have high priority groups on *opposite sides*.
- *Z-isomers* have both highest priority groups on the *same side*.

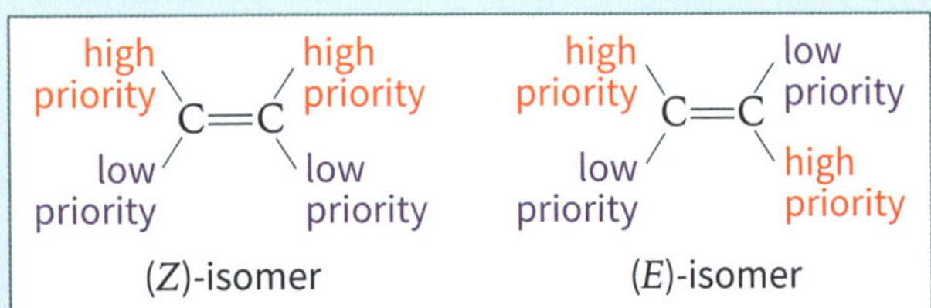


## *Cis–trans* isomerism

*Cis–trans* isomerism is a special case of *E-Z* isomerism when:

- two of the groups attached to each of the carbon atoms in the double bond are the same (e.g. in but-2-ene there are two $-CH_3$ groups)
- a hydrogen atom is attached to each carbon in the double bond.

The *cis*-isomer is the *Z*-isomer. The *trans*-isomer is the *E*-isomer.

$H_3C$ $CH_3$ / C=C / H H

*cis*-but-2-ene

$H_3C$ H / C=C / H $CH_3$

*trans*-but-2-ene

## Cahn-Ingold-Prelog

If there are four different groups around a C=C bond, the **Cahn-Ingold-Prelog** (CIP) priority rules are used to determine which two groups are the highest priority. The mass of the atoms that are bonded directly to each carbon within the double bond, determines the priority.

- The atoms with the highest mass have the highest priority: if they are on the same side of the double bonds it is $Z$-isomer; if they are on different sides of the double bond it is an $E$-isomer.
- If the atoms bonded directly to the carbon atom of the double bond are the same, then the atoms bonded to these atoms are examined to determine which has the highest priority.

## Addition reactions of alkenes

Compared to the $\sigma$-bond (347 kJ mol$^{-1}$), a $\pi$-bond (265 kJ mol$^{-1}$) has a relatively low bond enthalpy. The difference in strength of the $\sigma$- and $\pi$-bonds is the key to the reactivity of alkenes. The $\sigma$-bond *is stable and requires a large amount of energy to* break, the $\pi$-bond is much weaker and will break more readily. In an **addition reaction** the $\pi$-bond breaks, new bonds form, and a small molecule is added.

### Addition of hydrogen

The unsaturated double bonds in alkenes can be removed by the addition of hydrogen with a nickel catalyst at 423 K to form an alkane. This process is know as **hydrogenation**.

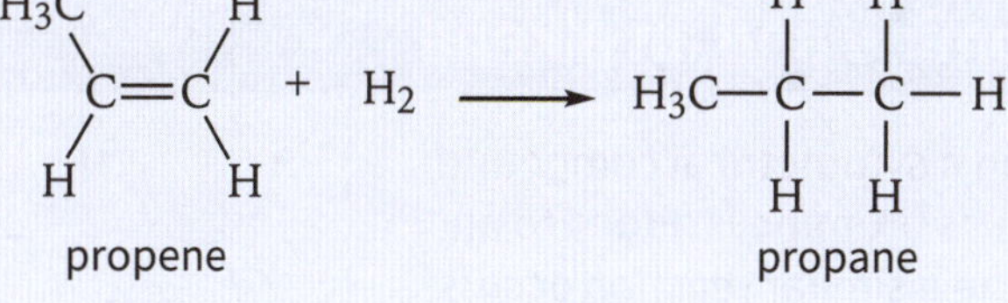

### Addition of halogens

$H_3C(H)C{=}CH_2$ + $Br_2$ ⟶ $H_3C{-}CH(Br){-}CH_2(Br)$

propene

1,2-dibromopropane

The addition of diatomic Halogens across a double bond can be used as a test to determine if a double bond is present in a compound. When bromine water is added to a compound with a double bond the bromine atoms add across the double bond. The solution will decolourise and a dibromoalkane forms.

### Addition of steam

Alcohols are formed when an alkene and steam react in the presence of phosphoric acid catalyst. The water molecule will add as a hydrogen group and a hydroxyl group. Steam is an *unsymmetrical molecule* and when it reacts with an *unsymmetrical alkene* two different alcohols are produced.

$CH_3{-}CH{=}CH_2$ + $H_2O(g)$ ⟶ $CH_3{-}CH_2{-}CH_2OH$ (propan-1-ol) or $CH_3{-}CH(OH){-}CH_3$ (propan-2-ol)

propan-1-ol

propan-2-ol

## Electrophilic addition of alkenes

An **electrophile** is an *electron pair acceptor*: it is attracted to electron-dense regions.

An **addition reaction** is where one compound or part of a compound is added to another.

Thus **electrophilic addition** is the adding of an electrophile to another compound.

## Hydrogen bromide and an alkene

1. Hydrogen bromide is an unsymmetrial compound and has a dipole. The slightly positive hydrogen is attacked by the electron-dense region around the double bond.
2. The $\pi$-bond breaks and a bond forms between one of the carbon atoms and the hydrogen atom.
3. The bond in H—Br breaks by heterolytic fission.
4. These two ions are attracted to each other to complete the reaction.

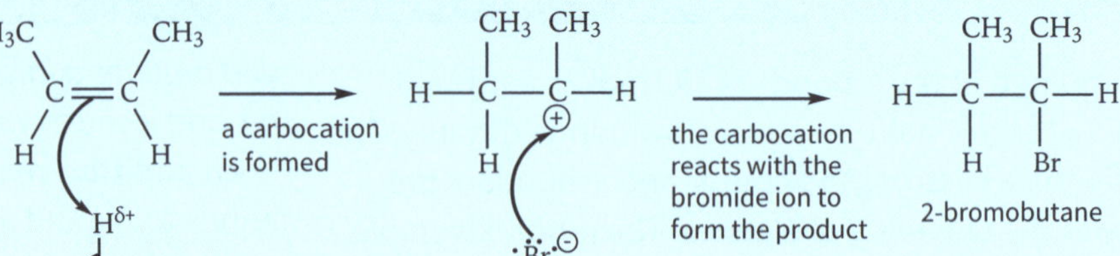

## Bromine and an alkene

When a symmetrical compound such as bromine is added to an alkene a similar reaction occurs.

induced dipole

a carbocation is formed

the carbocation reacts with the bromide ion to form the product

1,2-dibromopropane

### Markownikoff's rule

When an unsymmetrical alkene reacts with an unsymmetrical molecule a mixture of two different products is formed. Markownikoff's rule is used to predict the formation of the major and minor products. The hydrogen of the hydrogen halide will be attracted preferentially to the carbon that has the most hydrogen atoms and the least carbon atoms already attached to it.

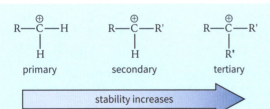

*primary* carbocation

1-bromopropane

***minor product***

*secondary* carbocation

2-bromopropane

***major product***

primary

secondary

tertiary

stability increases

The addition of a hydrogen halide across a double bond to give the major products happens via the most stable carbocation intermediate.

## Polymers from alkenes

Alkenes can undergo **addition polymerisation**: the double bond breaks allowing many alkenes (**monomers**) to join together to form a **polymer**. For example, monomers of propene can join together to polypropene.

You will need to be able to draw polymers from monomers and to identify repeating units in a polymer chain.

## Polymers and the environment

Polymers made from alkenes make plastics. The alkenes are from crude oil which is a finite resource, and the use of plastics has a number of environmental problems. Plastic polymers are cheap to produce and widely used for a range of purposes. Unfortunately these products often end up in landfill so there is an increasing move to re-use waste and find alternatives.

Polymers are made from crude oil and store a lot of energy within them. **Waste polymers** can be used as fuels to generate heat or to generate steam to drive a turbine which can produce electricity.

Polymers can be recycled as feedstock in the production of new products. Monomers can be reclaimed from unsorted and unwashed plastics waste.

## Polyvinylchloride (PVC)

Polyvinylchloride (PVC) or polychloroethene has a high chlorine content. When PVC is disposed of, chlorine is released, so dumping it in landfill is not suitable and burning PVC releases corrosive hydrogen chloride gas. This toxic waste product needs to be removed safely or the material needs to be recycled.

## Alternative polymers

As an alternative to traditional polymers, new polymers that are biodegradable or photodegradable are being developed. This is going to reduce our dependency on finite resources and reduce the problem of disposal of waste plastics.

**Biodegradable polymers** can be broken down by microorganisms into water, carbon dioxide, and biological compounds. These leave no toxic residue and are better for the environment. **Photodegradable polymers** are oil-based but have been treated so that light can be used as a trigger to start the degradation process.

# Retrieval

Learn the answers to the questions below, then cover the answers column with a piece of paper and write as many as you can. Check and repeat.

| | Questions | Answers |
|---|---|---|
| 1 | What is the shape around each carbon in an alkane? | tetrahedral |
| 2 | How does chain length affect boiling point? | increases as the length of the carbon chain increases |
| 3 | What triggers the breaking of a Cl—Cl bond? | UV light |
| 4 | What two types of bonding does a carbon–carbon double bond have? | a $\sigma$-bond and a $\pi$-bond |
| 5 | Why does *E*-*Z* isomerism occur? | different groups attached to each carbon around a double bond |
| 6 | What is the catalyst in hydrogenation? | nickel |
| 7 | Which type of carbocation is most stable? | tertiary |
| 8 | What are two uses of waste polymers? | feedstock or fuel |
| 9 | What is an electrophile? | an electron pair acceptor |
| 10 | What is stereoisomerism? | same structural formula but the atoms have a different arrangement in space |
| 11 | Which orbitals are overlapping in a carbon–carbon single bond? | two p-orbitals |
| 12 | What is incomplete combustion? | combustion in a limited supply of oxygen |
| 13 | What is homolytic fission? | when a bond breaks so that one electron goes to each of the atoms |
| 14 | What is a radical? | a compound or atom with an unpaired electron |
| 15 | What does a dashed wedge represent? | bonds that are going behind the page |
| 16 | Why does branching decrease boiling point? | the overall intermolecular forces decrease as there are fewer points of contact |
| 17 | What rule is used to determine if a compound is an *E*- or *Z*-isomer? | Cahn–Ingold–Prelog priority rules |
| 18 | What rule is used to determine the major or minor products when hydrogen halides react with an asymmetrical alkene? | Markownikoff's rule |
| 19 | Why is burning waste PVC problematic? | HCl gas given off |
| 20 | Which type of isomer, *E* or *Z*, is a *cis*-isomer equivalent to? | *Z*-isomer |

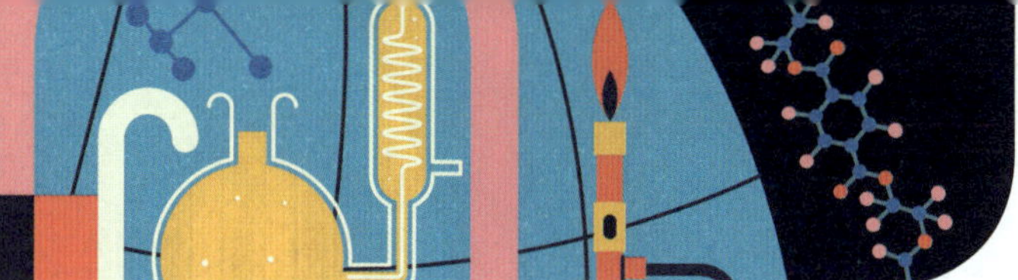



## Maths skills

Practise your maths skills using the worked example and practice questions below.

| Rearranging equations | Worked example | Practice |
|---|---|---|
| When rearranging an equation, it is important to think how to 'undo' what is happening to that variable.<br><br>For example, to make $c$ the subject of the equation: $a = bc + dc$ is multiplied by $b$ and then added to $d$. To make $c$ the subject you need to reverse the order of the operations ('undo'). First, subtract $d$ from both sides:<br><br>$a - d = bc$<br><br>Then divide by $b$:<br><br>$\frac{a-d}{b} = c$<br><br>Remember, $\log_{10}$ is the opposite of 10 to the power of a number and ln is the opposite of e to the power of a number. You can undo powers with power roots, for example, undo $^3$ (cubing) with $\sqrt[3]{}$ (cube root). | Make $[OH^-]$ the subject of each equation.<br><br>**Questions**<br><br>**1** rate = $k[OH^-][RCl]$<br><br>**2** pOH = $-\log_{10}[OH^-]$<br><br>**Answers**<br><br>**1** $[OH^-]$ is being multiplied by $k$ and $[RCl]$ so divide by both of these to 'undo' it.<br><br>rate = $k[OH^-][RCl] \div k[RCl]$<br><br>$[OH^-] = \frac{\text{rate}}{k[RCl]}$<br><br>**2** Undo the minus, by multiplying by minus one. Then undo log by ten to the power.<br><br>pOH = $-\log_{10}[OH^-]$<br><br>$-$pOH = $\log_{10}[OH^-]$<br><br>$[OH^-] = 10^{-pOH}$ | Rearrange the following to make the letter in curly brackets {…} the subject.<br><br>**1** $y = mx + c$ $\{m\}$<br><br>**2** $p = qr^2 + s$ $\{s\}$<br><br>**3** $K_c = \frac{[A][B]}{[C]\{[D]\}}$<br><br>**4** $K_p = \frac{p(X)^2p(Y)}{p(Z)}$ $\{p(Y)\}$<br><br>**5** $pK_a = -\log_{10}\left(\frac{[H^+][A^-]}{[HA]}\right)$ $\{[H^+]\}$ |

For answers and more practice questions visit www.oxfordrevise.com/scienceanswers

Even more practice and interactive revision quizzes are available on kerboodle

# Practice

## Exam-style questions

**1** **(a)** The C=C bond in alkenes consists of a $\sigma$-bond and a $\pi$-bond.

In the space below, give the two bonds between the two carbons.

C C

**[2]**

**Synoptic link**

6.2.2

**(b) (i)** Give the skeletal formulae for the two stereoisomers of 3-methylpent-2-ene and name each isomer.

**[3]**

**Exam tip**

Draw the backbone, insert the double bond, and then draw the methyl group.

**(ii)** Explain why these two isomers exhibit *E-Z* -isomerism isomerism but $(CH_3)_2{=}CH(CH_3)_2$ does not.

........................................................................................

........................................................................................

........................................................................................

........................................................................................**[3]**

**Exam tip**

There are three marks so one mark is necessary to explain why $(CH_3)_2C{=}CH(CH_3)_2$ does not exhibit *E-Z* isomerism.

**(c) (i)** State which of the compounds in **Figure 1.1** would exhibit *E-Z* isomerism and which would exhibit optical isomerism.

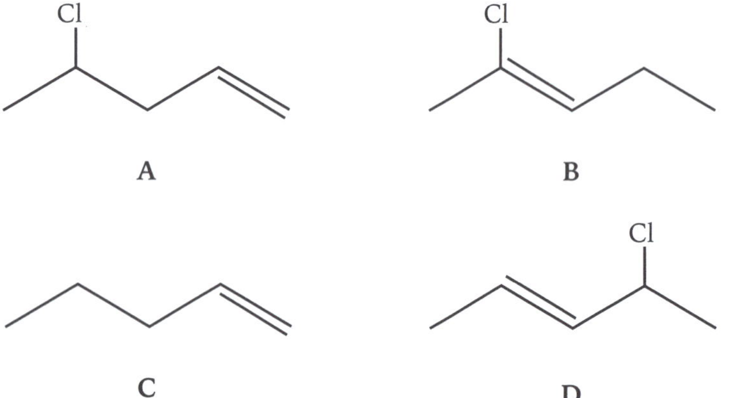


**Figure 1.1**

........................................................................................

........................................................................................

........................................................................................**[2]**

**(ii)** Give the systematic name of compound **D**.

........................................................................................**[2]**

**Exam tip**

The C=C has priority over the methyl group in terms of numbering.

**2** Cyclopropane can undergo a radical substitution with chlorine in the presence of UV light to give the halogenoalkane chlorocyclopropane, also shown in **Figure 2.1**.

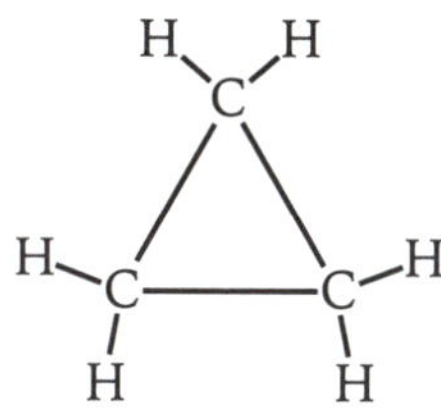

cyclopropane

chlorocyclopropane

1 product of the free-radical reaction

**Figure 2.1**

**(a)** Give the skeletal formula of cyclopropane.

**[1]**

**(b)** Using molecular formulae, write the balanced symbol equation for the reaction that gives chlorocyclopropane as the main product.

**[2]**

> **! Exam tip**
>
> Use information at the start of a question. Quite often the answer to a question is helped by referring to information already given.

**(c)** The mechanism for the reaction involves initiation, propagation, and termination reactions.

Write the equation for the initiation reaction including the conditions required.

**[2]**

> **! Exam tip**
>
> Write conditions on the arrow.

**(d) (i)** Describe the characteristics of a propagation reaction.

..............................................................................................

..............................................................................................**[1]**

**(ii)** Write the two equations for the propagation reactions using molecular formulae.

**[2]**

> **Exam tip**
>
> Your answer should include a free radical.

**(e) (i)** Describe the characteristics of a termination reaction.

..............................................................................................

..............................................................................................**[1]**

**(ii)** In one of the termination reactions, two of the hydrocarbon radicals combine to give a compound.

Draw the skeletal formula of this compound.

**[1]**

**(iii)** Give an equation for another termination reaction that could take place.

..............................................................................................

..............................................................................................**[1]**

**3** In this question we are assuming that jet-plane fuel consists of the alkane, decane.

In a typical short-haul jet-plane flight, 1000 kg of carbon dioxide is produced.

**(a)** Calculate the number of moles of carbon dioxide formed during the time of the flight.

number of moles = .....................**[2]**

**(b) (i)** Give the molecular formula of decane.

..............................................................................................

..............................................................................................**[1]**

**(ii)** Give the balanced symbol equation for the complete combustion of 1 mol of decane and use it to calculate the number of moles of decane burned in the flight.

**[2]**

**Synoptic links**

2.1.3a 3.2.1g

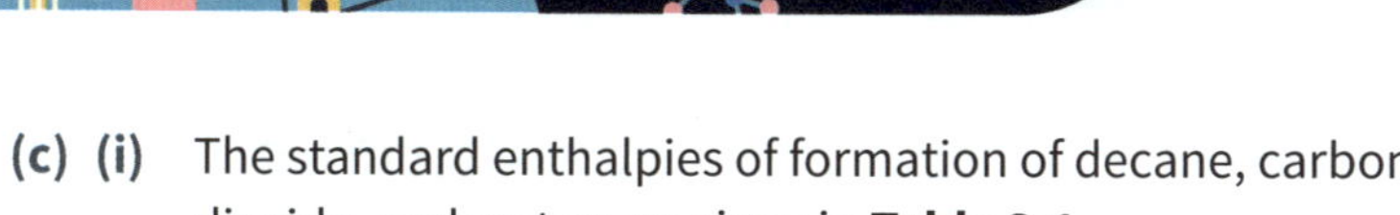



**(c) (i)** The standard enthalpies of formation of decane, carbon dioxide, and water are given in **Table 3.1**.

Use these values to find the standard enthalpy of combustion for decane.

| Compound | Standard enthalpy of formation ($kJ\,mol^{-1}$) |
|---|---|
| decane | –556.6 |
| carbon dioxide | –393.5 |
| water | –285.9 |

**Table 3.1**

standard enthalpy of combustion = .................... $kJ\,mol^{-1}$ **[3]**

**(iii)** Calculate the amount of heat generated by burning the decane in this flight.

Give your answer in MJ and to 3 significant figures. **[2]**

amount of heat generated = .................... MJ **[3]**

**Exam tip**

Use $\Delta_r H^\ominus$ = sum of $\Delta_f H^\ominus$ for the products – sum of $\Delta_f H^\ominus$ for the reactants.

**4** Alkenes are very important compounds in the chemical industry because they can be used to form addition polymers.

**(a) (i)** Draw the repeat unit for the addition polymer formed by 2-methylpropene.

**Exam tip**

You need to use square brackets and dotted lines.

**[2]**

**(ii)** Describe the results obtained when bromine water is added separately to 2-methylpropene and the addition polymer poly-2-methylpropene.

..........................................................................................

..........................................................................................

..........................................................................................

.......................................................................................... **[3]**

(iii) Explain the different results for the test and use a balanced symbol equation to support your explanation.

..........................................................................................

..........................................................................................

..........................................................................................

..........................................................................................

..........................................................................................

..........................................................................................

..........................................................................................

.......................................................................................... [3]

> **Exam tip**
>
> Always give the starting colour and resulting colour of the test reagent.

(b) The repeat unit for an addition polymer is shown in **Figure 4.1**.

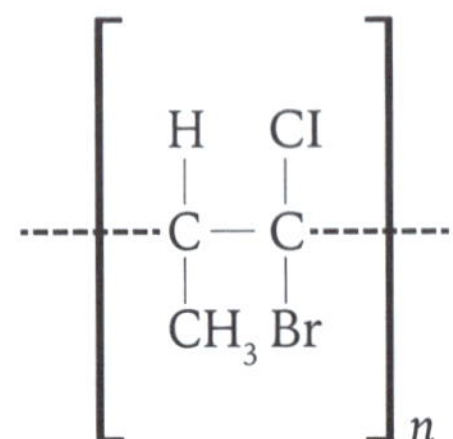


**Figure 4.1**

(i) Give the displayed formulae of the two monomers that could be used to form this addition polymer.

> **Exam tip**
>
> Draw all bonds and atoms.

[3]

(ii) Give the systematic name under each isomer. [2]

5 (a) Give the mechanism for the major product of the addition of hydrogen bromide to but-1-ene.

[5]

**(b) (i)** Give the structural formulae of the primary and secondary carbocations formed when HBr reacts with propene. **[2]**

**(ii)** Give the products of the reaction between 2-methylpropene and steam.

Identify which one is the major product and which one is the minor product. **[3]**

**6** This question concerns the use of alkanes as fuels for cars and the problems this causes for the environment.

**(a)** Carbon monoxide and carbon particulates are the products of the incomplete combustion of alkanes.

Give the balanced symbol equations for the formation of these two products when octane is burned:

**(i)** Carbon monoxide and water only. **[1]**

**(ii)** Carbon and water only. **[1]**

**(b)** Calculate the volume of carbon monoxide formed (at room temperature and pressure) when 1 kg of octane undergoes incomplete combustion. **[3]**

**(c)** Catalytic converters are now fitted to the exhausts of cars to remove carbon monoxide and nitrogen monoxide (NO). Nitrogen monoxide is formed by the reaction between nitrogen and oxygen in the car engine and exhaust.

**(i)** Give the balanced symbol equation for this reaction. **[1]**

**(ii)** Nitrogen monoxide and carbon monoxide react with the catalyst to produce carbon dioxide and an element.

Give the balanced symbol equation for this reaction. **[1]**

**(iii)** Using oxidation numbers, explain why this is a redox reaction. **[2]**

**(iv)** Suggest why catalytic converters do not work until the car engine has warmed up. **[2]**

**Exam tip**

Use $\frac{1}{2}$ values for numbers of oxygen molecules.

# Knowledge

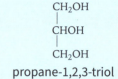

## 12 Alcohols and haloalkanes

### Naming alcohols

An alcohol is characterised by one or more –OH functional groups within the compound. An alcohol with two –OH groups is a -diol and one with three is a -triol.

butane-1,4-diol

$CH_2OH$
$CHOH$
$CH_2OH$

propane-1,2,3-triol

### Polarity

The oxygen in the –OH group gives the alcohols different properties compared to alkanes of a similar length. Alkanes are non-polar because the electronegativity of hydrogen and carbon are similar. However, alcohols are polar, due to the difference in electronegativity between oxygen and hydrogen $–O^{\delta-}–H^{\delta+}$. This polar bond explains the difference in solubility and volatility compared to alkanes.

- Alcohols have a lower volatility compared to alkanes with the same number of carbon atoms.
- Alcohols are soluble in water but become less soluble as the number of carbons increases. Alkanes are not soluble.

### Hydrogen bonding

The polarity of alcohols allows for hydrogen bonds to form. In non-polar molecules, intermolecular bonding is limited to weak London forces; hydrogen bonds are stronger.

- Hydrogen bonding allows a compound to be soluble in water, so alcohols are completely soluble.
- Hydrogen bonds require more energy to overcome intermolecular forces than London forces, so alcohols have higher boiling points than comparable alkanes, and are less volatile.

### Classification of alcohols

Depending on the location of the –OH group, alcohols can be classified as primary, secondary, or tertiary. Primary, secondary, and tertiary alcohols react differently in different situations.

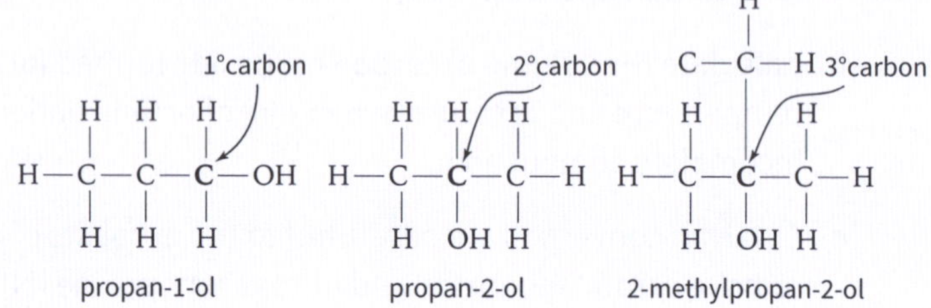


propan-1-ol propan-2-ol 2-methylpropan-2-ol

A **primary alcohol** has the –OH group attached to a carbon atom that has *two* hydrogen atoms and *one* alkyl group attached to it.

A **secondary alcohol** has the –OH group attached to a carbon atom that has *one* hydrogen atom and *two* alkyl groups attached to it.

A **tertiary alcohol** has the –OH group attached to a carbon atom that has *no* hydrogen atoms and *three* alkyl groups attached to it.

### Combustion of alcohols

When in a plentiful supply of oxygen, an alcohol will burn to give carbon dioxide and water. This reaction is exothermic, releasing a large amount of energy:

$$C_2H_5OH(l) + 3O_2(g) \rightarrow 2CO_2(g) + 3H_2O$$

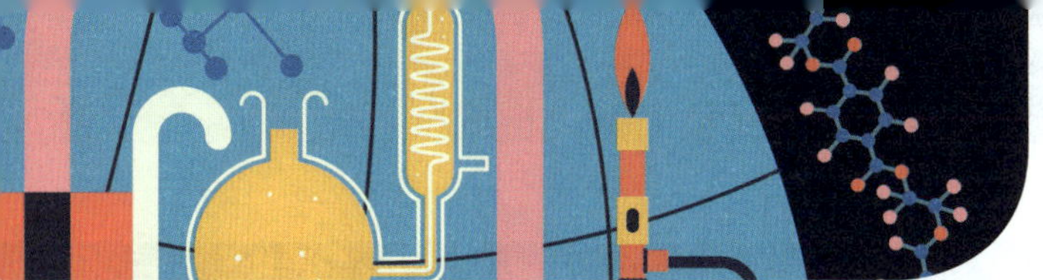



## Oxidation of alcohols

### Oxidation of primary alcohols

When a primary alcohol is heated with acidified potassium dichromate, an aldehyde will be formed. Further oxidation to a carboxylic acid is prevented by distilling off the aldehyde as it is made. To produce a carboxylic acid, the primary alcohol needs to be heated strongly under reflux. Then full oxidation of the aldehyde (formed initially in the reaction) occurs to form the carboxylic acid:

$$CH_3CH_2CH_2CH_2OH + [O] \xrightarrow[\text{distil}]{K_2Cr_2O_7/H_2SO_4} CH_3CH_2CH_2CHO + H_2O$$

butan-1-ol; [O] = oxidising agent; butanal

$$CH_3CH_2CH_2CH_2OH + 2\,[O] \xrightarrow[\text{reflux}]{K_2Cr_2O_7/H_2SO_4} CH_3CH_2CH_2COOH + H_2O$$

butan-1-ol; 2 [O] = oxidising agent; butanoic acid

### Oxidation of secondary alcohols

Secondary alcohols can only be oxidised to ketones. This is done under reflux to ensure the reaction goes to completion.

## Dehydration reaction of alcohols

$$CH_3CH_2OH \longrightarrow H_2C{=}CH_2 + H_2O$$

An alcohol can undergo a **dehydration** reaction to remove the –OH group and a hydrogen atom to form an alkene and water. This is done by heating an alcohol under reflux in the presence of an acid catalyst (e.g., $H_3PO_4$ or $H_2SO_4$).

## Substitution reactions of alcohols

The alcohol group can be substituted with halide ions when heated under reflux in the presence of acid and a sodium halide. The first part of the reaction is the formation of the hydrogen halide, which can then react with the alcohol to form a haloalkane.

$$NaBr(s) + H_2SO_4(aq) \rightarrow NaHSO_4(aq) + HBr(aq)$$

$$CH_3CH(OH)CH_3 + HBr \rightarrow CH_3CHBrCH_3 + H_2O$$

propan-2-ol → 2-bromopropane

# Haloalkanes

## Substitution reactions

Haloalkanes are organic compounds that contain at least one Halogen atom. Due to the difference in electronegativity between the Halogen atoms and carbon atoms, haloalkanes will contain polar bonds.

Haloalkanes will undergo **nucleophilic substitution** reactions. A **nucleophile** is an atom with a negative charge or an atom with a δ– charge; they have a lone pair on an electronegative atom that can be used to form a covalent bond. A nucleophile is an *electron pair donor*. The nucleophiles will replace the Halogen in a substitution reaction.

Depending of the position of the Halogen within the compound, haloalkanes can be characterised as primary, secondary, or tertiary.

1°carbon

1-chloropropane

2°carbon

2-chloropropane

3°carbon

2-chloro-2-methylpropane

When a haloalkane reacts with aqueous sodium hydroxide or potassium hydroxide the nucleophile will be the $OH^-$ ion and be substituted for the Halogen. The rate of the reaction will depend on the strength of the carbon–Halogen bond.

$$\text{H—C—C}^{\delta+}\text{—Br}^{\delta-} \;\; {}^-\text{OH} \longrightarrow \text{H—C—C—OH} + \text{Br}^-$$

# Measuring the rate of hydrolysis

The rate of hydrolysis of primary haloalkanes can be measured by carrying out the reaction in the presence of silver nitrate and ethanol. A coloured precipitate of the silver halide will form. The ethanol is a solvent in this reaction and the nucleophile is water from the silver nitrate solution. In this reaction X represents a Halogen:

$$RCH_2CH_3X + H_2O \rightarrow RCH_2CH_3OH + H^+ + X^-$$

$$Ag^+ + X^- \rightarrow AgX$$

- Chloroalkanes form a *white precipitate slowly*.
- Bromoalkanes form a *cream precipitate* faster than chloroalkanes.
- Iodoalkanes form a *yellow precipitate* rapidly.

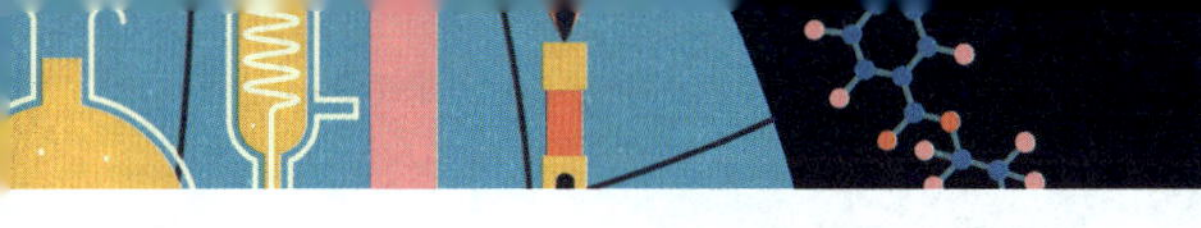



## Rates of hydrolysis and carbon–Halogen bond enthalpy

The ability of the carbon–Halogen bond to break in a reaction is determined by the bond polarity and the bond enthalpy.

Within a carbon–Halogen bond the Halogens are more electronegative. This allows the carbon atom to be attacked by nucleophiles. The C—F bond is the most polar, so the carbon atom in this bond is most readily attacked by a nucleophile. However, this does not make the C—F bond the most reactive.

Bond enthalpy is a more important determinant of reactivity. Moving down the group, the strength of the bond decreases:

- the C—F bond is the strongest bond, so is unreactive
- the C—I bond is the weakest bond, so is the most reactive.

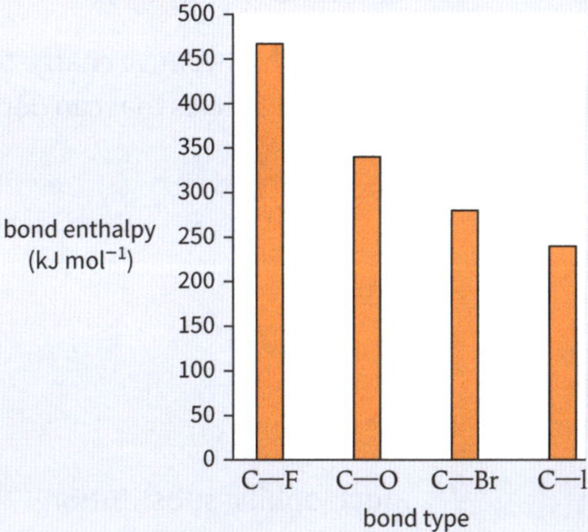


## Halogen radicals as pollutants

**Chlorofluorocarbons (CFCs)** are environmental pollutants that have accumulated in the atmosphere over years and are destroying the ozone layer.

Ozone, $O_3$, is found naturally in the upper atmosphere and is beneficial as it absorbs ultraviolet radiation. When CFCs enter the atmosphere the UV radiation causes the C—Cl bond to break, forming a chlorine **radical**. This is an **initiation reaction**. The C—Cl bond breaks as it has a lower bond enthalpy than the C—F bond. In a series of **propagation reactions** these radicals catalyse the breakdown of ozone and leave a hole in the protective ozone layer. Other radicals, such as •NO, have a similar effect.

**Initiation reaction:**

$CF_2Cl_2 \rightarrow Cl\bullet + CF_2Cl\bullet$

**Propagation reactions:**

$Cl\bullet + O_3 \rightarrow \bullet ClO + O_2$

$\bullet ClO + O_3 \rightarrow 2O_2 + Cl\bullet$

$\bullet ClO + O \rightarrow O_2 + Cl\bullet$

**Termination reaction:**

$O_3 + O \rightarrow 2O_2$

Wide-ranging research into the effect that the hole in the ozone layer has on humans and the effect that CFCs have on the ozone layer has led to the development of alternative compounds that are chlorine free.

# Retrieval

Learn the answers to the questions below, then cover the answers column with a piece of paper and write as many as you can. Check and repeat.

| | Questions | Answers |
|---|---|---|
| 1 | Of the Halogen–carbon bonds, which is the strongest? | C—F |
| 2 | How can the rate of hydrolysis be followed? | with the speed of the formation of a silver precipitate |
| 3 | What does CFC stand for? | chlorofluorocarbons |
| 4 | What is the definition of a nucleophile? | atom with a negative charge or an atom that has a $\delta-$ charge/ a species that can donate a lone pair of electrons |
| 5 | Is 2-chloropronae a primary, secondary, or tertiary haloalkane? | secondary |
| 6 | What colour precipitate will a bromoalkane form? | cream |
| 7 | What is the chemical formula for ozone? | $O_3$ |
| 8 | What type of reaction happens when a haloalkane reacts with aqueous sodium hydroxide or potassium hydroxide ? | nucleophilic substitution |
| 9 | Why do haloalkanes contain polar bonds? | due to the difference in electronegativity between the halogens and carbon |
| 10 | Which Halogen–carbon bond is the most reactive? | C—I |
| 11 | Why do alcohols have a higher boiling point than comparable alkanes? | due to their stronger intermolecular bonds or hydrogen bonds |
| 12 | Are alcohols more volatile than alkanes? | alcohols are less volatile |
| 13 | How many OH group are there in a -triol molecule? | three |
| 14 | Why are alcohols polar? | due to the difference in electronegativity between oxygen and hydrogen |
| 15 | What do primary alcohols partially oxidise to? | aldehydes |
| 16 | What do primary alcohols fully oxidise to? | carboxylic acids |
| 17 | What type of intermolecular bonding is found in alcohols but not in alkanes? | hydrogen bonding |
| 18 | What do secondary alcohols oxidise to? | ketones |
| 19 | What do tertiary alcohols oxidise to? | they do not oxidise |
| 20 | What colour change can be seen in the oxidation of alcohols by potassium dichromate(VII)? | orange acidified potassium dichromate(VII) ($Cr_2O_7^{2-}$) will reduce to green chromium ions ($Cr^{3+}$) in the reaction |

Put paper here

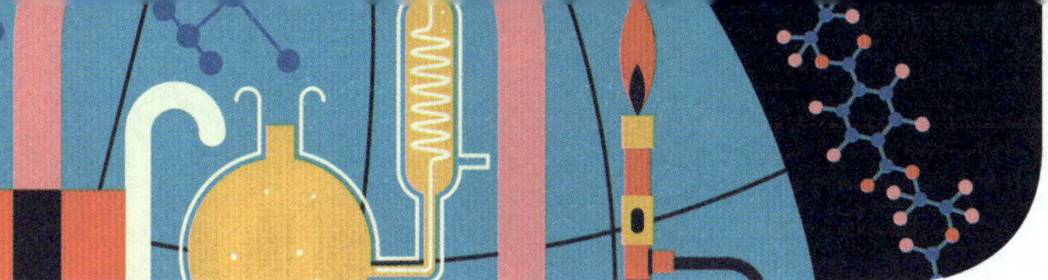



# Maths skills

Practise your maths skills using the worked example and practice questions below.

## Algebraic equations

Solving equations is a very important skill, especially when when calculating values in kinetics and equilibria questions.

There are two main steps to solving equations:

1. substitute values into the equation
2. re-arrange the equation.

These steps can happen in either order. If you are confident with algebra then do step 2 first, if you are less confident with algebra do step 1 first.

However, sometimes you cannot do step 1 first because you might be missing a quantity that will cancel out. So, it is important to practise doing step 2 first.

## Worked example

**Question**

Calculate the rate constant in the equation:

rate = $k[H^+]^2$

rate = $4.0 \times 10^{-3}$ mol dm$^{-3}$ s$^{-1}$

$[H^+] = 1.0 \times 10^{-3}$

**Answer**

Re-arrange, substitute then calculate.

Make $k$ the subject by dividing by $[H^+]^2$:

$$k = \frac{\text{rate}}{[H^+]^2}$$

substitute:

$$k = \frac{4.0 \times 10^{-3}}{(1.0 \times 10^{-3})^2}$$

calculate:

$k = 4.0 \times 10^{3}$ mol$^{-1}$ dm$^{3}$ s$^{-1}$

## Practice

Calculate $[H^+]$ in each case.

1 $K_a = \frac{[H^+]^2}{[HA]}$
- $K_a = 1.75 \times 10^{-5}$ mol dm$^{-3}$
- $[HA] = 0.500$ mol dm$^{-3}$

2 $K_a = \frac{[H^+]^2}{[HA]}$
- $K_a = 1.51 \times 10^{-5}$ mol dm$^{-3}$
- $[HA] = 0.200$ mol dm$^{-3}$

3 rate = $k[H^+[Br^-]$
- rate = $0.400 \times 10^{-4}$ mol dm$^{-3}$ s$^{-1}$
- $k = 1.60 \times 10^{-4}$ mol$^{-1}$ dm$^{3}$ s$^{-1}$
- $[Br^-] = 0.600$ mol dm$^{-3}$

4 rate = $k[H^+][CH_3COOCH_3]$
- rate = $8.2 \times 10^{-6}$ mol dm$^{-3}$ s$^{-1}$,
- $k = 1.50 \times 10^{-3}$
- $[CH_3COOCH_3 = 0.02$ mol dm$^{-3}$

# Practice

## Exam-style questions

1 This question concerns the reactivity of the carbon–Halogen bond in haloalkanes and how quickly it breaks in substitution reactions. 

The relevant data is given in **Table 1.1**:

| Electronegativity values of carbon and the halogens | |
|---|---|
| **Element** | **Electronegativity** |
| carbon | 2.5 |
| fluorine | 4.0 |
| chlorine | 3.5 |
| bromine | 2.8 |
| iodine | 2.6 |

| Carbon–Halogen bond enthalpies | |
|---|---|
| **Bond** | **Bond enthalpy ($kJ\,mol^{-1}$)** |
| C—F | 467 |
| C—H | 413 |
| C—Cl | 346 |
| C—Br | 290 |
| C—I | 228 |

**Table 1.1**

**(a)** If difference in electronegativity is the factor that determines the relative reactivity of the carbon—Halogen bond, predict the trend in reactivity of the compounds as you go from C—Cl to C—I. Explain your answer.

..........................................................................................

..........................................................................................

..........................................................................................

..........................................................................................

..........................................................................................

..........................................................................................

..........................................................................................

..........................................................................................

.......................................................................................... **[3]**

**(b)** Haloalkanes react very slowly with water to form halide ions and the corresponding alcohol.

The general equation, where **X** represents a Halogen, is given below:

$$C_4H_9X(l) + H_2O(l) \rightarrow C_4H_9OH(l) + \mathbf{X}^-(aq) + H^+(aq)$$

Two students decided to test which of the two factors, electronegativity or bond-enthalpy, determined the relative reactivities of the carbon—Halogen bonds.

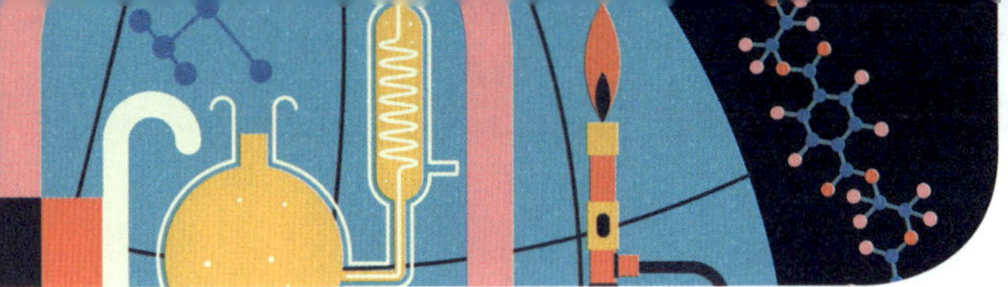



(i) Describe an experiment they could use to investigate this question. The students can be provided with any apparatus you feel necessary and the following chemicals:

Silver nitrate solution; ethanol; haloalkanes of your choice.

You must also explain how your method shows the rate of the reaction and explain the practical method where necessary.

........................................................................................

........................................................................................

........................................................................................

........................................................................................

........................................................................................

........................................................................................

........................................................................................

........................................................................................

........................................................................................[6]

(ii) Describe the results that should be obtained by the students and explain what they show.

........................................................................................

........................................................................................

........................................................................................[2]

**Exam tip**

Do not go into great detail, just give the main points, including important explanations.

**2** Haloalkanes undergo nucleophilic substitution reactions.

**(a)** Define the term nucleophile.

........................................................................................

........................................................................................[1]

**Synoptic link**

3.3.11.1

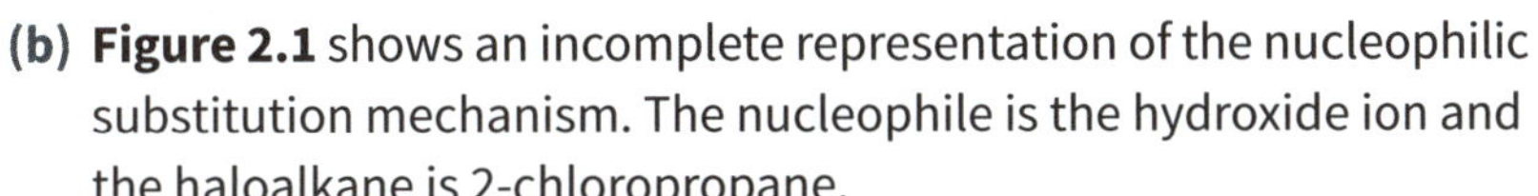

**(b)** **Figure 2.1** shows an incomplete representation of the nucleophilic substitution mechanism. The nucleophile is the hydroxide ion and the haloalkane is 2-chloropropane.

Adapt the diagram to show a correct representation of the mechanism.

You need to add lone-pair electrons, charges, dipoles, curly arrows, and the products of the reaction.

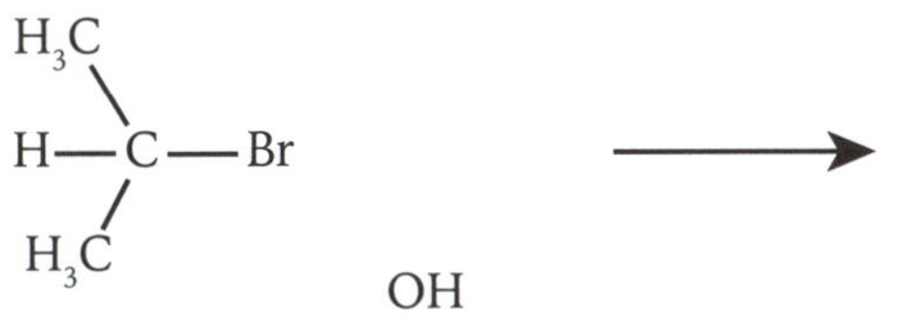


[6]

**Figure 2.1**

**Exam tip**

Make sure it is clear where the curly arrows are coming from and going to.

(c) Adapt the following equations to show the full balanced equation. You must give the structural formulae of the organic product and the other products, and balance the equation.

$CH_2BrCH_2Br + \ldots NaOH \rightarrow \ldots CH_3CHBrCH_2Br + \ldots NaOHNH_3 \rightarrow \ldots$ **[4]**

**3** A group of students was given the task of preparing 1-bromobutane from an alcohol.

The equation for the reaction is:

$$C_4H_9OH(l) + HBr(aq) \rightarrow C_4H_9Br(l) + H_2O(l)$$

A summary of the procedure is given below:

- 4.80 g of the alcohol was refluxed with a mixture of sodium bromide and concentrated sulfuric acid (hydrogen bromide was made in situ in this reaction).
- After refluxing for several minutes, they changed the apparatus and distilled the remaining mixture.
- The resulting distillate was shaken with sodium carbonate solution.
- The resulting mixture was then added to a separating funnel and the lower layer was run off, leaving the aqueous layer behind.

**(a)** Give the systematic name of the alcohol the students must use in the preparation and give its structural formula.

.................................................................................................**[2]**

**(b) (i)** Name the impurity removed using sodium carbonate and give the balanced symbol equation for the reaction.

**[2]**

**(ii)** Define the term reflux.

.................................................................................................

.................................................................................................**[1]**

**(iii)** Give two reasons why a separating funnel can be used to separate the 1-bromobutane from the aqueous layer.

.................................................................................................

.................................................................................................

.................................................................................................**[2]**

**(c)** After purification, 5.75 g of pure 1-bromobutane was obtained. Calculate the percentage yield from the preparation.

[Relevant relative atomic masses: C = 12; H = 1; O = 16; Br = 79.9]

percentage yield = ......................% **[4]**

4 **(a)** This question concerns the three alcohols shown in **Figure 4.1**.

| A | B | C |
|---|---|---|
| $H_3C{-}C(CH_3)(CH_3){-}OH$ | $H_3C{-}C(CH_3)(CH_3){-}CH_2{-}OH$ | $H_3C{-}C(CH_3)(CH_3){-}CH_2{-}OH$ |

**Figure 4.1**

**(i)** Give the systematic name of each alcohol. Classify them as either primary (1°), secondary (2°), or tertiary (3°).

........................................................................................

........................................................................................

........................................................................................

........................................................................................ **[3]**

**(ii)** Two of the alcohols undergo dehydration to give the same product.

Identify the alcohols by their letters and give the structural formula of the dehydration product.

........................................................................................

........................................................................................ **[1]**

**(iii)** Identify the alcohol that cannot undergo the dehydration. Explain why it cannot be dehydrated.

........................................................................................

........................................................................................

........................................................................................ **[2]**

**(iv)** Give the structural formula and systematic name of one more alcohol that is isomeric with the alcohols **A** and **C**.

........................................................................................ **[2]**

**(b)** **Figure 4.2** shows a simplified structure of an alcohol. Complete the diagram to show the hydrogen bonding between two molecules of the alcohol. **[3]**

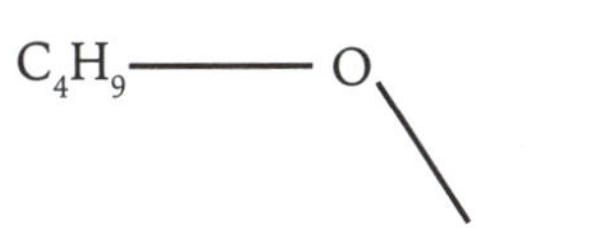


**Figure 4.2**

**Exam tip**

There are three marks so there are three features you need to include in your diagram.

5 **(a)** Give the details of a chemical test you could use to distinguish between the two isomeric alcohols shown in **Figure 5.1**.

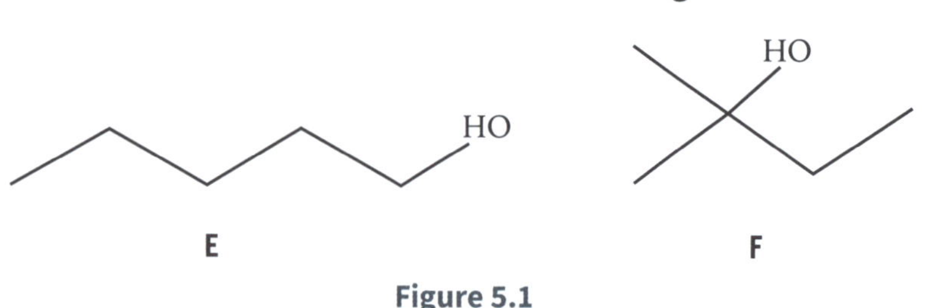


**Figure 5.1**

..............................................................................................

..............................................................................................

..............................................................................................

..............................................................................................

..............................................................................................[3]

**(b)** Two groups of students (group I and group II) were given the task of preparing two compounds from pentan-1-ol.

Group I was asked to prepare pentanal and group II to prepare pentanoic acid.

**(i)** State the name of the reagent they both used.

..............................................................................................

..............................................................................................[1]

**(ii)** Explain why group I did not use refluxing in their method.

..............................................................................................

..............................................................................................

..............................................................................................[2]

**(iii)** Explain why group II did not use distillation in their method.

..............................................................................................

..............................................................................................

..............................................................................................[2]

**(c)** Give a balanced symbol equation for both reactions.

[2]

6 The formulae below show the structures of two esters:

| $CH_3CH_2COOCH_2CH_2CH(CH_3)_2$ | $CH_3COOCH_2CH_2CH(CH_3)_2$ |
|---|---|
| **G** | **H** |

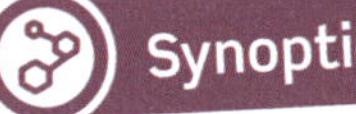

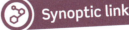

**Synoptic link**

6.1.3c

**Exam tip**

The part of the ester with –CO– comes from the carboxylic acid.

**(a)** Give the structural formulae of the carboxylic acid and the alcohol used to prepare these two esters.

**(i)** Ester **G**.

[1]

(ii) Ester **H**.

[1]

(b) The equation below shows the reaction used to prepare an ester. The masses of reactants used, the mass of the product formed, and the relative molecular masses are also given in **Table 6.1**.

$$CH_3COOH(l) + CH_3CH_2CH(OH)CH_3(l) \rightleftharpoons CH_3COOCH(CH_3)CH_2CH_3(l) + H_2O(l)$$

| Mass used or formed (g) | 9.00 | 14.8 | 14.5 |
|---|---|---|---|
| $M_r$ | 60 | 74 | 116 |

**Table 6.1**

(i) Name a reagent that could be used as a catalyst for the reaction.

..............................................................................

..............................................................................[1]

(ii) State which reactant is the limiting reactant.

Explain your answer.

..............................................................................

..............................................................................

..............................................................................

..............................................................................[3]

(iii) Calculate the percentage yield for the preparation.

percentage yield = ....................[2]

(iv) Suggest why the yield is below 100%.

..............................................................................

..............................................................................

..............................................................................[2]

# 13 Organic synthesis

## Quickfit apparatus

Distillation and reflux are two common techniques in chemistry, that can be set up using **Quickfit apparatus**.

Similar equipment is used for both **distillation** and **reflux** but the equipment is arranged in different ways so the solution can be heated gently (under distillation) and products removed, or heated strongly (under reflux).

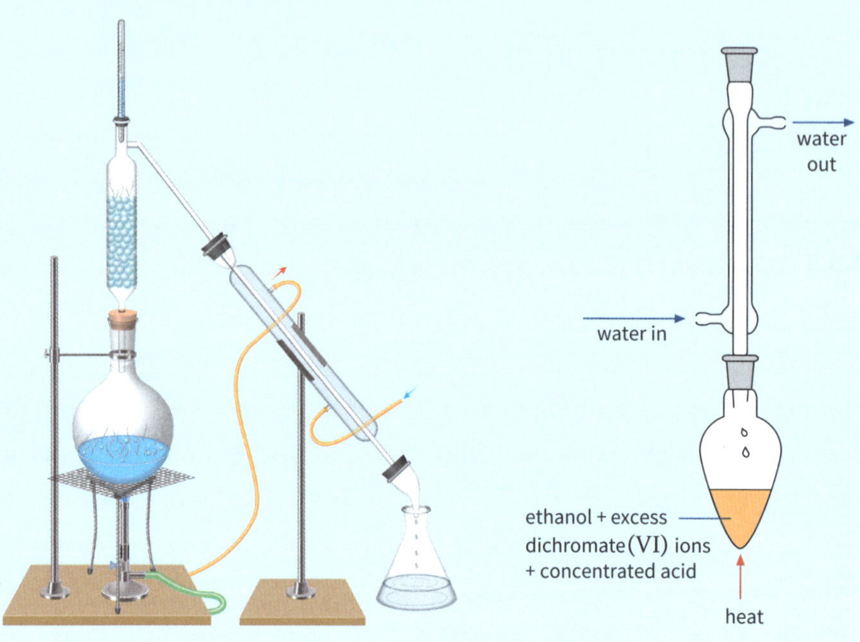


## Preparation and purification of organic liquids

The preparation and purification of an organic liquid, such as ethyl ethanoate, requires several steps.

If the organic liquid is made under reflux, it is separated out from the unreacted reactants with a separating funnel. The **organic layer** is separated from an **aqueous layer**.

If there is any water left in the organic product, it can be dried with an anhydrous salt ($MgSO_4$ or $CaCl_2$).

The sample may contain two organic liquids with similar boiling points: this mixture should undergo **redistillation** to separate out the desired product.

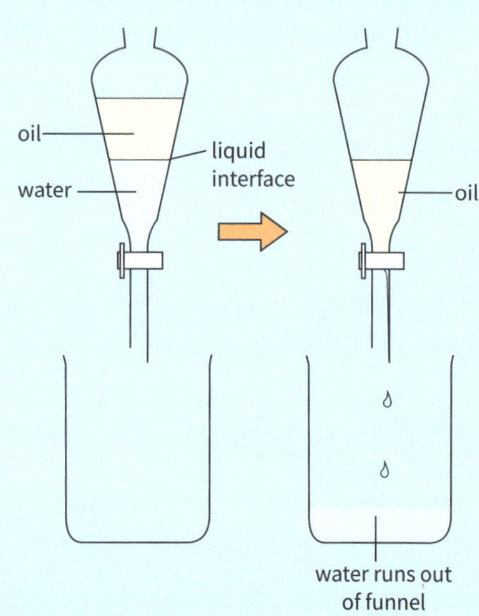

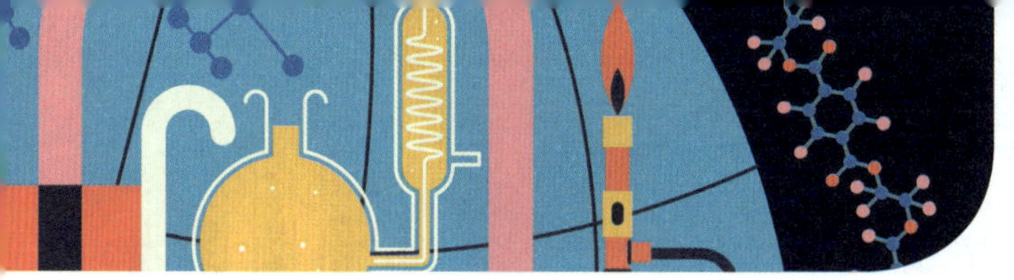



| Homologous series | Functional group | Homologous series | Functional group |
|---|---|---|---|
| alkene | >C=C< | carboxylic acid | —C(=O)—OH |
| alcohol | —OH | ester | —C(=O)—O—C |
| haloalkane | —Cl—Br—F | amine | $—NH_2$ |
| aldehyde | —C(=O)—H | acyl chloride | —C(=O)—Cl |
| ketone | >C=O | nitrile | —CN |

## Synthetic routes

In chemistry a compound is rarely made in a single reaction and only has one functional group.

You need to be able to identify multiple functional groups within a compound and predict the properties of this compound.

A flowchart is useful for planning multi-step organic synthesis routes that an organic compound can have. Then you will be able to predict its properties and reactions.

## Flowchart of reactions

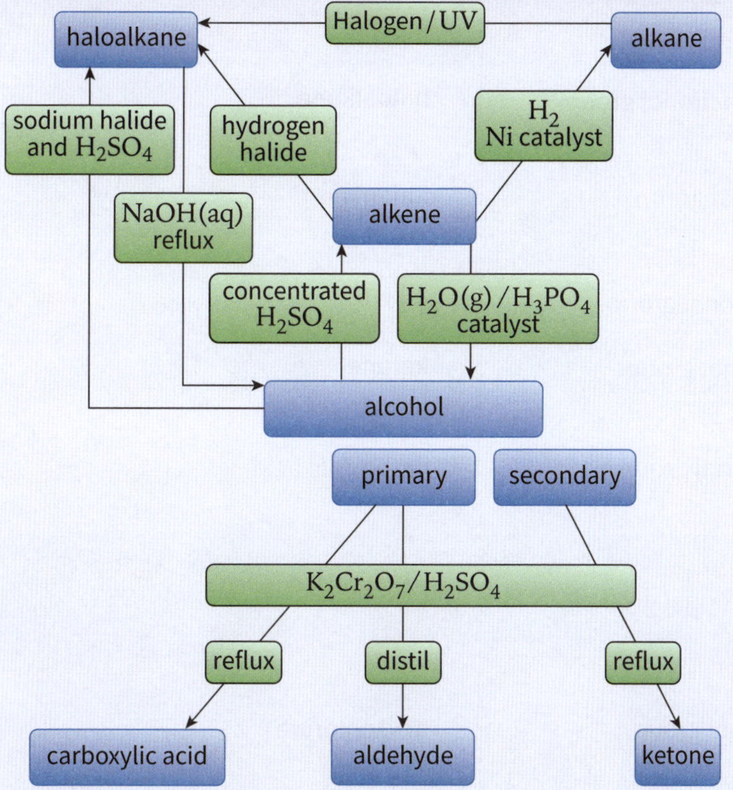

# Retrieval

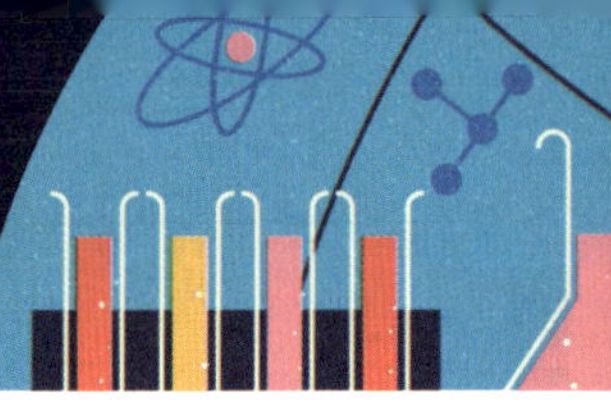

Learn the answers to the questions below, then cover the answers column with a piece of paper and write as many as you can. Check and repeat.

| | Questions | Answers |
|---|---|---|
| 1 | How can a solution be heated gently? | under distillation |
| 2 | What catalyst is used for the synthesis of an alcohol from an alkene? | phosphoric acid |
| 3 | What apparatus is used for the synthesis of a carboxylic acid from a primary alcohol? | reflux |
| 4 | Which type of alcohol can be converted into a ketone? | secondary |
| 5 | How can a mixture of an organic liquid and aqueous liquid be separated? | with a separating funnel |
| 6 | Name the $-NH_2$ functional group | amine |
| 7 | What is the catalyst for the reaction that removes double bonds from alkenes? | Ni / nickel |
| 8 | Name the $-CN$ functional group | nitrile |
| 9 | What apparatus is used for the synthesis of an alcohol from a haloalkane? | reflux |
| 10 | How can a solution be heated strongly? | reflux |
| 11 | Name the $-OH$ functional group. | alcohol |
| 12 | Name the $-Cl/-Br/-F$ functional group. | haloalkane |
| 13 | Name the $-C(=O)H$ functional group. | aldehyde |
| 14 | Name the $>C=C<$ functional group. | alkene |
| 15 | Name the $>C=O$ functional group. | ketone |
| 16 | Name the $-C(=O)OH$ functional group. | carboxylic acid |
| 17 | Name the $-C(=O)O-C$ functional group. | ester |
| 18 | Name the $-C(=O)Cl$ functional group. | acyl chloride |

Put paper here

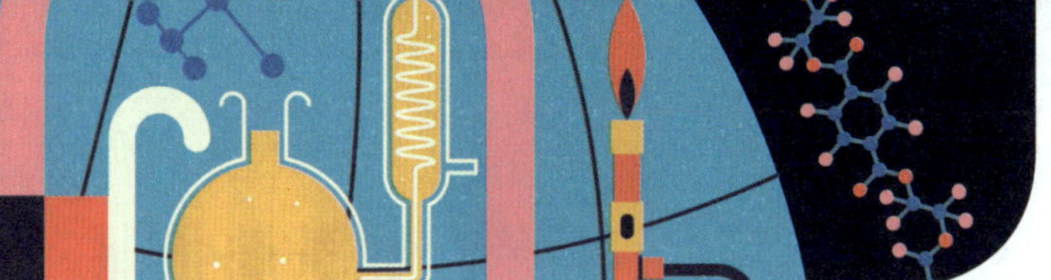



| | | |
|---|---|---|
| 19 | How can any water left after separating be removed from an organic product? | by drying the product with $MgSO_4$ or $CaCl_2$ |
| 20 | How can a sample containing two organic liquids with similar boiling points be separated? | by redistillation |

Put paper here

## Practical skills

Practise your practical skills using the worked example and practice questions below.

### Preparing an organic liquid

When preparing a pure sample of an organic liquid there are many different processes to consider. As well as focusing on the chemical reaction, it is equally important to focus on the purification stages as well. Common purification steps include:

- adding $Na_2CO_3$ to neutralise any acid that could include a catalyst
- leaving the liquid to stand in a separating funnel to separate out the organic and aqueous layer
- adding anhydrous $MgSO_4$ or $CaCl_2$ to dehydrate (remove water from) the solution.

### Worked example

A student is trying to prepare a sample of ethyl hexanoate, which smells of pineapple.

**Questions**

1. Name the two organic reagents and catalyst the student will require.
2. After the reaction is complete, the student adds sodium carbonate. Explain why.
3. The student leaves the liquid in a separating funnel until two layers form. Which layer (top or bottom) is the organic layer?

**Answers**

1. Ethanol, hexanoic acid, *concentrated* sulfuric acid.
2. To neutralise any excess acid (hexanoic and sulfuric).
3. The top layer is the organic layer as it is less dense.

### Practice

1. When adding the sodium carbonate, the student needs to invert the separating funnel and open the funnel from time to time.
   - a Why does the funnel need to be inverted?
   - b Why does the funnel need to be opened?
2. Some anhydrous calcium chloride is added in small amounts until the liquid becomes clear. It is then filtered.
   - a Why is the solution filtered?
   - b Why is anhydrous calcium chloride added until the liquid goes clear?

# Practice

## Exam-style questions

1 Cyclohexene can be prepared by the elimination of water from cyclohexanol. The equation for the reaction is given in **Figure 1.1**, along with data concerning relevant physical properties.

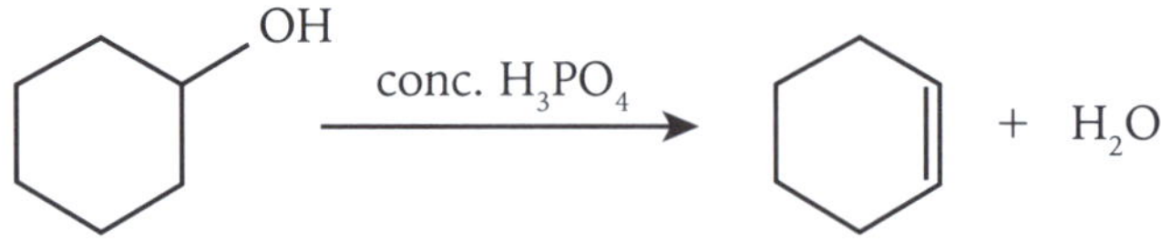


**Figure 1.1**

| Molecule | cyclohexanol | cyclohexene |
|---|---|---|
| Boiling point (°C) | 162.000 | 83.000 |
| Solubility in water ($g\,dm^{-3}$) | 43.000 | insoluble |
| Density ($g\,cm^{-3}$) | 0.962 | 0.779 |
| Molar mass ($g\,mol^{-1}$) | 100.000 | 82.000 |

**Table 1.1**

**(a)** Calculate the atom economy of this reaction.

atom economy = .................... **[1]**

**(b)** A description of the method used is shown below:

20.0 g of cyclohexane and 5 $cm^3$ of concentrated phosphoric acid were placed in a round-bottomed flask. The apparatus was set up for distillation and the mixture was gently heated.

The distillate obtained was then shaken with an aqueous solution of sodium chloride in a tap-funnel. The two layers were separated by opening the tap. Anhydrous calcium chloride was added to the organic layer.

The mixture was then filtered and the filtrate redistilled. The liquid boiling at 81–85 °C was collected and retained.

**(i)** Suggest why concentrated phosphoric acid is added to the reaction mixture.

..........................................................................................

.......................................................................................... **[1]**

**(ii)** Using the data above, suggest why the reaction mixture is gently heated.

..........................................................................................

.......................................................................................... **[1]**

**(iii)** Give the reason for adding anhydrous calcium chloride.

..........................................................................................

.......................................................................................... **[1]**

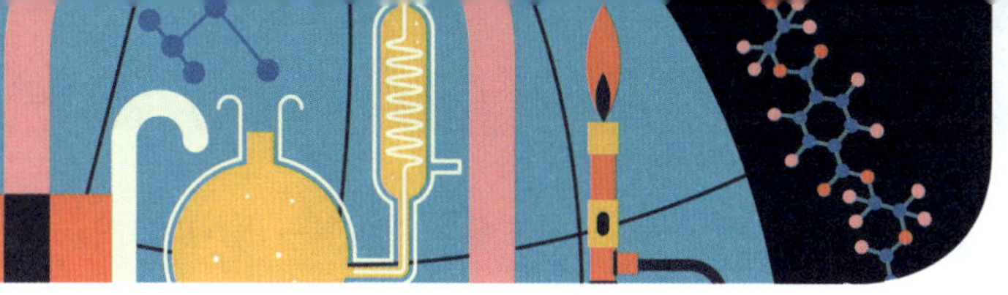



**(iv)** Explain the choice of temperature range over which the distillate is collected.

............................................................................................

............................................................................................ **[1]**

**(c)** 9.50 $cm^3$ of the final distillate were collected. Calculate the percentage yield for the preparation.

Use density to find mass.

percentage yield = ..................... **[4]**

**2** When haloalkanes are refluxed with aqueous sodium hydroxide, alcohols are formed. For example, for a bromoalkane, RBr, the reaction can be written as:

$$\mathrm{RBr(l) + NaOH(aq) \rightarrow NaBr(aq) + ROH(l)}$$

The extent of the reaction can be measured by titrating the sodium hydroxide remaining against standard hydrochloric acid.

In one experiment, 0.548 g of an unknown monobromoalkane, **A**, was refluxed with 50.0 $cm^3$ of 0.200 mol $dm^{-3}$ aqueous sodium hydroxide solution. When the reaction was complete, the remaining sodium hydroxide solution required 15.0 $cm^3$ of 0.400 mol $dm^{-3}$ hydrochloric acid for complete neutralisation. The alcohol formed did not react with acidified potassium dichromate solution.

**Synoptic links**

2.1.3a 4.2.3a 2.13a
2.13b 2.1.3g

**(a)** Describe briefly how you would set up the apparatus for refluxing.

............................................................................................

............................................................................................

............................................................................................ **[2]**

**(b) (i)** Calculate the number of moles of sodium hydroxide present at the beginning of the experiment.

..................... moles **[1]**

**(ii)** Calculate the number of moles of sodium hydroxide that reacted with the bromoalkane.

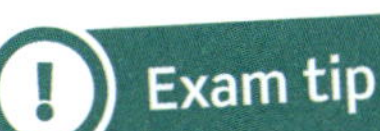

Calculate the number of moles of sodium hydroxide remaining and from this the number that reacted.

..................... moles **[2]**

**(iii)** Calculate the number of moles of the monobromoalkane present (assuming complete reaction) and its relative molecular mass.

**[2]**

**(iv)** Give the formula of the bromoalkane and its structure. Explain how you arrived at your answer.

[Relative atomic masses: Br = 79.9; C = 12; H = 1]

..........................................................................................

..........................................................................................

..........................................................................................

..........................................................................................

..........................................................................................

.......................................................................................... **[5]**

> **Exam tip**
>
> Mono-bromoalkanes have one bromine atom present. Work out the number of carbons and hydrogens present and then read the description at the beginning of the question.

**(c)** Give the balanced symbol equation for the reaction of **A** with aqueous sodium hydroxide.

**[1]**

**3** Lactic acid is a biological compound and is the end-product of anaerobic respiration. Its structural formula is $CH_3CH(OH)(COOH)$.

> **Synoptic link**
>
> 6.2.2c

**(a) (i)** Give the systematic name of lactic acid.

..........................................................................................

.......................................................................................... **[1]**

**(ii)** Name the functional groups present in a molecule of lactic acid.

..........................................................................................

..........................................................................................

.......................................................................................... **[2]**

**(iii)** Give the structural formula of a structural isomer of lactic acid that contains the same functional groups.

**[1]**

(iv) Lactic acid exhibits stereoisomerism.

Name this type of stereoisomerism and explain why it is possible with lactic acid.

..............................................................................................

..............................................................................................

..............................................................................................[2]

(b) **Figure 3.1** shows a set of reactions using lactic acid as the starting compound along with reagents and conditions, but the products are missing.

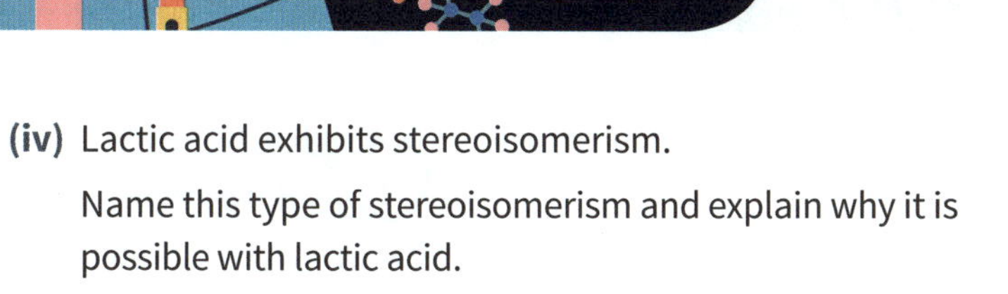


**Figure 3.1**

(i) Give the skeletal formulae of **B**, **C**, and **D**.

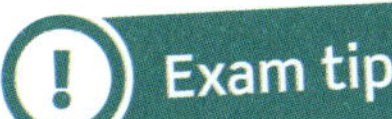


Concentrate on the functional groups present.

[3]

(ii) Name the type of reaction taking place when:

- **B** is converted to **C**
- **C** is converted to **D**

..............................................................................................

..............................................................................................

..............................................................................................[2]

**4** 2-methylpropene is an alkene.

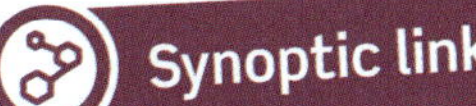


4.1.3f

(a) Describe a chemical test to prove it is an alkene and give the expected observations.

..............................................................................................

..............................................................................................

..............................................................................................[2]

**Exam tip**

This question stipulates that the observations should be given but even if they are not and more than one mark is awarded, give the observations.

(b) A two-step synthetic pathway starting with 2-methylpropene is shown in **Figure 4.1**.

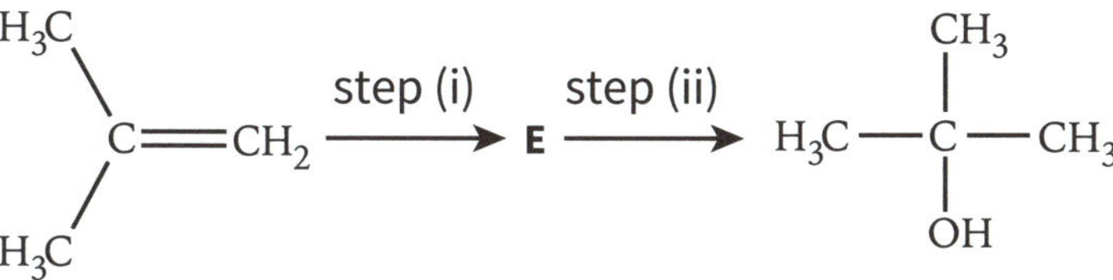


**Figure 4.1**

Give the systematic name or structural formula of **E**.

..............................................................................................

..............................................................................................[1]

**(c)** Give the reagents and conditions needed for steps (i) and (ii) in the reaction pathway shown in **Figure 4.1**.

...........................................................................................

...........................................................................................

...........................................................................................

...........................................................................................

.......................................................................................**[4]**

> **Exam tip**
>
> You have to look at both the starting compound and the end-product and decide what reactions of the starter give a compound from which the end product can be formed.

**(d)** Give the balanced symbol equations for steps (i) and (ii) in the reaction pathway shown in **Figure 4.1**.

...........................................................................................

...........................................................................................

.......................................................................................**[2]**

**5** Compound **F**, shown in **Figure 5.1** below, has three functional groups.

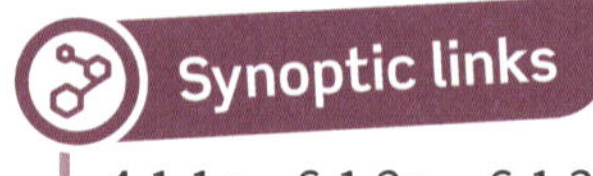

> **Synoptic links**
>
> 4.1.1c 6.1.2a 6.1.3c

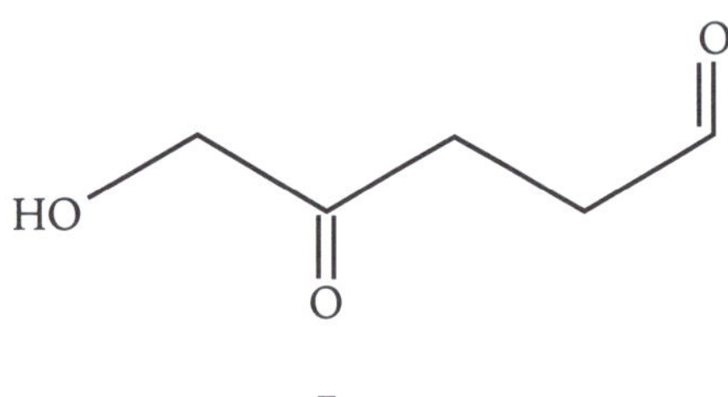


**F**

**Figure 5.1**

**(a) (i)** Name the three functional groups present in compound **F**.

...........................................................................................

...........................................................................................

...........................................................................................

.......................................................................................**[3]**

**(ii)** Give the molecular and structural formulae of **F**.

.......................................................................................**[2]**

**(b) (i)** Use a drawing to give the skeletal formula of the compound formed when **F** is refluxed with excess acidified potassium dichromate.

**[1]**

(ii) Describe any colour changes taking place in the reaction.

..............................................................................................

..............................................................................................[1]

> **Exam tip**
> Do not forget the other product formed.

(iii) Using structural formulae, give the balanced symbol equation for the oxidation reaction.

You may represent the oxidising agent as [O].

[3]

**6** 2-chlorobutane can be prepared in a number of ways. Two of these are shown below:

**Reaction I** $CH_3CH_2CH{=}CH_3 + HCl \rightarrow CH_3CH_2CHClCH_3$

**Reaction II** $CH_3CH_2CH_2CH_3 + Cl_2 \xrightarrow{\text{UV light}} CH_3CH_2CHClCH_3 + HCl$

**(a)** State which of the two reactions has the higher atom economy?

..............................................................................................

..............................................................................................[1]

> **Exam tip**
> What intermediates are formed?

**(b)** Explain why **reaction I** will not give 100% yield.

..............................................................................................

..............................................................................................

..............................................................................................[2]

**(c)** **Reaction II** gives a low percentage yield.

Explain why.

> **Exam tip**
> What other products might form?

..............................................................................................

..............................................................................................

..............................................................................................[2]

**(d)** 2-chlorobutane can be used to prepare butan-2-one ($CH_3COCH_2CH_3$) by a two-step synthetic pathway.

Describe the pathway, including details of the reagents and conditions.

..............................................................................................

..............................................................................................

..............................................................................................

..............................................................................................

..............................................................................................

..............................................................................................[5]

# Knowledge

## 14 Analytical techniques (IR and MS)

### Infrared spectroscopy

Covalent bonds vibrate as they absorb infrared (IR) radiation, which makes them bend and stretch. The stretching is a rhythmic movement in line with the bond, so the distance between the atoms increases and decreases; while the bending changes the angle between atoms.

The degree of bend and stretching depends on:

- the *mass* of the atoms
- the *strength* of the bond.

These properties can be used to identify compounds, as each bond absorbs IR at a certain frequency and this gives a *fingerprint*, which can be analysed to determine identity. The wavenumber (this is proportional to the frequency) is used to identify the type of bond in a molecule.

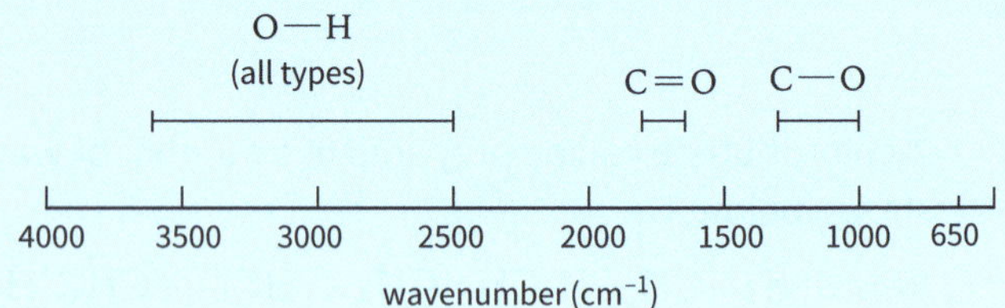


### IR spectra of common functional groups

The *Data Sheet* has values to show which peaks correlate to which bonds, but you should be familiar with the common ones. Most compounds have a peak around 3000 $cm^{-1}$ due to absorption by C—H bonds.

The O—H bond in an alcohol is identified by the peak produced between 3200 $cm^{-1}$ and 3600 $cm^{-1}$.

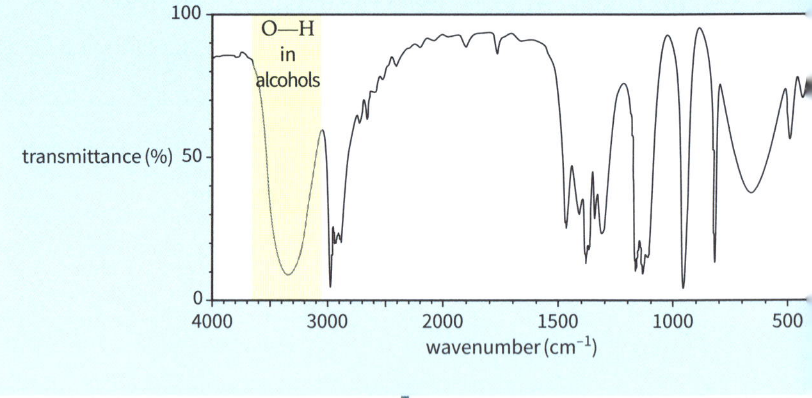


The C=O bond in an aldehyde or a ketone is identified by the peak produced within the region 1630 $cm^{-1}$ to 1820 $cm^{-1}$. It is characteristically very narrow.

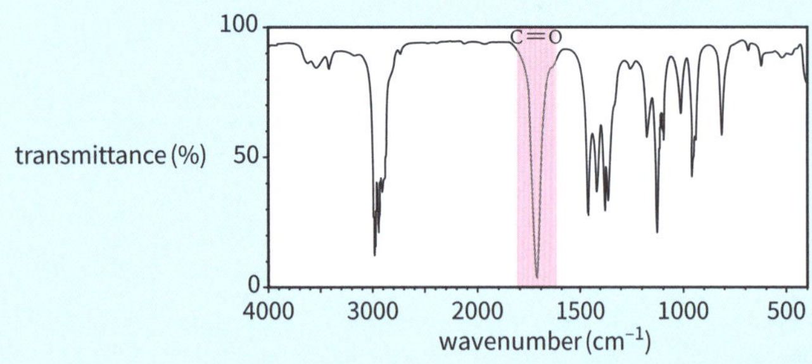


A carboxylic acid has both the O—H group and the C=O group, so shows a characteristic pair of peaks.

- The C=O bond produces a peak between 1630 $cm^{-1}$ and 1820 $cm^{-1}$
- O—H bond produces a broad peak between 2500 $cm^{-1}$ and 3330 $cm^{-1}$.

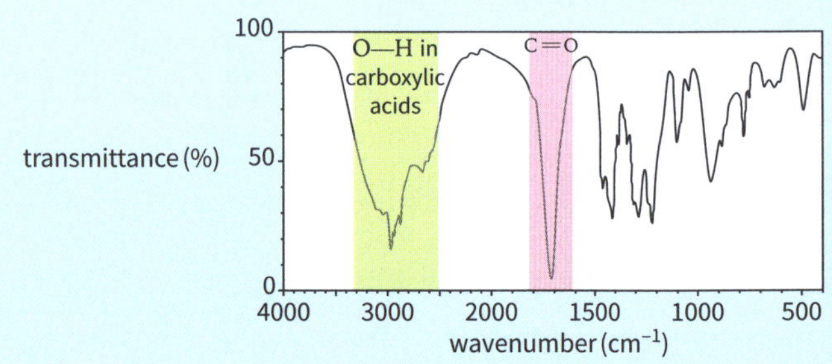

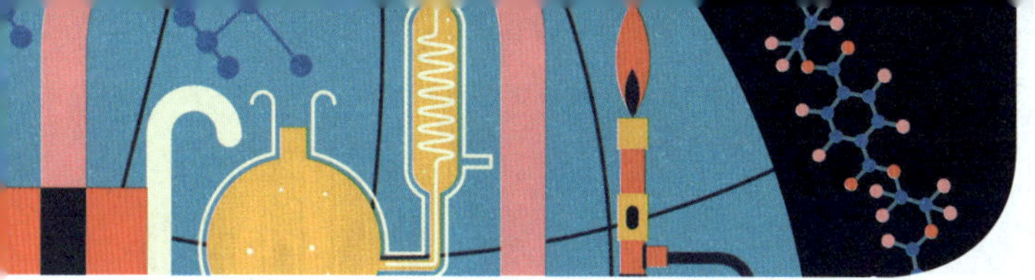

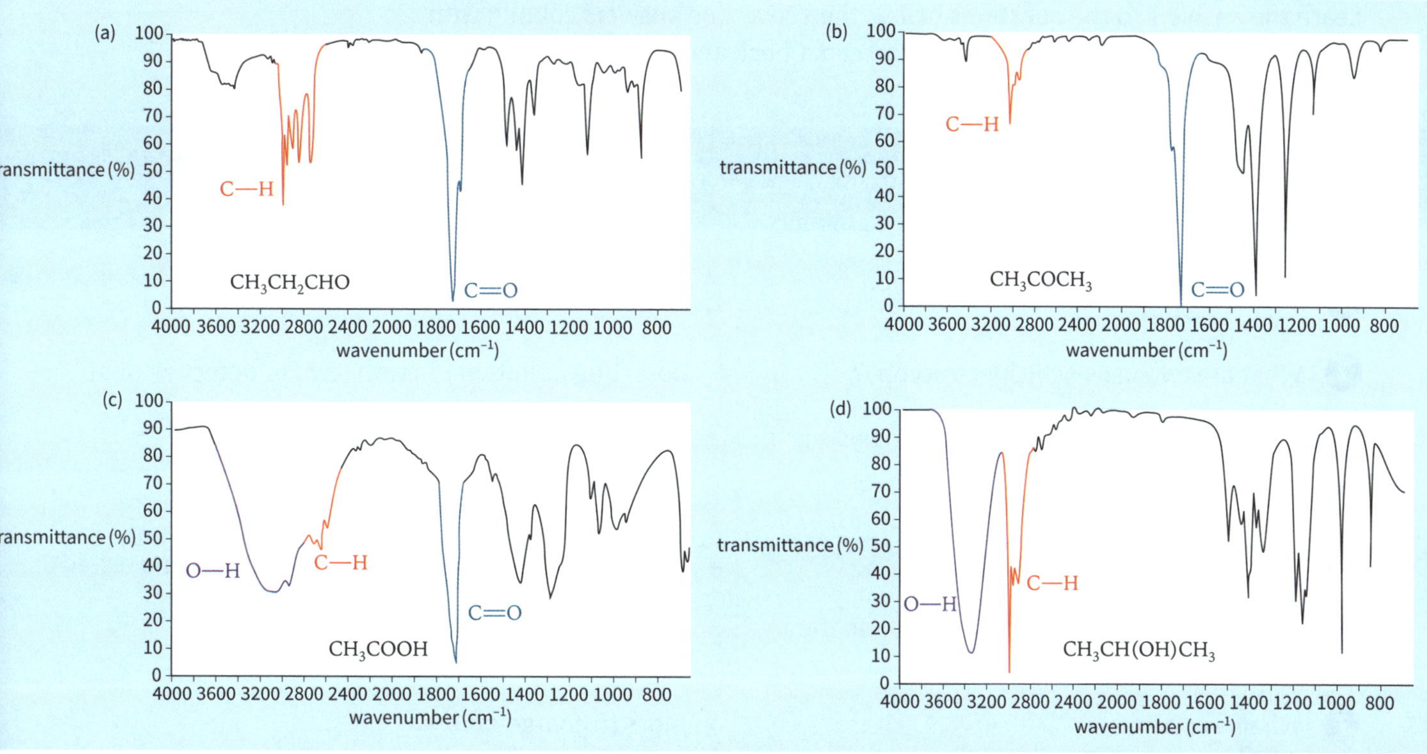


## Uses of IR spectroscopy

The applications of IR spectroscopy are wide ranging.

They can be used to determine the level of pollutants in cities, and in breath tests to detect the presence of ethanol.

## Mass spectrometry

Mass spectrometry can be used to determine the identity of organic compounds and differentiate between isomers of the same mass.

The compound is ionised and positive ions are accelerated towards a detector. The time of flight through the instrument is measured and the mass to charge ratio ($m/z$) of the ion is plotted on a graph showing the relative abundance.

The largest peak will be the molecule as a positive ion; this is the **molecular ion** and gives the mass of the whole compound. Peaks of lower mass will show how a compound has been fragmented.

There may be a small M + 1 peak due to a compound having a small percentage of carbon-13.

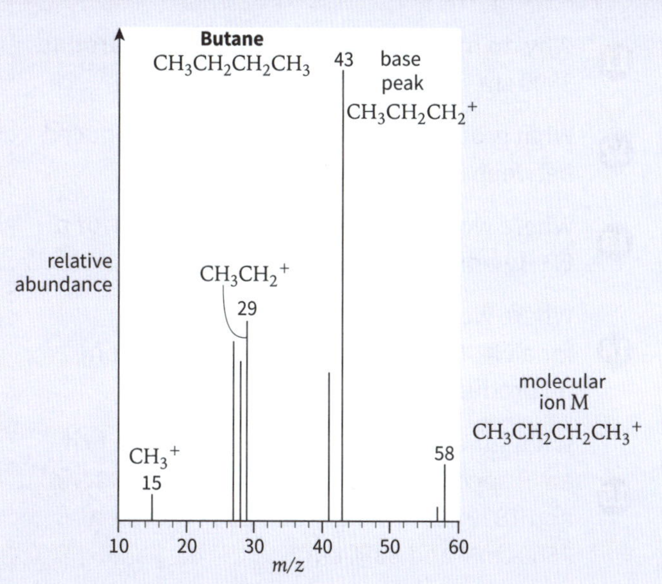

# Retrieval

Learn the answers to the questions below, then cover the answers column with a piece of paper and write as many as you can. Check and repeat.

| | Questions | Answers |
|---|---|---|
| 1 | How can isomers be differentiated on a mass spectra? | $m/z$ of fragments |
| 2 | What are two uses of IR spectroscopy? | detecting pollution / breath tests to detect alcohol |
| 3 | What information can elemental analysis give? | the empirical formula |
| 4 | Why is methane a greenhouse gas? | it re-emits radiation that warms the Earth |
| 5 | What is a molecular ion? | positively charged ion of the compound being tested |
| 6 | What happens to covalent bonds when they absorb IR radiation? | they bend or stretch |
| 7 | What is $m/z$? | mass to charge ratio |
| 8 | What order should you look at data from a range of analyses when trying to determine the identity of a compound? | elemental analysis, mass spectra, IR spectra |
| 9 | What two things determine the degree of bending and stretching of a bond after absorbing IR radiation? | the mass of the atoms and the strength of the bond |
| 10 | What is the name of the region that shows the characteristic of a functional group on an IR spectrum? | fingerprint region |
| 11 | Why do most compounds have a peak around 3000 $cm^{-1}$? | this is due to absorption by C—H bonds |
| 12 | What group is identified by the peak produced between 3200 $cm^{-1}$ and 3600 $cm^{-1}$? | the O—H bond in an alcohol |
| 13 | Where would a peak in IR spectroscopy for a C—O bond be produced? | between 1000 and 1300 $cm^{-1}$ |
| 14 | Where would a peak in IR spectroscopy for a C=O bond in a aldehyde or a ketone be produced? | in the region 1630 $cm^{-1}$ to 1820 $cm^{-1}$ |
| 15 | What functional group would you identify if an IR spectrum had a peak between 1630 $cm^{-1}$ and 1820 $cm^{-1}$ and a broad beak between 2500 $cm^{-1}$ and 3330 $cm^{-1}$ | a carboxylic acid |
| 16 | What is the largest peak in a mass spectrum due to? | the molecular ion |

Put paper here

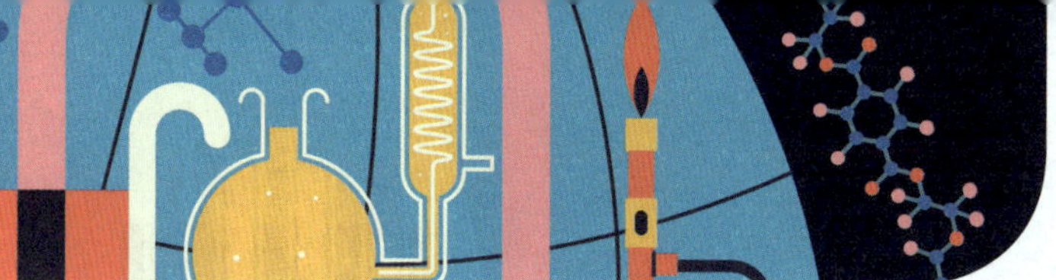



| | Question | | Answer |
|---|---|---|---|
| 17 | Why might there be a small $M + 1$ peak in a mass spectrum? | Put paper here | because of a small percentage of carbon-13 |
| 18 | What is measured by a mass spectrometer? | | time of flight of positive ions through the instrument |
| 19 | Why is IR spectroscopy useful when identifying a compound? | Put paper here | it can be used to identify bonds and functional groups |
| 20 | Why is mass spectroscopy useful when identifying a compound? | | it can be used to identify molecular mass |

## Maths skills

Practise your maths skills using the worked example and practice questions below.

| Standard form | Worked example | Practice |
|---|---|---|
| Very often in chemistry you encounter numbers that are very big or very small. For example, the size of atoms and molecules is very small, but the number of molecules in a sample is very big. Standard form helps you write numbers effectively.<br><br>Standard form is expressed as $a \times 10^n$ where $1 \leq a < 10$ and $n$ is a positive or negative whole number. For example, the following numbers are in standard form:<br><br>$6.02 \times 10^{23}$ or $1.00 \times 10^{-14}$<br><br>These numbers are not in standard form:<br><br>0.005 or 96 500<br><br>You need to be able to key in expressions in standard form into a calculator. Look for the EXP or $\times 10^x$ button. | **Questions**<br><br>Calculate the following and give your answers in standard form.<br><br>**a** $(6 \times 10^7) \div (5 \times 10^{-2})$<br><br>**b** $0.09 \div 0.000621$<br><br>**c** $\sqrt{6\,000\,000}$<br><br>**d** $\frac{R}{N_A}$<br>(where $R$ is the gas constant and $N_A$ is Avogadro's constant)<br><br>**Answers**<br><br>**a** $1.2 \times 10^9$<br><br>**b** $0.09 \div 0.000621 = 145$<br>In standard form this is $1.45 \times 10^2$.<br><br>**c** $\sqrt{6\,000\,000} = 2449$<br>In standard form this is $2.449 \times 10^3$.<br><br>**d** $\frac{R}{N_A} = \frac{8.31}{6.02 \times 10^{23}} = 1.38 \times 10^{-23}$ | Calculate the following and give your answers in standard form.<br><br>**1** $3 \times 10^4 \times 2 \times 10^5$<br><br>**2** $(3.6 \times 10^5) \div (4.5 \times 10^{-7})$<br><br>**3** $6.25 \times 10^8 + 5.4 \times 10^9$<br><br>**4** $36^2 \times 5^3$<br><br>**5** $N_A{}^3$ |

## Exam-style questions

1 The infrared spectra shown below are for three unbranched compounds with the same number of carbon atoms. They are, in random order, a primary alcohol, an aldehyde, and a carboxylic acid.

**(a)** Identify the type of compound from the spectrum.

Explain how you made your choice.

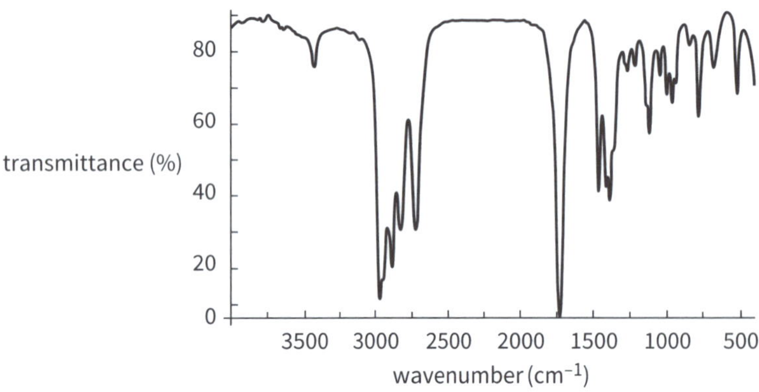


**Figure 1.1**

..........................................................................................

..........................................................................................

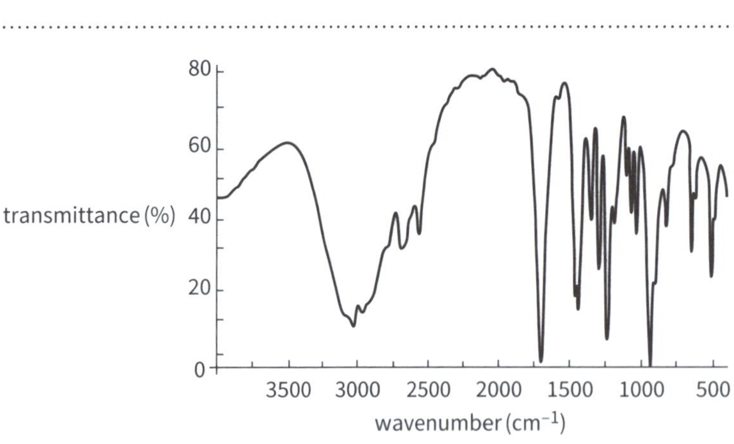


**Figure 1.2**

..........................................................................................

..........................................................................................

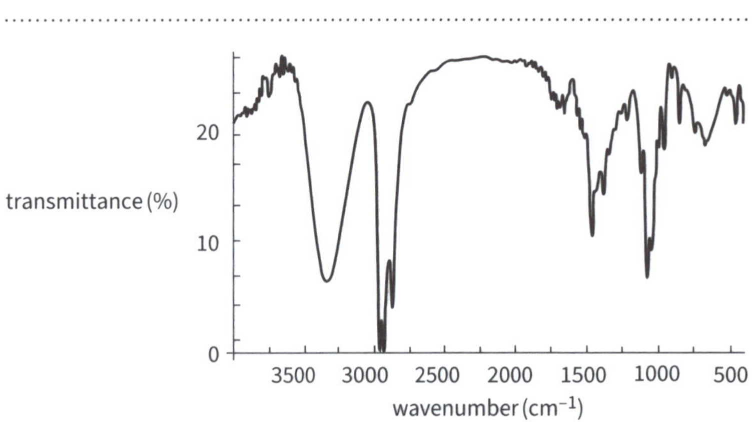


**Figure 1.3**

..........................................................................................

..........................................................................................**[6]**

> **Exam tip**
>
> Use the datasheet and use wavenumbers.

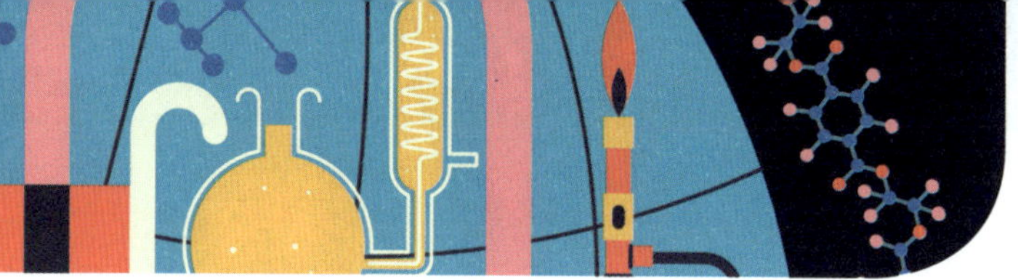



**(b)** Describe the reagents and conditions required for converting the primary alcohol (R—$CH_2OH$) into the carboxylic acid (R—COOH).

..............................................................................................

..............................................................................................

.............................................................................................. **[2]**

**(c)** Describe and explain how the spectra of the products would differ from the spectra of the reactants.

..............................................................................................

..............................................................................................

..............................................................................................

..............................................................................................

..............................................................................................

.............................................................................................. **[5]**

**2** Compound **A** contains carbon, hydrogen, and oxygen. On analysis, the compound contained 62.1% carbon and 10.3% hydrogen.

**(a)** Calculate the empirical formula of compound **A**.

empirical formula = .................... **[2]**

**(b)** The infrared and mass spectra of compound **A** are shown in **Figures 2.1** and **2.2** below.

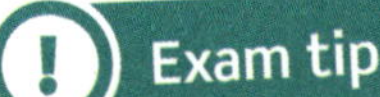



When answering questions on IR spectra, simply quote the numbers on the datasheet.

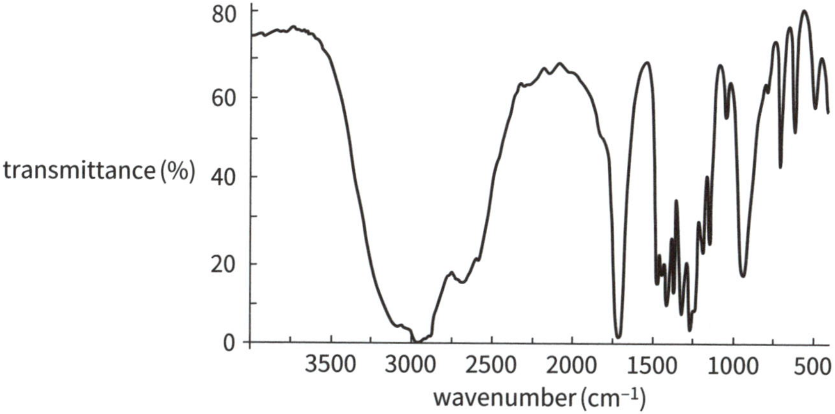



**Figure 2.1**

For answers and more practice questions visit www.oxfordrevise.com/scienceanswers

Even more practice and interactive revision quizzes are available on kerboodle

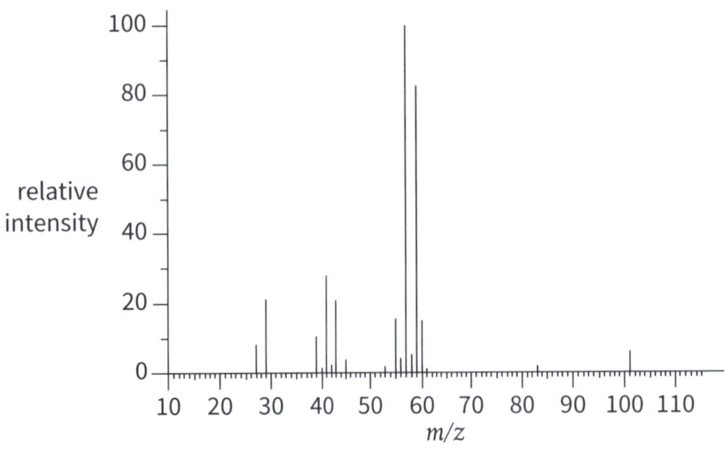


**Figure 2.2**

Identify the functional group present in compound **A** and explain your answer.

........................................................................................

........................................................................................

........................................................................................ **[2]**

**(c)** Determine the molecular formula of **A**.

Show your reasoning.

........................................................................................

........................................................................................

........................................................................................ **[2]**

**(d)** On the mass spectrum of **A**, the peak at $m/z = 57$ is caused by a tertiary carbocation. The peak at $m/z = 59$ is due to a cation that makes up the rest of the molecule.

**(i)** Give the structures of both carbocations and explain how you arrived at your answers.

........................................................................................

........................................................................................

........................................................................................

........................................................................................

........................................................................................ **[4]**

**(ii)** Give the skeletal or structural formula of compound **A**.

**[1]**

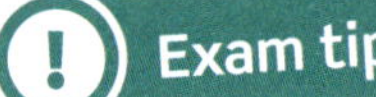

**Exam tip**

For the second carbocation, find the mass of the functional group and its contribution to the mass.

**3** The structural formulae of the isomeric compounds butanoic acid and ethyl ethanoate are given below:

| | |
|---|---|
| butanoic acid | $CH_3CH_2CH_2COOH$ |
| ethyl ethanoate | $CH_3COOCH_2CH_3$ |

**Synoptic link**

6.1.3c

**(a) (i)** Draw the displayed formulae of both compounds. **[2]**

**(ii)** Suggest any similarities and differences in their infrared spectra. **[2]**

**(b)** Ethyl ethanoate can be formed from the reaction of ethanol with ethanoic acid.

Outline how you could prepare ethyl ethanoate in the laboratory using ethanol as the only organic compound available (you would need a catalyst).

Give balanced symbol equations for any reactions that take place in your method. **[5]**

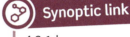

**Exam tip**

Consider the functional groups and where their absorptions occur in the IR spectrum.

**4** Compound **B** is an unbranched carbon compound that has the following percentage composition: carbon 68.2%; hydrogen 13.63%; and oxygen 18.2%. On its mass spectrum, the molecular ion peak is at $m/z = 88$ and its most prominent peak is at $m/z = 59$.

**Synoptic link**

4.2.1d

**(a)** Calculate the empirical and molecular formulae of **B**. **[3]**

**(b)** The infrared spectrum of **B** is shown in **Figure 4.1**.

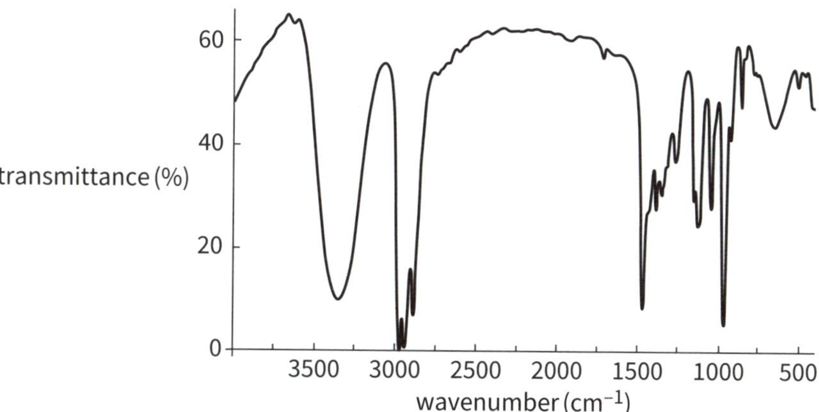


**Figure 4.1**

**(i)** Identify the functional group present in **B**. **[1]**

**(ii)** Give the names and skeletal formulae of the three compounds with the molecular formula of **B** and the same functional group. **[3]**

**(c)** When heated with concentrated phosphoric acid, **B** is dehydrated to give just two products, that are *E-Z* isomers.

**(i)** Identify these two *E-Z* isomers using skeletal formulae and name them. **[2]**

**(ii)** State which of the three compounds given in your answer to **4 (b) (ii)** is **B**.

Explain why you chose this compound and not the other two.

In your answer make it clear which isomer you are referring to. **[3]**

**Exam tip**

There are only two *E-Z* isomers with the formula $C_5H_{10}$.

**(d)** Identify the fragment responsible for the peak at $m/z = 59$ in the mass spectrum of **B**. **[1]**

**5** Compound **C** is an unbranched compound containing four carbons. **C** has no stereoisomers. When analysis was carried out on **C** the following results were obtained.

- When heated gently with alcoholic silver nitrate solution, **C** gave a pale cream precipitate that was partially soluble in ammonia solution.
- When **C** was refluxed with aqueous NaOH, compound **D** was obtained.
- When **D** was refluxed with excess acidified dichromate solution, compound **E** was formed.
- Compounds **C**, **D**, and **E** decolourised bromine water.
- The infrared spectrum of **E** is shown in **Figure 5.1**.

**Synoptic links**

4.1.3c 4.1.3f 4.2.1c 4.2.2a

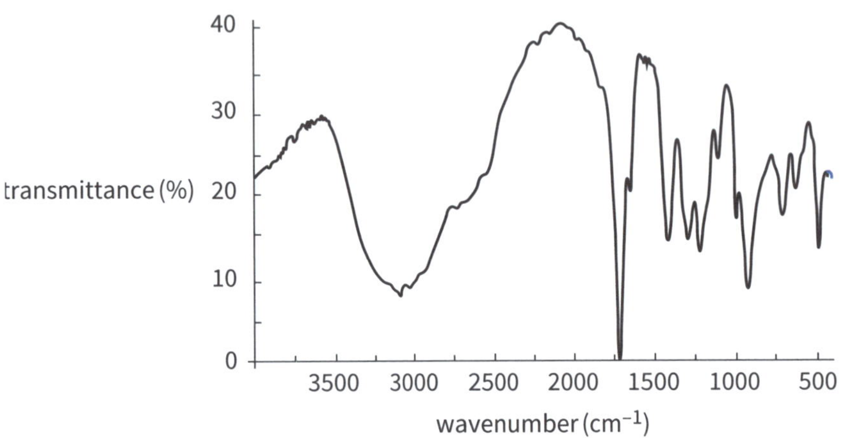


**Figure 5.1**

**(a)** Identify the functional group in compound **E** that is responsible for this infrared spectrum.

Explain your answer. **[2]**

**Exam tip**

Work backwards from **E**.

**(b)** Identify the functional groups in compounds **C** and **D**.

Explain your reasoning. **[3]**

**(c)** Identify the structures of **C**, **D**, and **E** from the information given and give their skeletal formulae.

Explain your answers. **[4]**

**(d)** Give the balanced symbol equation for the reaction of **C** with bromine. **[1]**

**6** Ozonolysis is a technique used to deduce the groups attached to a C=C bond in alkenes. A simplified description is shown in **Figure 6.1**.

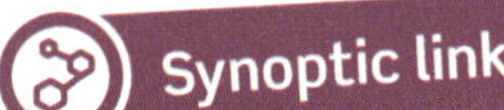

**Synoptic link**

4.2.1c

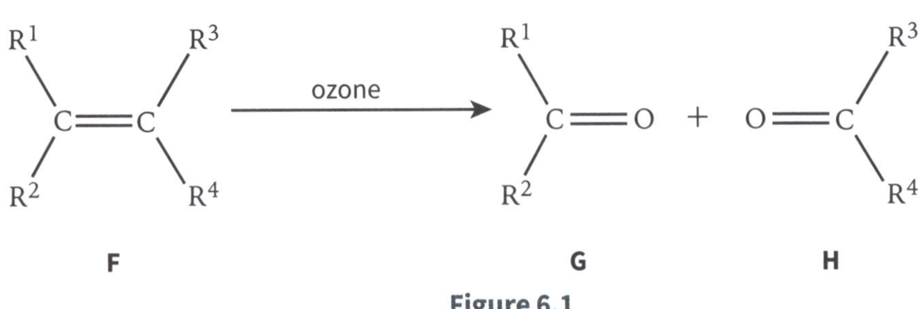


**Figure 6.1**

If any one of the substituents $R_1$ to $R_4$ are hydrogen, then the resulting compounds will be aldehydes. If both $R_1$ and $R_2$ are alkyl groups then the resulting compounds will be ketones.

- In the mass spectrum of an alkene, **F**, the molecular ion peak was at $m/z = 84$.
- After ozonolysis, **F** gave two products, **G** and **H**.
- The mass spectra for **G** and **H** gave molecular ion peaks at $m/z = 44$ for **G** and $m/z = 72$ for **H**.

**(a)** Calculate the molecular formula of **F**. Show how you arrived at your answer. **[2]**

**Exam tip**

How many carbons? Divide by 12.

**(b)** The infrared spectra of the two products, **G** and **H**, are shown below (**Figures 6.2** and **6.3**). **Figures 6.4** and **6.5** show along with the spectra of what is formed after each of them is refluxed with acidified potassium dichromate solution.

Infrared spectrum for **G**

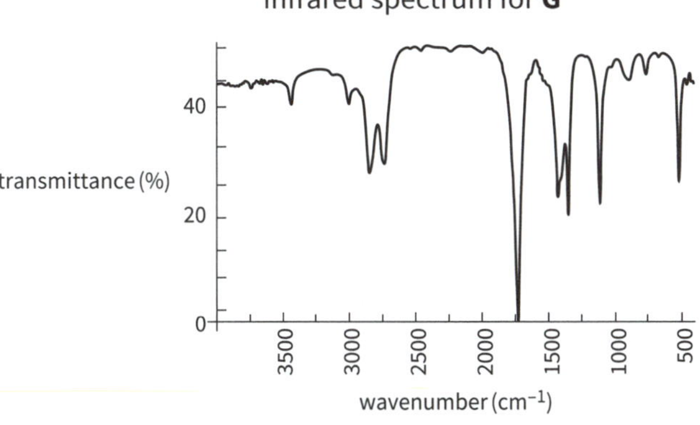


**Figure 6.2**

Infrared spectrum for **H**

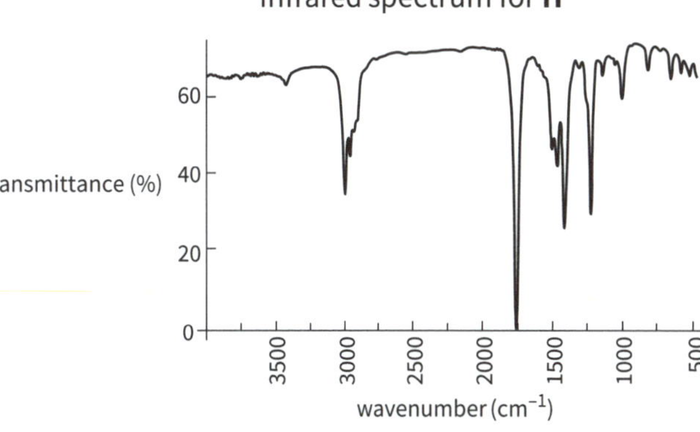


**Figure 6.3**

Infrared spectrum for **G** after reflux with $H^+/Cr_2O_7^{2-}$

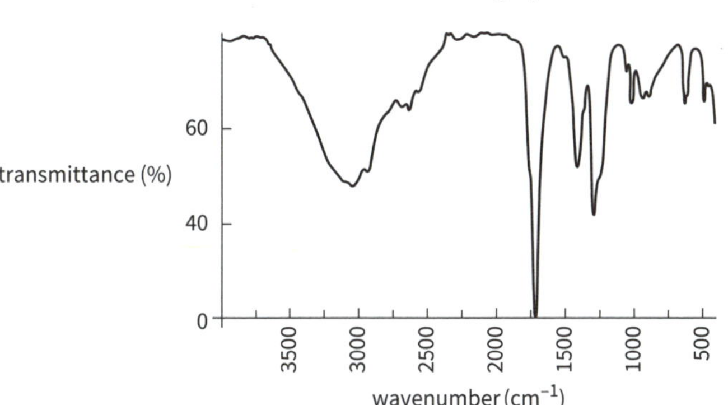


**Figure 6.4**

Infrared spectrum for **H** after reflux with $H^+/Cr_2O_7^{2-}$

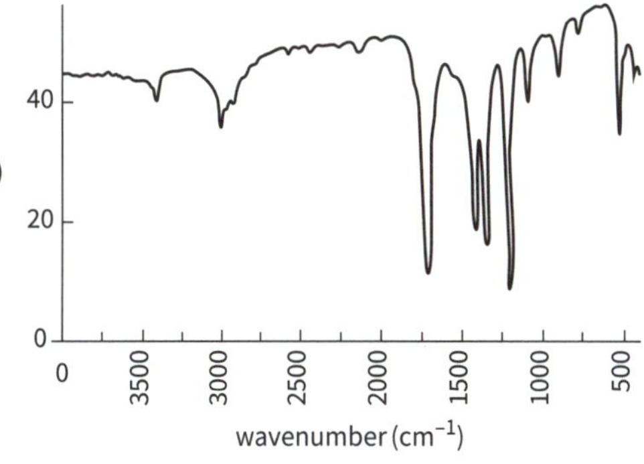


**Figure 6.5**

**(i)** Using the information from the infrared and mass spectra, identify the functional group in **G** and its structural formula. Explain your reasoning. **[4]**

**(ii)** Using the information from the infrared and mass spectra, identify the functional group in **H** and its structural formula and name.

Explain your reasoning. **[4]**

**(c)** Give the structural formula of the alkene **F**. **[1]**

# Knowledge

## 15 Reaction rates and equilibrium (quantitative)

### Measuring the rate of reaction

The rate of a reaction can be measured in a range of different ways: by looking at the reactants being used up (e.g. by loss of mass or change in pH) or looking at the production of products (e.g. by gas being released or change in colour or turbidity).

The rate of reaction at a specific time can be found by measuring the gradient of the tangent at that time on a concentration–time graph. The rate at different times can compared by comparing the gradients.

### The rate equation and order of reaction

For any reaction a **rate equation** can be produced. Based on experimental data, this shows how reactants influence the rate. Reactions can be zero, first, or second order with respect to certain reactants. The overall order of a reaction is the sum of all the individual orders. The order of a reaction can be determined from graphs of rate experiments.

For two reactants [A] and [B], the rate equation is:

rate = $k[A]^m[B]^n$

overall order = $m + n$

$k$ is the **rate constant**; it is specific for every reaction and varies with temperature. $m$ and $n$ are the **orders of reaction** with respect to each reactant.

For a reaction A + B → C + D

| Orders of the reaction | Rate equation | Overall order | Affect on rate |
|---|---|---|---|
| A (or B) is first order | rate = $k[A]$ | first | doubling the concentration of A would double the rate of reaction |
| A and B are first order | rate = $k[A][B]$ | second | doubling the concentration of A or B individually would double the rate of reaction; doubling the concentration of both would quadruple the rate of reaction |
| A (or B) is second order | rate = $k[A]^2$ | second | doubling the concentration of A would quadruple the rate of reaction |
| A and B are second order | rate = $k[A]^2[B]^2$ | fourth | doubling the concentration of A or B individually would quadruple the rate of reaction |
| A is first order and B is second order | rate = $k[A][B]^2$ | third | doubling the concentration of A individually would double the rate of reaction. Doubling the concentration of B individually would quadruple the rate of reaction |

### Zero order

When the rate of a reaction is not affected by concentration, the reaction is a **zero order reaction** with respect to the reactant. The rate expression is:

rate = $k$.

The units for $k$ for this reaction is **$mol\,dm^{-3}\,s^{-1}$**. A concentration–time graph will be a straight line with a negative gradient. A rate–concentration graph will be a horizontal line.

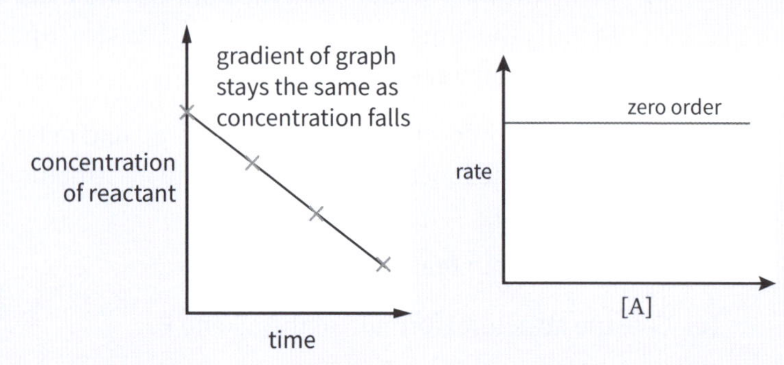

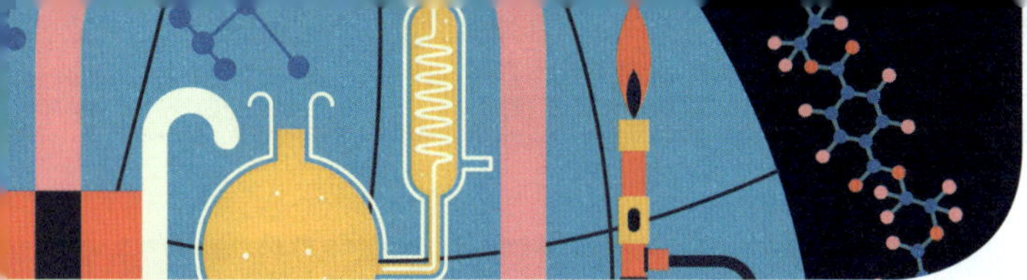



## First order

For **first order reactions** the rate of reaction will be directly proportional to the concentration. If the reaction is first order with respect to the concentration of reactant A:

$$\text{rate} = k[\text{A}]$$

The unit for $k$ for this reaction is $\mathbf{s^{-1}}$.

For a first-order reaction, the **half-life** ($t_{1/2}$), the time for the concentration of a reactant to half, is constant.

$$k = \ln \frac{2}{t_{1/2}}$$

A rate–concentration graph for a first order reaction is a straight line through the origin.

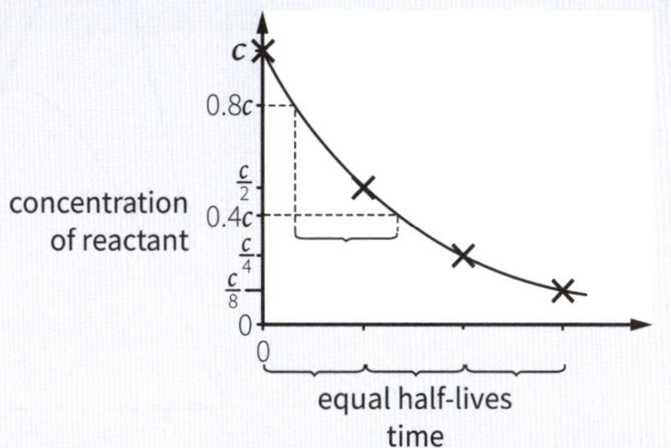


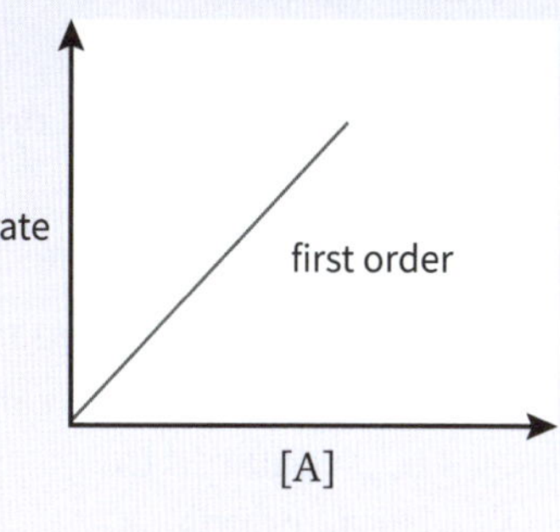


## Second order

For **second order reactions** the rate of reaction is proportional to the square of the concentration. A *rate–concentration graph* will be an upward curve with increasing gradient, but $k$ cannot be determined from this graph. The rate constant is determined from a graph of *rate–concentration*$^2$. This is a straight line. For a reaction where the rate is dependent on the concentration of both reactants A and B the rate equation is:

$$\text{rate} = k[\text{A}][\text{B}]$$

The units for $k$ for this reaction are $\mathbf{mol^{-1}\ dm^{3}\ s^{-1}}$.

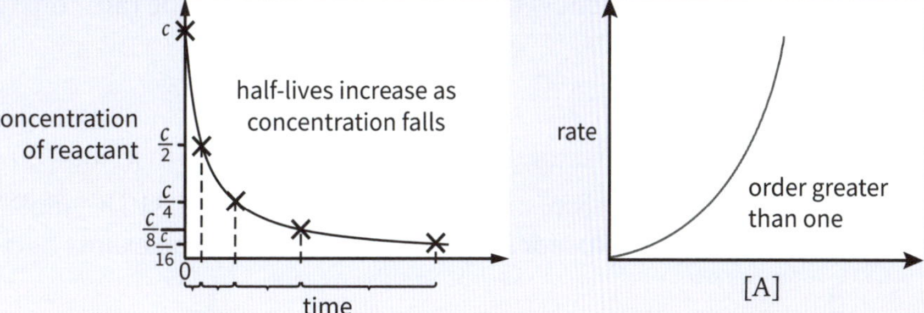


## The initial rate method

The **initial rate method** can also be used to determine the order of a reaction. At a given temperature a range of experiments are set up to test which concentrations will affect the initial rate of reaction. The initial rate is the gradient when time = 0 on a *concentration–time* graph. The initial rates are compared to work out the order of reaction with respect to each reactant. For example, to determine the orders of reactants for $\text{A} + \text{B} \rightarrow \text{C} + \text{D}$ the following applies:

| [A] (mol dm$^{-3}$) | [B] (mol dm$^{-3}$) | Initial rate (mol dm$^{-3}$ s$^{-1}$) |
|---|---|---|
| $1.0 \times 10^{-3}$ | $1.0 \times 10^{-3}$ | $3.2 \times 10^{-4}$ |
| $2.0 \times 10^{-3}$ | $1.0 \times 10^{-3}$ | $6.4 \times 10^{-4}$ |
| $2.0 \times 10^{-3}$ | $2.0 \times 10^{-3}$ | $6.4 \times 10^{-4}$ |

[A] stays the same but [B] is doubled and rate is the same so [B] is zero order

[B] stays the same but [A] is doubled and rate is doubled so [A] is first order

## The rate constant and temperature

If the rate equation for a reaction is:

rate = $k[A][B]^2$

only the value of $k$ changes with temperature, not the order of the reaction.

At higher temperatures:

- more particles have the energy to overcome the activation energy ($E_a$) required to break bonds,
- particles move faster, collide more frequently with the correct orientation.

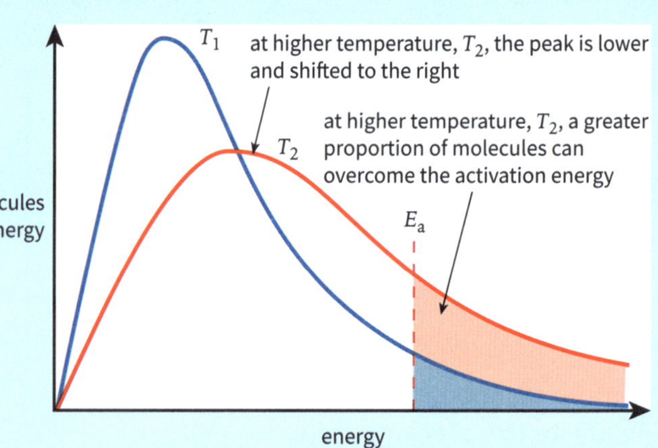


The rate constant, $k$, increases as temperature increases.

The higher the value for $k$, the faster the rate of reaction.

## The Arrhenius equation

The **Arrhenius equation** show the relationship between the rate constant and temperature.

$$k = Ae^{-\frac{E_a}{RT}}$$

$A$ is the pre-exponential factor, $E_a$ is the activation energy in kJ mol$^{-1}$, $T$ is the temperature in kelvin, and $R$ is the gas constant, 8.31 J K$^{-1}$ mol$^{-1}$. This equation and the value for $R$ will be given to you in the exam.

The logarithmic form of the equation is:

$$\text{In}\,k = -\frac{E_a}{RT} + \text{In}\,A$$

Plotting In $k$ against $\frac{1}{T}$ gives a straight-line graph with the equation:

$$ym = mx + c$$

By comparing the two equations $E_a$ and $A$ can be calculated because:

gradient $(m) = \frac{-E_a}{R}$

$c = y$-intercept = In $A$

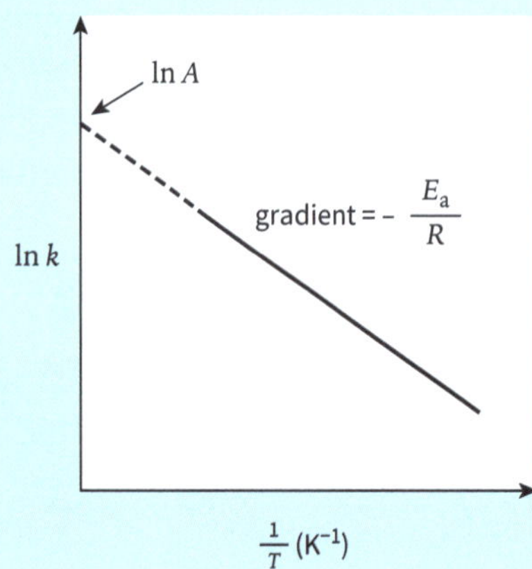

## The rate-determining step

The majority of reactions that happen are not simple and involve a number of steps; some of these are fast steps and some are slow steps.

It is like people queuing to leave a building through a revolving door: it does not matter how quickly they reach the exit if the door can only take one person at a time. The time taken to get through the door limits the speed at which people can leave. The slowest step is is the **rate-determining step**.

The rate-determining step gives information about possible mechanisms. For example, for $A + B \rightarrow C + D$, if rate = $k[A]$, then molecule B cannot be involved in the rate-determining step. If rate = $k[A][B]$, then the slowest step in the reaction mechanism probably involves both A and B, whereas a reaction rate proportional to $[A]^2$ suggests a rate-determining step in which two molecules of A react to give an intermediate.

## The equilibrium constant $K_p$

For a reversible reaction that occurs in a gaseous state, the equilibrium constant is dependent on pressure as well as temperature.

$K_p$ is the equilibrium constant for a system that contains gaseous molecules at a constant temperature, calculated from partial pressures.

The equilibrium constant is not affected by a catalyst, only the rate at which equilibrium is achieved is affected by a catalyst.

## Mole fraction and partial pressure

For a system that contains a mixture of gaseous molecules the **partial pressure**, $p$, of a gas is the contribution that the gas makes to the total pressure of the system. The sum of the partial pressures of each gas is equal to the total pressure of the system.

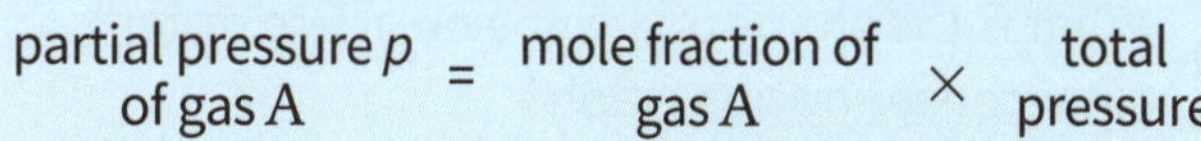

partial pressure $p$ of gas A = mole fraction of gas A × total pressure

$$\text{mole fraction of gas A} = \frac{\text{number of moles of gas A in mixture}}{\text{total number of moles of gas in the mixture}}$$

$K_p$ is the equilibrium constant for a reaction involving gases at a constant temperature. Changing temperature will affect the value of $K_p$ for endothermic and exothermic reactions, following the principles of Le Châtelier.

## The rate constant $K_p$

For a gaseous reaction in equilibrium:

$$aA(g) + bB(g) \rightleftharpoons cC(g) + dD(g)$$

$$K_p = \frac{p(C)^c\, p(D)^d}{p(A)^a\, p(B)^b}$$

This is in a similar format to the expression for $K_c$, used for equilibrium reactions where concentration affects the equilibrium constant of a reaction.

$$K_c = \frac{[C]^c[D]^d}{[A]^a[B]^b}$$

$K_p$ and $K_c$ expressions do not include terms for any reactants or products in the solid phase.

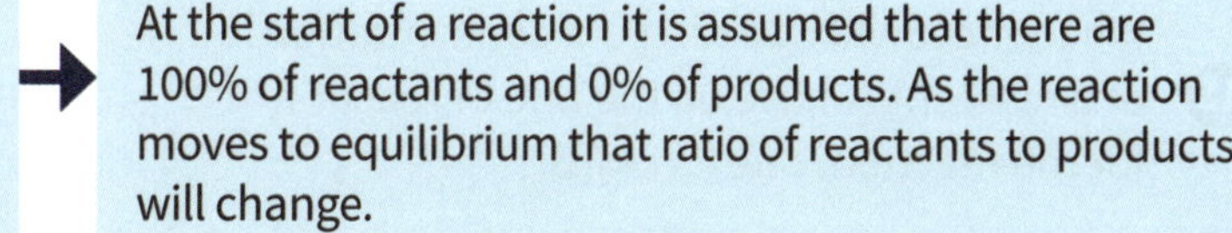

At the start of a reaction it is assumed that there are 100% of reactants and 0% of products. As the reaction moves to equilibrium that ratio of reactants to products will change.

# Retrieval

Learn the answers to the questions below, then cover the answers column with a piece of paper and write as many as you can. Check and repeat.

| | Questions | Answers |
|---|---|---|
| 1 | What is the overall order of this reaction: rate = $k$[A][B]? | second |
| 2 | What are the units for $k$ in a zero order reaction? | mol dm$^{-3}$ s$^{-1}$ |
| 3 | What type of graph will give a straight line for a second order reaction? | rate–concentration$^2$ |
| 4 | Which part of the Arrhenius equation, $\ln k = -\frac{E_a}{RT} + \ln A$ is equal to the gradient of a straight line graph of $\ln k$ against $\frac{1}{T}$? | $-\frac{E_a}{R}$ |
| 5 | What is partial pressure? | the contribution of each individual gas to the total pressure |
| 6 | What is the order with respect to reactant A for this reaction: rate = $k$[A][B]? | first |
| 7 | What is the shape of the line on a zero order rate–concentration graph? | flat line |
| 8 | How can the partial pressure of a gas be determined? | partial pressure $p$ of gas = mole fraction of gas × total pressure |
| 9 | What state do some of the reactants need to be in for partial pressure to be calculated? | gaseous |
| 10 | What are the units for $k$ for a second order reaction? | mol dm$^3$ s$^{-1}$ |
| 11 | What is the order of a reaction if the units for $k$ are s$^{-1}$? | first |
| 12 | What is the order of a reaction if doubling the concentration has no effect on rate? | zero |
| 13 | Which part of the Arrhenius equation, $\ln k = -\frac{E_a}{RT} + \ln A$ is equal to the $y$-intercept of a straight-line graph of $\ln k$ against $\frac{1}{T}$? | $\ln A$ |
| 14 | How can the mole fraction of a gas A be calculated? | mole fraction of gas A = $\frac{\text{number of moles of gas A in mixture}}{\text{total number of moles of gas in the mixture}}$ |
| 15 | What effect will doubling the concentration of a reactant A have on the rate of reaction for the reaction, rate = $k$[A][B]? | it would double |
| 16 | What is the rate determining step? | the slowest step in a reaction |

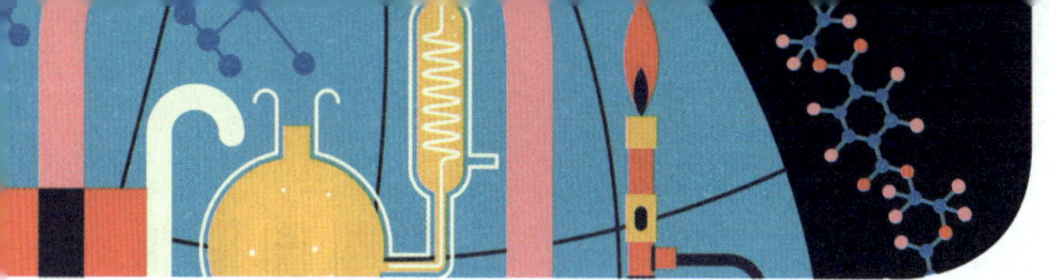



| | Question | | Answer |
|---|---|---|---|
| 17 | What is the relationship between rate and concentration for a first order reaction? | Put paper here | they are directly proportional to each other |
| 18 | What would happen to the rate of reaction if the value of the rate constant $k$ was increased? | Put paper here | it would increase |
| 19 | How does the shape of a Maxwell–Boltzman distribution curve change when the temperature is increased? | Put paper here | it shifts it to the right and lowers the peak of the curve |
| 20 | What time value is the gradient of the tangent measured on on a initial rate graph? | Put paper here | $T = 0$ |

# Practical skills

Practise your practical skills using the worked example and practice questions below.

### Initial rate method

Clock reactions are a very good way to measure the rate of a reaction by an initial rate method. For a reaction to be used in an initial rate method there needs to be a sudden clear colour change. Two common reactions are:

- the iodine clock, which produces a very dark blue iodine starch complex
- thiosulfate and acid, in which elemental sulfur is produced, which makes the solution opaque.

For these experiments it is acceptable to approximate the rate as $\frac{1}{\text{time}}$ instead of $\frac{\Delta(\text{concentration})}{\text{time}}$.

If the temperature is varied, the activation energy can be determined from plotting a graph of $\ln k$ against $\frac{1}{\text{time}}$.

### Worked example

The rate constant for a chemical reaction at 20 °C is $9.00 \times 10^{-3}$ mol dm$^{-3}$ s$^{-1}$.

The activation energy for the reaction is 36.0 kJ mol$^{-1}$.

**Questions**

1 Calculate the pre-exponential factor, $A$.

2 Calculate the rate constant, $k$ when $T = 30$ °C.

**Answers**

1 $k = Ae^{-\frac{E_a}{RT}}$

$\therefore A = \frac{k}{e^{-\frac{E_a}{RT}}}$

$= \frac{9 \times 10^{-3}}{e^{-\frac{36 \times 10^3}{8.31 \times 293}}}$

$= 2.37 \times 10^4$

2 $k = 23\,740 \times e^{-\frac{36 \times 10^3}{8.31 \times 2303}}$

$= 0.0147$ mol dm$^{-3}$ s$^{-1}$

So, increasing the temperature by 10 °C, means the reaction happens about 63% faster!

### Practice

A student investigates the iodine clock reaction to measure the activation energy, $E_a$, of a reaction. The student measures the time of the reaction at different temperatures, but keeps all reactant concentrations the same. The following measurements are recorded.

| Temperature (°C) | Time (s) |
|---|---|
| 20 | 88 |
| 30 | 52 |
| 40 | 32 |
| 50 | 20 |
| 60 | 13 |

1 Calculate $\ln\left(\frac{1}{\text{time}}\right)$ and $\frac{1}{T\text{ (K)}}$ for each set of data.

2 Plot a suitable graph.

3 Determine the gradient of the graph.

4 Calculate the activation energy, $E_a$ in kJ mol$^{-1}$

# Practice

## Exam-style questions

**1** A student was investigating the effect that varying the concentration of two compounds **X** and **Y** had on the rate of reaction.

They determined the rate equation to be:

$$\text{rate} = k[\text{X}]^2$$

**(a) (i)** State the order of reaction with respect to **Y**.

................................................................................ **[1]**

**(ii)** At 25 °C the initial rate was $4.3 \times 10^{-3}$ mol dm$^{-3}$ s$^{-1}$ when the concentration of **X** was 0.38 mol dm$^{-3}$ and the concentration of **Y** was 0.43 mol dm$^{-3}$.

Calculate the value for the rate constant at this temperature.

Give its units.

rate constant = ......................

units = ...................... **[3]**

**(b)** The Arrhenius equation can be used to explore the relationship between the rate constant, $k$, the activation energy, $E_a$, and the temperature, $T$.

This can be expressed as:

$$\ln k = \frac{E_a}{RT} + \ln A$$

Use your answer to **1(a)(ii)** and this equation to calculate the activation energy at 25 °C given that $\ln A = 19.8$ and the gas constant $R = 8.31$ J K$^{-1}$ mol$^{-1}$.

(If you were unable to obtain a value for **1(a)(ii)**, you should use $3.3 \times 10^{-3}$ as the rate constant. This is *not* the correct value.)

activation energy = ......................kJ mol$^{-1}$ **[4]**

**(c)** Two possible mechanisms are proposed for the reaction.

**Mechanism 1**

$2\text{X} \rightarrow \text{J}$ **SLOW**

$\text{J} + \text{Y} \rightarrow \text{A} + \text{B}$

**Mechanism 2**

$\text{X} + \text{Y} \rightarrow \text{A} + \text{B}$

Which mechanism is correct? Give a reason for your answer.

................................................................................

................................................................................

................................................................................ **[2]**

**Exam tip**

Go back to the rate equation to check your answer.

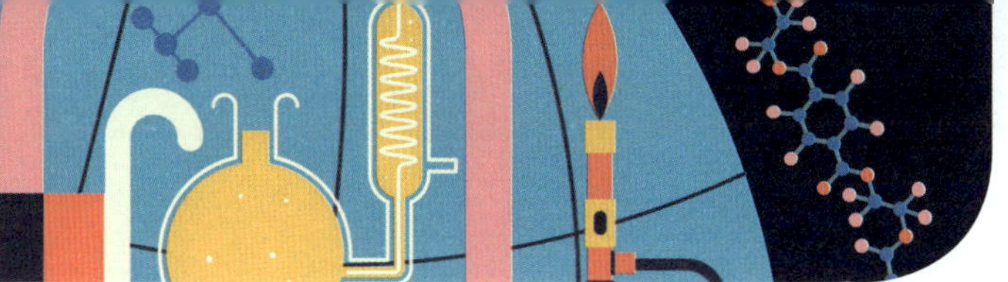



2 The results in **Table 2.1** were obtained for the reaction:

$$A + 3B \rightarrow C + D$$

| Experiment | [A] (mol dm$^{-3}$) | [B] (mol dm$^{-3}$) | Initial rate (mol dm$^{-3}$ s$^{-1}$) |
|---|---|---|---|
| 1 | 0.50 | 1.00 | $2.50 \times 10^{-3}$ |
| 2 | 0.50 | 2.00 | $1.00 \times 10^{-2}$ |
| 3 | 1.00 | 0.50 | $1.25 \times 10^{-3}$ |

**Table 2.1**

**(a)** Give the order of the reaction with respect to **B**.

........................................................................................ **[1]**

**(b)** Give the order of the reaction with respect to **A**.

........................................................................................ **[1]**

**(c)** Determine the rate equation for this reaction.

Calculate the rate constant and give its units.

rate constant = ....................

units = .................... **[4]**

**Exam tip**

Pay close attention to the standard form. It is easy to miss when a rate is faster.

3 The following reaction happens between three compounds **X**, **Y**, and **Z**:

$$X(aq) + 2Y(aq) + Z(aq) \rightarrow D(aq) + E(aq)$$

**Table 3.1** shows some rates at varying concentrations of each reactant.

| Experiment | [X] (mol dm$^{-3}$) | [Y] (mol dm$^{-3}$) | [Z] (mol dm$^{-3}$) | Rate (mol dm$^{-3}$ s$^{-1}$) |
|---|---|---|---|---|
| 1 | $1 \times 10^{-3}$ | $1 \times 10^{-3}$ | $2 \times 10^{-3}$ | $7.50 \times 10^{-3}$ |
| 2 | $2 \times 10^{-3}$ | $2 \times 10^{-3}$ | $2 \times 10^{-3}$ | $3.00 \times 10^{-2}$ |
| 3 | $5 \times 10^{-4}$ | $2 \times 10^{-3}$ | $4 \times 10^{-3}$ | $3.75 \times 10^{-3}$ |

**Table 3.1**

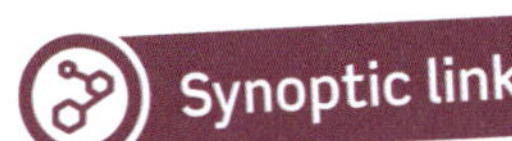


3.2.3

**(a) (i)** Give the order of the reaction with respect to **X**.

........................................................................................ **[1]**

**(ii)** Give the order of the reaction with respect to **Y**.

........................................................................................ **[1]**

**(iii)** Give the order of the reaction with respect to **Z**.

........................................................................................ **[1]**

**(iv)** Determine the rate equation for this reaction.

Calculate the rate constant and give its units.

rate equation = ....................

rate constant = ....................

units of rate constant = .................... **[4]**

**(b) (i)** In reality, this reaction can form an equilibrium if allowed to occur within a closed system.

$$X(aq) + 2Y(aq) + Z(aq) \rightleftharpoons D(aq) + E(aq) \quad \Delta H = -15\,kJ\,mol^{-1}$$

1.50 moles of **X**, 2.00 moles of **Y**, and 0.80 moles of **Z** were placed in 500 $cm^3$ of distilled water and allowed to reach equilibrium. At equilibrium, 0.60 moles of **E** was present.

Give an expression for the equilibrium constant $K_c$.
Calculate its value and give its units.

$K_c$ = ........................................

units = ........................................ **[5]**

**(ii)** State the effect on the magnitude of $K_c$ if the solution at equilibrium is heated.

Explain your answer.

...............................................................................

...............................................................................

............................................................................... **[3]**

**Exam tip**

All explanations for changes to equilibrium must be linked to Le Châtelier's principle.

**4** A student was investigating the rate of reaction between magnesium and hydrochloric acid.

**Figure 4.1** shows the student's graph of their results.

**Synoptic links**

2.1.3 5.1.3

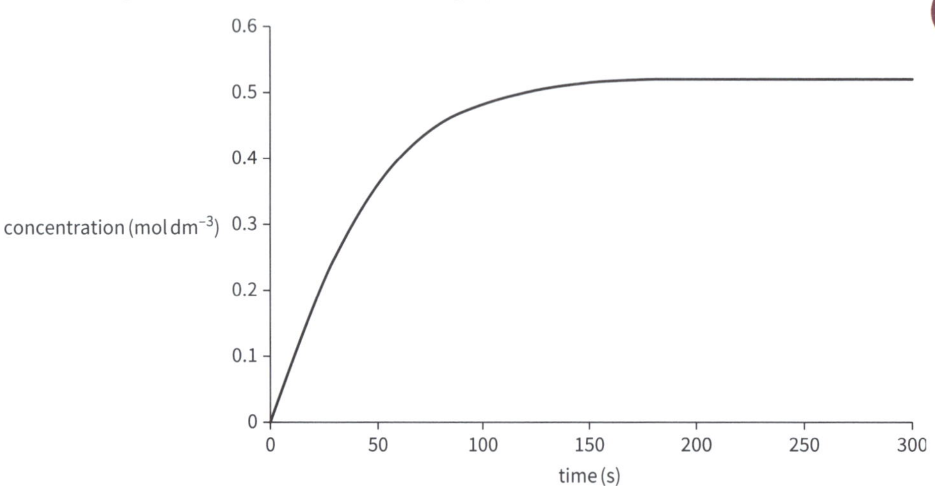


**Figure 4.1**

**(a)** Determine the rate of reaction at 50 seconds. **[2]**

**(b)** Another way of measuring this reaction is to collect the volume of gas produced.

Use a diagram to show suitable apparatus to record the volume of gas produced in a certain time limit.

Make sure you label all the equipment involved. **[2]**

**(c) (i)** The student used 1.9 g of magnesium ribbon and 100 $cm^3$ of 2.0 mol $dm^{-3}$ hydrochloric acid.

Calculate the maximum amount in moles of hydrogen gas produced by the reaction. **[3]**

**(ii)** Calculate the pH of the solution after the reaction has finished. **[4]**

To find the gradient of a curve, you need to draw a tangent.

**5** Hydrogen peroxide, when acidified, reacts with iodide ions to form iodine. When carried out in the presence of starch and sodium thiosulfate, a sudden dark blue colour determines the endpoint of the reaction.

This reaction is often known as an 'iodine clock' reaction.

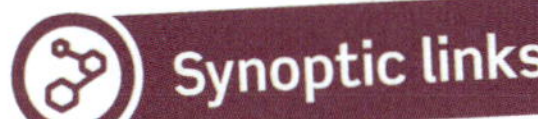

3.1.5.4 3.1.2.5 3.1.12.2

**(a) (i)** A sample set of results is shown in **Table 5.1**.

Complete the table and plot a graph with $\frac{1}{T}$ on the $y$-axis and $\log_{10}$ (volume of KI(aq)) on the $x$-axis. Draw a straight line of best fit. **[6]**

| Experiment | Volume of KI(aq) ($cm^3$) | $\log_{10}$ (volume of KI(aq)) | Time (s) | $\log10\left(\frac{1}{T}\right)$ |
|---|---|---|---|---|
| 1 | 10 | | 142 | |
| 2 | 16 | | 92 | |
| 3 | 20 | | 64 | |
| 4 | 30 | | 50 | |
| 5 | 40 | | 40 | |
| 6 | 50 | | 28 | |

**Table 5.1**

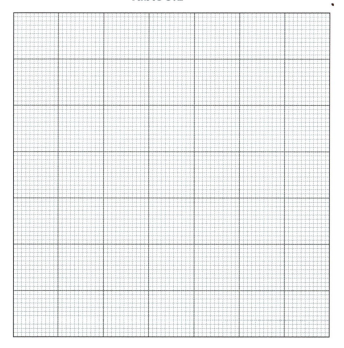

**Exam tip**

When drawing a graph pick a sensible scale. The points should cover over half the paper but do not choose a scale that is hard to read. Avoid fractions and threes.

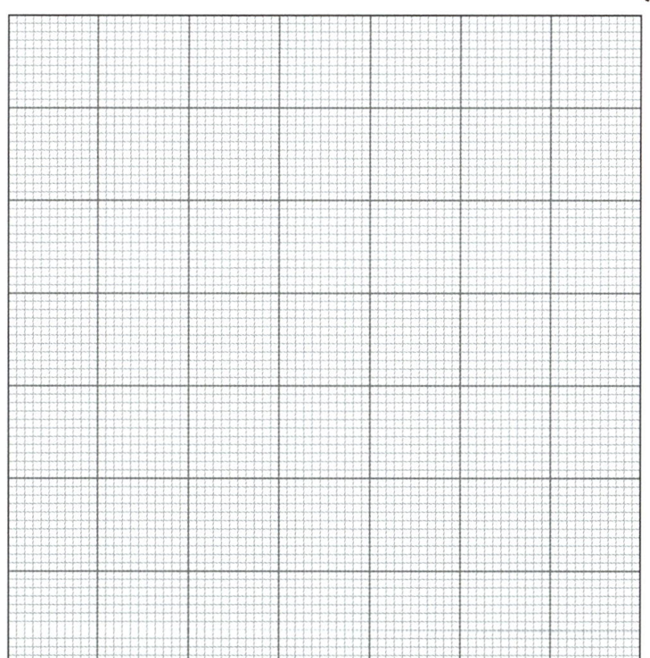

(ii) The gradient of the straight line of the graph in **5(a)(i)** is equal to the order of reaction.

Determine the gradient to 2 decimal places and deduce the order of reaction. **[4]**

(b) The student judged the time for the colour to appear by observing it on a bench with a stopwatch.

Give one way the method could be improved to provide a more accurate time. **[2]**

**Exam tip**

Do not mention repeats or getting a more precise timer or measuring cylinder.

6 Propanone reacts with iodine in the presence of an acid catalyst:

$$CH_3COCH_3(aq) + I_2(aq) \rightarrow CH_3COCH_2I(aq) + HI(aq)$$

(a) Define the term catalyst. **[2]**

**Synoptic link**

3.2.2

(b) (i) Use the data in **Table 6.1** to determine the order of reaction with regard to propanone, iodine, and hydrogen ions. **[3]**

| Experiment | $[CH_3COCH_3]$ (mol dm$^{-3}$) | $[I_2]$ (mol dm$^{-3}$) | $[H^+]$ (mol dm$^{-3}$) | Initial rate (mol dm$^{-3}$ s$^{-1}$) |
|---|---|---|---|---|
| 1 | 0.30 | 0.30 | 0.15 | $1.60 \times 10^{-3}$ |
| 2 | 0.30 | 0.15 | 0.15 | $1.60 \times 10^{-2}$ |
| 3 | 0.15 | 0.30 | 0.15 | $8.00 \times 10^{-3}$ |
| 4 | 0.15 | 0.30 | 0.075 | $4.00 \times 10^{-3}$ |

**Table 6.1**

(ii) Give an equation to represent the rate for the reaction, calculate the value of the rate constant, $k$, and give its units. **[4]**

(c) (i) **Table 6.2** shows how the value for $k$ varies with temperature for the same reaction at different concentrations.

Complete the table and plot a graph of ln $k$ on the $y$-axis and $\frac{1}{T}$ on the $x$-axis.

Draw a straight line of best fit. **[6]**

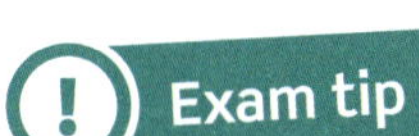

**Exam tip**

This plot is an important skill to practise. Make sure your $y$-axis is a negative axis with negative numbers increasing as you go down the page.

| Temperature $T$ / K | $\frac{1}{T}$ / K$^{-1}$ | Rate constant $k$ | ln $k$ |
|---|---|---|---|
| 293 | | 0.0030 | |
| 313 | | 0.0216 | |
| 333 | | 0.122 | |
| 353 | | 0.567 | |

**Table 6.2**

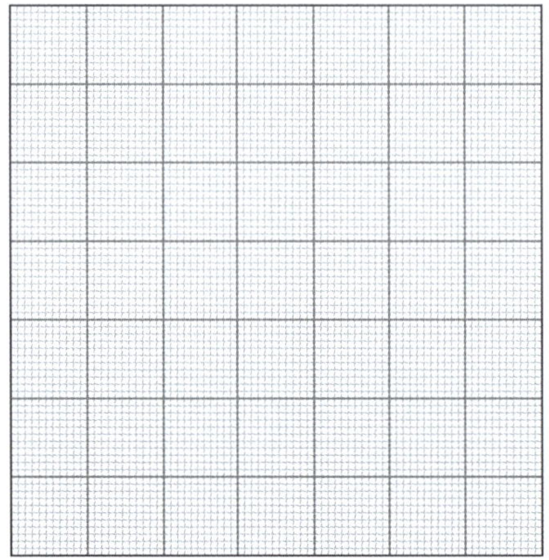

(ii) Use your line of best fit to calculate the activation energy of the reaction in $kJ\,mol^{-1}$. **[4]**

**7** An important step in the production of sulfuric acid is the oxidation of sulfur dioxide to sulfur trioxide:

$$2SO_2(g) + O_2(g) \rightleftharpoons 2SO_3 \qquad \Delta H = -197\ kJ\,mol^{-1}$$

A 3.8 mol sample of sulfur dioxide was mixed with a 2.0 mol sample of oxygen in a 50 $dm^3$ container. At equilibrium it was found that 0.8 moles of sulfur dioxide remained and the total pressure was 210 kPa at 298 K.

**(a) (i)** Give the amount, in moles, of each species at equilibrium. **[2]**

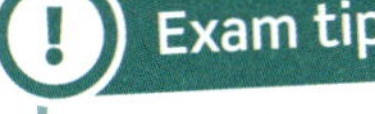


Square brackets will not be permitted.

**(ii)** Give the expression for $K_p$.

Calculate its value and give the units. **[5]**

**(iii)** State the effect on the position of equilibrium if the total pressure of the container was reduced to 105 kPa.

Explain your answer. **[3]**

**(iv)** The container is cooled to 200 K. State the effect this will have on $K_p$ and give a reason. **[3]**

**(b)** The reaction can also be expressed in reverse:

$$2SO_3 \rightleftharpoons 2SO_2(g) + O_2(g)$$

Determine the value of $K_p$ for this reaction and its unit. Give your answer to 3 significant figures. **[2]**

**8** Sulfuryl chloride, $SO_2Cl_2$, undergoes decomposition to form sulfur dioxide and chlorine gas:

2.2.2

$$SO_2Cl_2(g) \rightleftharpoons SO_2(g) + Cl_2(g)$$

A 500 $cm^3$ vessel was filled with 2.70 g of sulfuryl chloride and heated to 300 K. At equilibrium, 0.67 g had decomposed. The pressure was 125 kPa.

**(a) (i)** Give the amount, in moles, of each species at equilibrium. **[3]**

**(ii)** State an expression for $K_p$.

Calculate its value and give the units. **[5]**

**(b)** Draw the shape of a molecule of sulfuryl chloride.

Include all bonds and lone pairs.

Estimate the bond angle. **[3]**

Remember sulfur and oxygen have double bonds.

# Knowledge

## 16 pH and buffers

### Brønsted–Lowry acids and bases

A **Brønsted–Lowry acid** is a species that *donates a proton*. A hydrogen ion, $H^+$, is a proton, so an acid will donate hydrogen ions.

A **Brønsted–Lowry base** will *accept protons*. **Conjugate acid–base pairs** differ by a proton. A reaction can have two pairs that have donated and accepted protons.

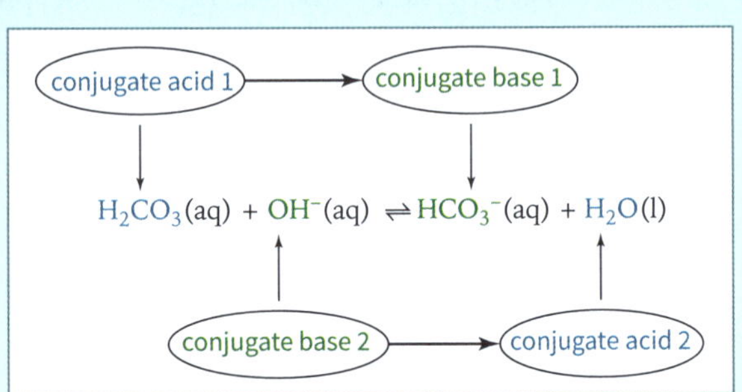


**Monobasic acids** have *one proton* to donate:

- hydrochloric acid, HCl
- nitric acid, $HNO_3$

**Dibasic acids** have *two protons* to donate:

- sulfuric acid, $H_2SO_4$
- carbonic acid, $H_2CO_3$

**Tribasic acids** have *three protons* to donate:

- boric acid, $H_3BO_3$

### Models of acid–base behaviour

Early ideas: acids and alkalis
Acids are sour (Latin acidus = sour)
Alkalis from plant ashes
(Arabic al-quili = plant ashes)

↓

1754, Guillaume François Rouelle (French):
A 'base' reacts with an acid to give a solid form.

↓ role of hydrogen

1832, Justus Liebig (German):
Acids contain hydrogen that can be replaced by a metal.

↓ ions in water

Arrhenius model 1884, Svante Arrhenius (Swedish):
- Acids form $H^+$ ions.
- Bases form $OH^-$ ions.

↓ central role of protons

Brønsted–Lowry model 1923, Johannes Brønsted (Danish) and Thomas Lowry (English):
Acid–base reactions transfer protons, $H^+$.
- An acid donates a proton.
- A base accepts a proton.

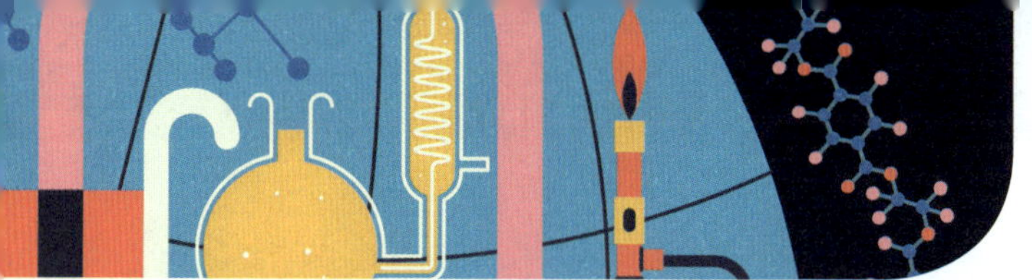



## pH expressed as a logarithm

The pH is a value of the concentration of hydrogen ions in a solution.

$pH = -\log_{10}[H^+(aq)]$

The lower the value of pH, the higher the concentration of hydrogen ions.

An increase of one number on the pH scale is equal to a tenfold increase in concentration of hydrogen ions, so a change from pH 4 to pH 5 will increase the concentration of hydrogen ions ten times.

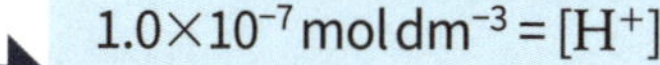

Practise using the log buttons on your calculator.

For *pure water* at 298 K:

$1.0\times10^{-7}\,mol\,dm^{-3} = [H^+]$

$pH = -\log_{10}[1.0\times10^{-7}] = pH\ 7$

$[H^+] = 10^{-pH}$

## pH of strong acids

The pH of strong acids can be determined using $pH = -\log_{10}[H^+(aq)]$.

In an aqueous solution, hydrogen ions fully dissociate from the acid so the concentration of the acid is equal to the concentration of hydrogen ions:

$$[acid(aq)] = [H^+(aq)].$$

## Calculating the concentration of hydroxide ions

The concentration of hydroxide ions can be calculated from pH using:

$$[H^+](aq)[OH^-](aq) = 1.0\times10^{-14}\,mol^2\,dm^{-6}\ (at\ 298\,K)$$

## The role of $H^+$ in reactions

When an acid reacts with metals, metal carbonates, metal oxides, and alkalis to produce salts, the hydrogen ions will react in the same way. They will be replaced by the metal ion to form a salt:

acid + metal → salt + hydrogen

acid + metal carbonate → salt + water + carbon dioxide

acid + metal oxide → salt + water

acid + metal hydroxide → salt + water

For the reactions between an acid and an alkali, the **spectator ions** can be removed to give:

$$H^+(aq) + OH^-(aq) \rightleftharpoons H_2O(l)$$

## The ionic product of water $K_w$

Hydrogen ions partially dissociate from water, setting up an equilibrium:

$$H_2O(l) \rightleftharpoons H^+(aq) + OH^-(aq)$$

Water can act both as an acid and as a base, with one molecule donating a proton and another accepting that proton.

$$\underset{acid}{H_2O(l)} + \underset{base}{H_2O(l)} \rightleftharpoons OH^-(aq) + H_3O^+(aq)$$

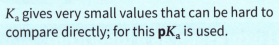

## The pH of weak acids

A weak acid, HA, partially dissociates:

$$HA(aq) \rightleftharpoons H^+(aq) + A^-(aq)$$

As this is an equilibrium:

$$K_a\ (\text{mol dm}^{-3}) = \frac{[H^+(aq)][A^-(aq)]}{[HA(aq)]}$$

$K_a$ is dependent on temperature.

$K_a$ gives very small values that can be hard to compare directly; for this **p$K_a$** is used.

p$K_a$ makes it easier to compare values.

$$pK_a = -\log K_a$$

$$K_a = 10^{-pKa}$$

The *larger* the value of $K_a$, the *smaller* the value of p$K_a$, the stronger the acid.

The *lower* the value of $K_a$, the *larger* the value of p$K_a$, the weaker the acid.

## Approximations used in the calculations for weak acids

**Approximation 1**

The acid will dissociate to give an equal ratio of acid and hydrogen ions. So at equilibrium the concentrations are approximately equal:

$$[H^+(aq)]_{eqm} \cong [A^-(aq)]_{eqm}$$

There will be a small contribution of $H^+$ from the dissociation of water, but this is minimal.

This assumption can sometimes break down for stronger weak acids where $[H^+(aq)]_{eqm} \cong [A^-(aq)]_{eqm}$ is no longer valid.

**Approximation 2**

The dissociation of weak acid is very small so any decrease in the concentration of the acid due to dissociation can be ignored.

$$[HA(aq)]_{eqm} = [HA(aq)]_{start}$$

Using these approximations the expression for $K_a$ for a weak acid is:

$$K_a = \frac{[H^+(aq)]^2}{[HA(aq)]}$$

$[H^+(aq)]$ can be calculated from this equation and the pH of a weak acid can be determined using:

$$pH = -\log_{10}[H^+(aq)]$$

## Neutralisation

A titration can be used to follow the progress of a a neutralisation reaction and find the endpoint of the reaction. Readings taken after each addition from the burette will show that the pH does not change in a linear fashion.

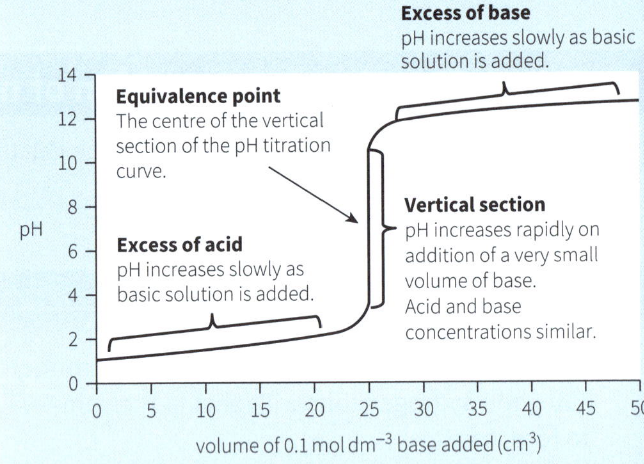
Excess of base
pH increases slowly as basic solution is added.
Equivalence point
The centre of the vertical section of the pH titration curve.
Vertical section
pH increases rapidly on addition of a very small volume of base.
Acid and base concentrations similar.
Excess of acid
pH increases slowly as basic solution is added.
pH
0
2
4
6
8
10
12
14
0
5
10
15
20
25
30
35
40
45
50
volume of 0.1 mol dm⁻³ base added (cm³)

## Choice of indicator

The choice of an indicator to find the endpoint of a titration needs to be carefully considered, as not all indicators are suitable for use with all combinations. The use of a pH probe will allow you to follow any reaction.

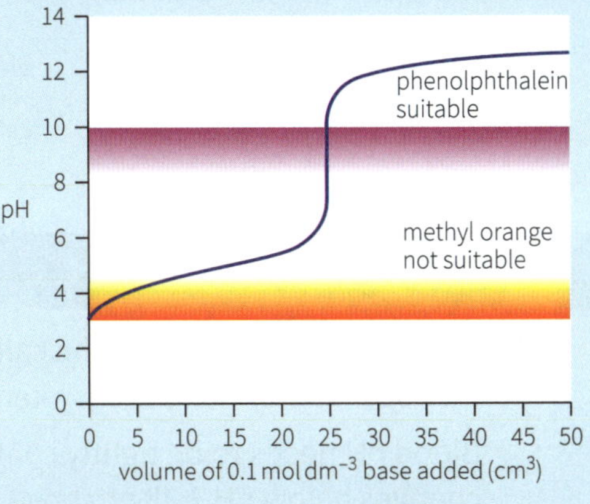


## Titration curves

These are typical shapes for **titration curves** for different combinations of acids and bases. For each of these curves, the base is added to a monoprotic acid.

The **equivalence point** (where moles of acid are equal to moles of base) is not always at pH 7, but the pH will change rapidly around this point.

**Strong acid and strong base**

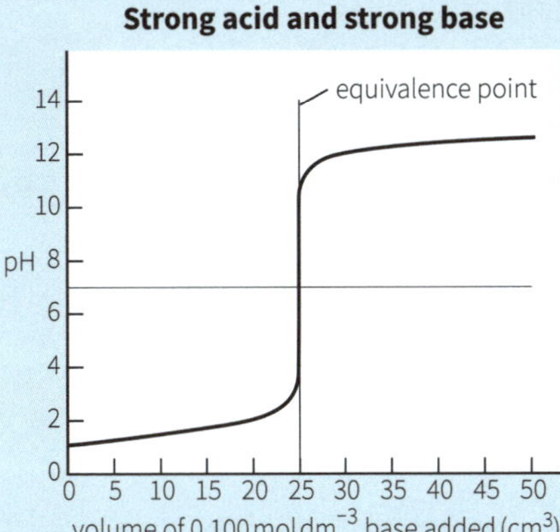


**Strong acid and weak base**

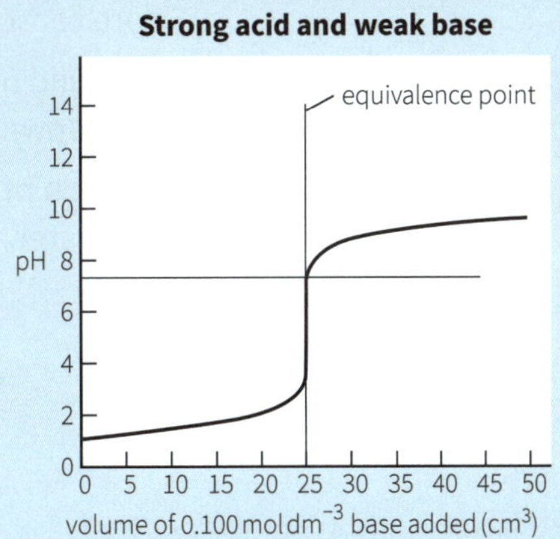


**Weak acid and weak base**

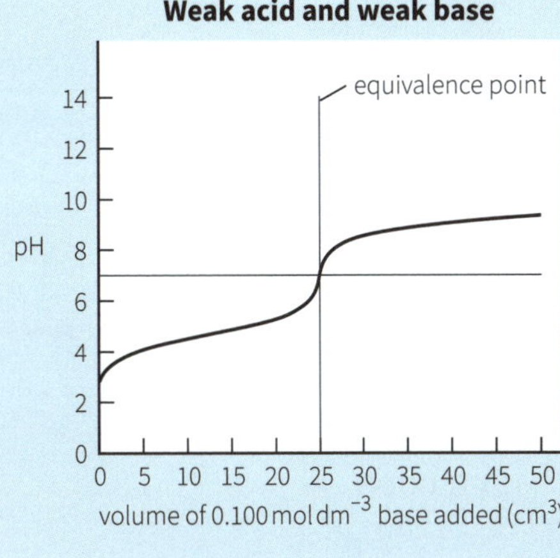


**Weak acid and strong base**

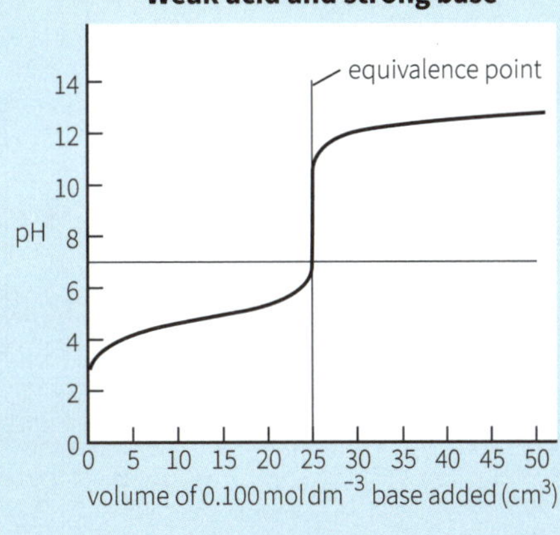

## Buffers

**Buffer solutions** are designed to resist changes in pH from the addition of small volumes of other solutions. They work by keeping the concentration of hydrogen ions and hydroxide ions constant by shifting the direction of equilibrium.

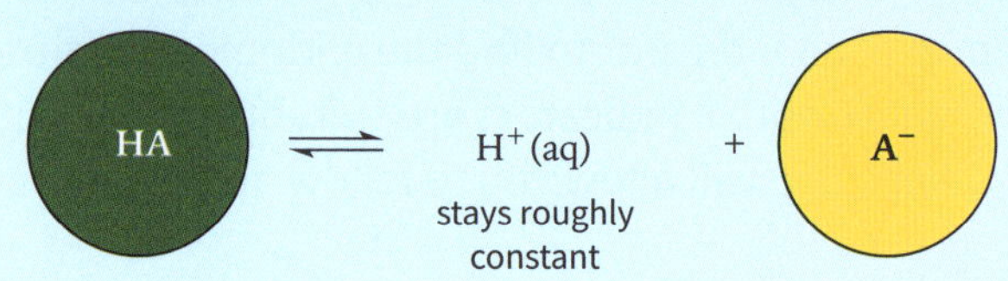


plenty of HA to make more $H^+(aq)$ if some is used up by alkali that gets added

plenty of $A^-$ to combine with any $H^+(aq)$ that gets added

## Acidic buffers

Acidic buffers are made from a mixture of a weak acid and a soluble salt of that acid; the pH is maintained below pH 7. For a weak acid the following equilibrium is set up:

$$HA(aq) \rightleftharpoons H^+(aq) + A^-(aq)$$

If an alkali is added, the $OH^-$ ions will react with the HA, removing the $OH^-$, and shifting to give water and $A^-$:

$$HA(aq) + OH^- \rightarrow H_2O(aq) + A^-(aq)$$

If an acid is added, the equilibrium will shift to the left and the new $H^+$ ions will be combined with the $A^-$ ions to give HA:

$$H^+(aq) + A^-(aq) \rightarrow HA(aq)$$

As the supplies of $A^-$ will soon be depleted by addition of large volume of acid, the salt of the acid is a source of extra $A^-$.

Conjugate acid–base pairs can be interconverted by transferring a proton; this is an equilibrium seen in buffers.

## Control of blood pH

Control of blood pH is via the carbonic acid–hydrogencarbonate buffer system.

Blood pH needs to be tightly controlled as changes below pH 7.35 or above pH 7.45 can cause health issues.

This range does not seem large, but it means a large difference in the concentration of hydrogen ions.

$$H_2CO_3(aq) \rightleftharpoons H^+(aq) + HCO_3^-(aq)$$

Addition of acid will shift the equilibrium to the left as $H^+$ reacts with the conjugate base to produce more carbonic acid.

The build up of carbonic acid is prevented by its breakdown into carbon dioxide.

Addition of hydroxide ions will react with the $H^+$ to give water, removing the hydroxide ions.

An acid buffer can also be produced by the partial neutralisation of a weak acid.

The addition of a small volume of alkali to an excess of acid will give a solution that is a mixture of the unreacted acid and its salt; for example, excess $CH_3COOH$ and NaOH.

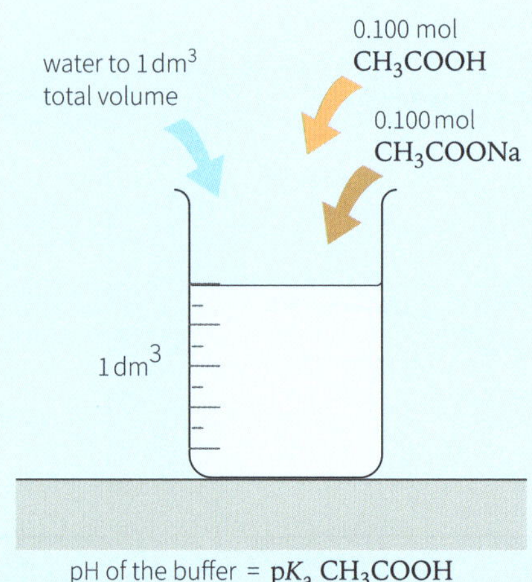


pH of the buffer = $pK_a$ $CH_3COOH$ = 4.77

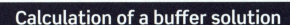

## Calculation of a buffer solution

To calculate the pH of a buffer solution use:

$$K_a = \frac{[H^+(aq)][A^-(aq)]}{[HA(aq)]}$$

assuming that $[A^-(aq)]$ is equal to the concentration of the salt.

However, in a buffer solution $[H^+(aq)] \neq [A^-(aq)]$ so $[H^+(aq)]\,[A^-(aq)]$ cannot be replaced with $[H^+(aq)]^2$. Rearranging the equation:

$$[H^+(aq)] = K_a \times \frac{[HA(aq)]}{[A^-(aq)]}$$

Then use: $pH = -\log_{10}[H^+(aq)]$.

## How to calculate the pH of a buffer made from a weak acid and its salt

Assuming the volumes and concentrations of each solution is known, use the following method to calculate the pH of the buffer:

**Step 1** Calculate the number of moles of HA and $A^-$ present:

$$\text{number of moles} = \frac{\text{concentration} \times \text{volume}}{1000}$$

**Step 2** Determine the concentration of HA and $A^-$ present in the buffer solution.

**Step 3** Calculate $[H^+]$:

$$[H^+(aq)] = K_a \times \frac{[HA(aq)]}{[A^-(aq)]}$$

**Step 4** Use: $pH = -\log_{10}[H^+(aq)]$

## To calculate the pH of an acid buffer produced by the partial neutralising of a weak acid, where the acid is in excess

**Step 1** Calculate the number of moles of $A^-$, $n(A^-)$, from the initial concentration of the base. $HA(aq) + OH^- \rightarrow H_2O(aq) + A^-(aq)$, so the moles of $A^-$ in the buffer will be equal to number of moles in the base at the start.

**Step 2** Calculate the number of moles of HA at the start, $n(HA)_{start}$

$$\text{number of moles} = \frac{\text{concentration} \times \text{volume}}{1000}$$

**Step 3** Calculate the number of moles of HA, $n(HA)_{end}$ in the buffer at the end:

$n(HA)_{end} = n(HA)_{start} - n(A^-)$

**Step 4** Determine the concentration of HA and $A^-$ present in the buffer solution.

**Step 5** Calculate $[H^+]$:

$$[H^+(aq)] = K_a \times \frac{[HA(aq)]}{[A^-(aq)]}$$

**Step 6** Use: $pH = -\log_{10}[H^+(aq)]$

# Retrieval

Learn the answers to the questions below, then cover the answers column with a piece of paper and write as many as you can. Check and repeat.

| | Questions | Answers |
|---|---|---|
| 1 | What is the formula for ethanoic acid? | $CH_3COOH$ |
| 2 | What is the equation to determine the pH of strong acids? | $pH = -\log_{10}[H^+(aq)]$ |
| 3 | What type of acid is nitric acid: strong or weak? | strong |
| 4 | What are the values and units for $K_w$ at 298 K? | $1.0 \times 10^{-14}\ mol^2\,dm^{-6}$ |
| 5 | What is the relationship between $pK_a$ and $K_a$? | $pK_a = -\log K_a$<br>$K_a = 10^{-pK_a}$ |
| 6 | What is a strong acid? | one whose hydrogen ions fully dissociate when in an aqueous solution |
| 7 | Which element always has a −1 oxidation state in compounds? | fluorine |
| 8 | What is an alkali? | a base that dissolves in water and releases aqueous hydroxide ions |
| 9 | Is an acid a *strong* or *weak* acid if it has a large value of $K_a$, and a small value of $pK_a$? | strong |
| 10 | What approximation is made about the concentration of hydrogen ions when working out the pH of a weak acid? | $[H^+(aq)]_{eqm} = [A^-(aq)]_{eqm}$ |
| 11 | Is ammonia an alkali? | yes |
| 12 | What is $K_w$? | the ionic product of water |
| 13 | What is the neutralisation equation? | $H^+(aq) + OH^-(aq) \rightleftharpoons H_2O(l)$ |
| 14 | What approximation is made about the concentration of HA when working out the pH of a weak acid? | $[HA(aq)]_{eqm} = [HA(aq)]_{start}$ |
| 15 | What is the shape of the part of a titration graph that shows the equivalence point? | a vertical line |
| 16 | What type of salt will be made from a reaction of ethanoic acid? | ethanoate salts |
| 17 | What is the assumption used to calculate the pH of a weak acid that cannot be used for buffers? | $[H^+(aq)] \neq [A^-(aq)]$ |
| 18 | How do buffers work? | they keep the concentration of hydrogen ions and hydroxide ions constant by shifting the direction of equilibrium |

Put paper here

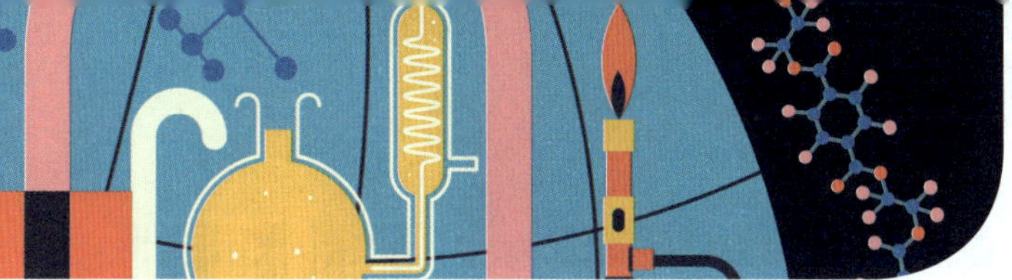



| | | |
|---|---|---|
| 19 | What needs to be taken into account when choosing an indicator for a titration? | the range of pH covered |
| 20 | How can the concentration of hydroxide ions in a solution be determined? | $[H^+(aq)][OH^-(aq)] = 1.0 \times 10^{-14}\, mol^2\, dm^{-6}$ |

Put paper here

## Practical skills

Practise your practical skills using the worked example and practice questions below.

### Measuring pH

A graph of pH against volume of acid or base added (a titration curve) can reveal important information about the acid or base.

Before carrying out this experiment it is important to calibrate the pH meter that you use. Do this by placing the pH probe in buffer solutions of known pH and allowing the readings to settle.

When drawing titration curves, draw a smooth curve between points. This requires practice and patience. You do *not* use a ruler to connect between the points because there is an underlying mathematical relationship between the data.

### Worked example

**Question**

Describe how the data is collected when monitoring the pH changes when NaOH is added to 25.00 $cm^3$ of a weak acid.

**Answer**

This would usually be an extended response question. The key points to include would be:

- calibrate the pH meter using buffer solutions of known concentration
- rinse the burette with NaOH before titration
- use a pipette to add acid to a clean conical flask
- add NaOH in small volumes (e.g. 1 $cm^3$)
- near the equivalence point, add NaOH in smaller volumes
- keep adding NaOH until NaOH is in excess
- swirl the flask and then measure pH after each addition.

### Practice

The data below shows how the pH changed when NaOH was added to 25.00 $cm^3$ of a weak acid.

| NaOH added ($cm^3$) | pH |
|---|---|
| 0.00 | 2.5 |
| 1.00 | 3.4 |
| 10.00 | 4.6 |
| 20.00 | 5.4 |
| 24.00 | 6.1 |
| 24.25 | 6.3 |
| 24.50 | 6.4 |
| 24.75 | 6.8 |
| 25.25 | 11.4 |
| 25.50 | 11.7 |
| 25.75 | 11.9 |
| 26.00 | 12.0 |
| 30.00 | 12.7 |
| 40.00 | 13.1 |
| 50.00 | 13.2 |

**1 a** Plot the titration curve for the data above.

**b** From your graph, determine the pH of the equivalence point.

**c** Which indicator would be a suitable indicator for this titration?

**2** The NaOH and weak acid both had concentrations of 0.500 mol $dm^{-3}$. Determine the $K_a$ of the weak acid.

# Practice

## Exam-style questions

2.1.3 2.1.4

1 Sulfuric acid is a strong Brønsted–Lowry acid. A student made up a solution of sulfuric acid by measuring 50.0 $cm^3$ of 3.00 mol $dm^{-3}$ into a 250 $cm^3$ volumetric flask. Distilled water was added until the volume was exactly 250 $cm^3$.

**(a)** Explain what is meant by the term strong Brønsted–Lowry acid.

..............................................................................................

..............................................................................................

.............................................................................................. **[2]**

**(b)** Outline the important practical steps that need to be taken to ensure all the acid is transferred to the flask and the volume is accurate at the end.

..............................................................................................

..............................................................................................

..............................................................................................

..............................................................................................

.............................................................................................. **[3]**

Know your definitions! No matter how hard you find chemistry, anyone can learn the definitions and score full marks.

**(c)** State the definition of pH as a mathematical relationship.

.............................................................................................. **[1]**

**(d)** Calculate the pH of the sulfuric acid in the 250 $cm^3$ volumetric flask.

pH = .................... **[4]**

2 Barium hydroxide is used as a reagent for checking the concentrations of organic acids, due to its lack of carbonate impurities.

A student weighs out 3.50 g of barium hydroxide and dissolves it into 500 $cm^3$ of distilled water.

The ionic product of water, $K_w = [H^+][OH^-] = 1 \times 10^{-14}$

**(a) (i)** State the units for $K_w$.

.............................................................................................. **[1]**

**(ii)** Calculate the pH of the barium hydroxide solution the student made.

**Exam tip**

Think about the solubility of carbonates.

pH = .................... **[3]**

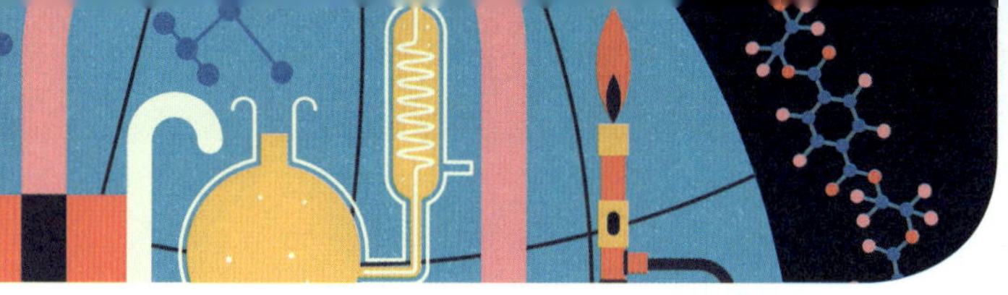



**(b)** A 100 $cm^3$ sample of the barium hydroxide solution was added to 250 $cm^3$ of 0.1 mol $dm^3$ hydrochloric acid. Calculate the pH of the final solution.

**Exam tip**

Write an equation.

pH = .................... **[5]**

**3** A student was investigating how the pH of a strong base changed when a weak acid was added. They poured 25 $cm^3$ of 0.10 mol $dm^{-3}$ sodium hydroxide solution into a beaker and filled a burette with 0.10 mol $dm^{-3}$ ethanoic acid ($K_a = 1.76 \times 10^{-5}$ mol $dm^{-3}$). They titrated the acid against the base and recorded the pH at regular intervals using a pH probe.

**(a)** Describe how the student ensured the pH probe correctly calibrated before use.

Make sure you include how to reduce the chance of contamination.

..........................................................................................

..........................................................................................

..........................................................................................

..........................................................................................

..........................................................................................

.................................................................................... **[4]**

**Exam tip**

This is a key aspect of the pH curve Required Practical. It is very likely that pH probes do not give the true value.

**(b)** Using the information given calculate the pH of the ethanoic acid in the burette.

pH = .................... **[4]**

(c) Sketch a graph of the pH curve that would be produced by this experiment. **[5]**

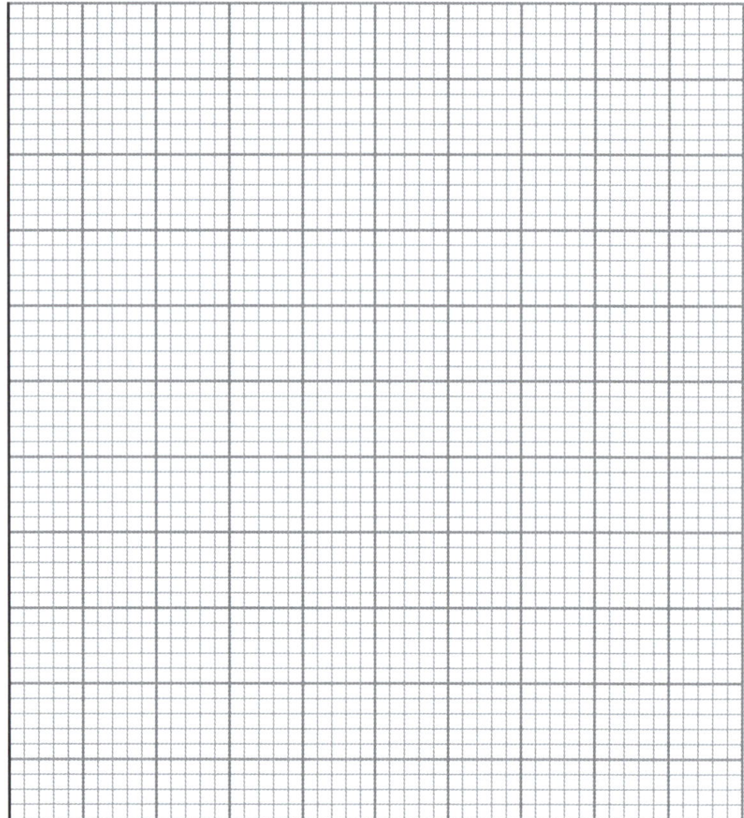

(d) **Table 3.1** lists some indicators and their pH ranges. Tick the indicator that is best used for this titration. **[1]**

| Indicator | pH range | Indicator to use |
|---|---|---|
| dichlorofluorescein | 4.0–6.6 | |
| thymolphthalein | 8.8–10.5 | |
| cresol purple | 1.2–7.4 | |
| indigo carmine | 12.6–14.0 | |

**Table 3.1**

4 A student made a buffer solution by dissolving 4.65 g of sodium benzoate ($M_r$ = 144.1 g mol$^{-1}$) into 250 cm$^3$ of 0.20 mol dm$^{-3}$ benzoic acid (p$K_a$ = 4.19).

(a) (i) Define buffer solution.

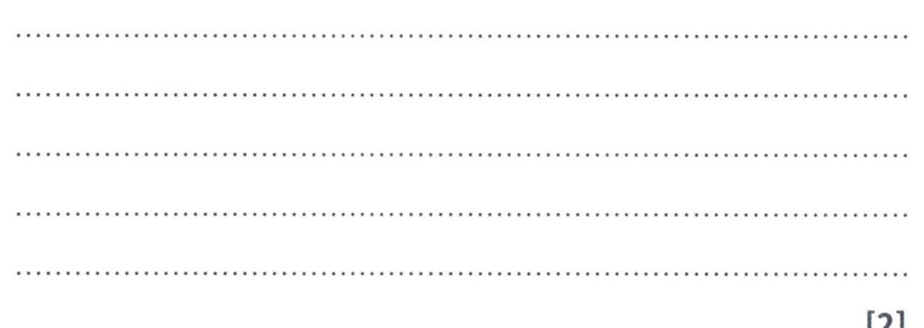

......................................................................... **[2]**

**Exam tip**

You need to know how to describe any required practical process.

(ii) Calculate the pH of the resulting buffer solution.

pH = .................... **[6]**

**(b)** Outline the practical procedure needed to obtain a pure sample of benzoic acid from an impure sample.

..........................................................................................

..........................................................................................

..........................................................................................

..........................................................................................

..........................................................................................

..........................................................................................

..........................................................................................

.................................................................................... **[5]**

**5** Formic acid is a monoprotic organic acid that is found in ant bites. A student completed a titration on a 25 $cm^3$ sample of 0.10 $mol\,dm^{-3}$ formic acid against a 0.10 $mol\,dm^{-3}$ sodium hydroxide solution. The student measured the pH, using a pH probe, as volumes of base were added.

2.1.3 4.1.1

**(a)** Sodium hydroxide is a strong Brønsted–Lowry base. Define strong Brønsted–Lowry base.

..........................................................................................

..........................................................................................

..........................................................................................

..........................................................................................

..........................................................................................

.................................................................................... **[2]**

**(b)** The student's results are shown in **Table 5.1**.

| Volume of formic acid added ($cm^3$) | pH |
|---|---|
| 0 | 2.40 |
| 2 | 3.20 |
| 4 | 3.30 |
| 6 | 3.40 |
| 8 | 3.50 |
| 10 | 3.60 |
| 12 | 3.70 |
| 14 | 3.80 |
| 16 | 3.90 |
| 18 | 4.00 |
| 20 | 4.20 |
| 22 | 4.40 |
| 24 | 5.00 |
| 26 | 12.50 |
| 28 | 12.80 |
| 30 | 12.90 |
| 32 | 12.95 |
| 34 | 12.98 |
| 36 | 13.00 |

**Table 5.1**

Plot the pH curve for this experiment. **[4]**

**(c) (i)** A sample of formic acid was found to have 26% carbon and 4.3% hydrogen by mass.

Determine the empirical formula of formic acid.

empirical formula = .................... **[4]**

**(ii)** The $M_r$ of formic acid is 46.0 g mol$^{-1}$. Give the molecular formula of formic acid.

..........................................................................................

.......................................................................................... **[1]**

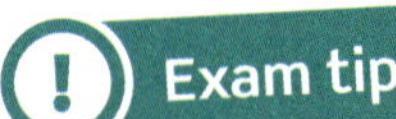


Draw out the structural formula before naming the molecule.

**(iii)** State the IUPAC name of formic acid.

..........................................................................................

.......................................................................................... **[1]**

**6** This question is focused on the ionic product of water, $K_w$. At 323 K the value for $K_w$ is $5.84 \times 10^{-14}$ mol$^2$ dm$^{-3}$.

**(a)** State the equation for $K_w$. **[1]**

**(b) (i)** Calculate the pH of water at 323 K.

pH = .................... **[4]**

**(ii)** Explain why water is still neutral at 323 K.

..........................................................................................

..........................................................................................**[1]**

**Exam tip**

Remember what 'neutral' actually means.

**(c)** The dissociation of water can represented by the following equation:

$$2H_2O(l) \rightleftharpoons H_3O^+(aq) + OH^-(aq)$$

Use the enthalpy data in **Table 6.1** to calculate the enthalpy change for the forward reaction.

| Compound | Enthalpy of formation of (kJ mol$^{-1}$) |
|---|---|
| $H_2O$ (l) | −293.0 |
| $H_3O^+$(aq) | −22.1 |
| $OH^-$(aq) | −300.0 |

**Table 6.1**

enthalpy change = .................... **[2]**

# Knowledge

## 17 Enthalpy, entropy, and free energy

### Born-Haber and enthalpy cycles

Construction of Born–Haber cycles involves lots of numbers and changes, so it is important you know the key terms to prevent confusion.

**Standard conditions** are 100 kPa and a stated temperature.

**Mean bond enthalpy** is the average enthalpy change when one mole of a given bond is broken in a range of molecules.

**Standard enthalpy of formation** ($\Delta_f H^\ominus$) is the enthalpy change when one mole of a substance is formed from elements (in their standard states) under standard conditions.

**Standard enthalpy of combustion** ($\Delta_c H^\ominus$) is the enthalpy change when one mole of a substance is completely burnt in oxygen.

**Standard enthalpy of atomisation** ($\Delta_{at} H^\ominus$) is the enthalpy change when one mole of gaseous atoms forms from the element in the standard state.

**First ionisation energy** ($\Delta_i H^\ominus$) is the enthalpy change when one mole of atoms in the gaseous state is converted to one mole of gaseous ions with a positive charge. The **second ionisation energy** relates to the removal of the second electron.

**First electron affinity** ($\Delta_{ea} H^\ominus$) is the enthalpy change when one mole of gaseous atoms forms one mole of gaseous ions with a negative charge. The **second electron affinity** relates to the gain of the second electron.

**Lattice enthalpy of formation** ($\Delta_{lat} H^\ominus$) is the enthalpy change when one mole of a solid ionic compound is formed from gaseous ions.

**Lattice enthalpy of dissociation** is the enthalpy change when one mole of a solid ionic compound is dissociated into gaseous ions.

**Standard enthalpy of hydration** ($\Delta_{hyd} H^\ominus$) is the enthalpy change when one mole of gaseous ions forms aqueous ions.

**Standard enthalpy of solution** ($\Delta_{sol} H^\ominus$) is the enthalpy change when one mole of solute forms a solution.

### Born-Haber cycles

**Born-Haber cycles** are used to determine unknown enthalpy changes. Drawing Born–Haber cycles needs to be done carefully. Check if the enthalpy value is positive (endothermic) or negative (exothermic), if a change is increasing or decreasing, and the order that the arrows go in.

1 Start with a clear list of all of the values that you have.
2 Start at zero, with everything in their standard states.
3 Sketch the diagram out first.
4 Ensure you have included every change and add state symbols. This will help if you add in equations, and you will be able to see if you skipped a step.
5 Work out the missing value, using Hess' law.

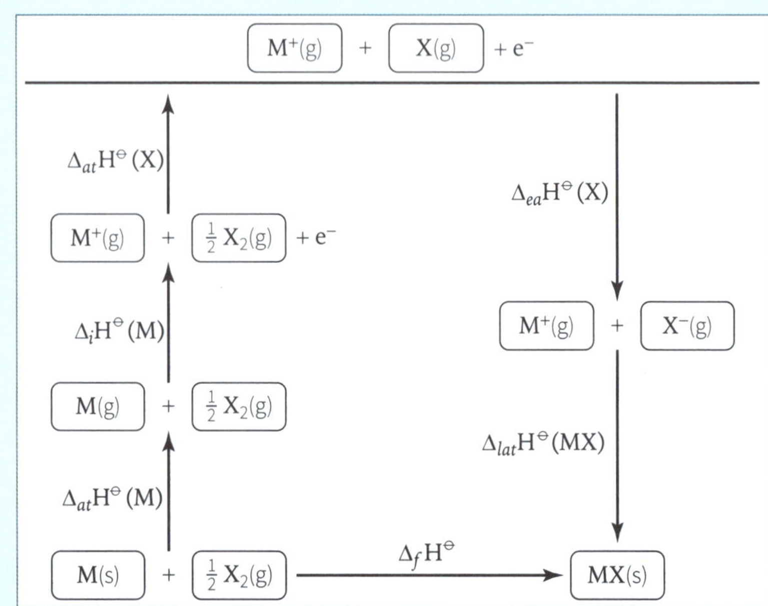

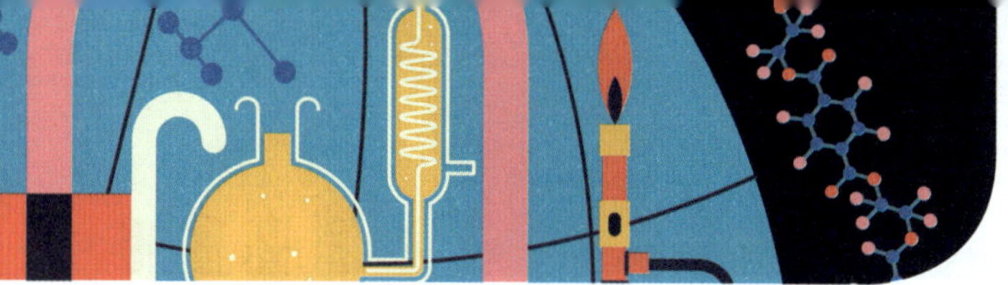



## Entropy and feasibility

Some reactions are energetically **feasible** and can happen spontaneously; this feasibility varies according to temperature. Being feasible does not tell you anything about the rate of a reaction; a feasible reaction could still be very slow. Enthalpy change ($\Delta H$) will tell you if a reaction is endothermic or exothermic but it cannot tell you if a reaction is feasible.

A reaction will favour randomness, gas over a liquid, and spreading out of particles. Randomness can be measured as **entropy** ($\Delta S$).

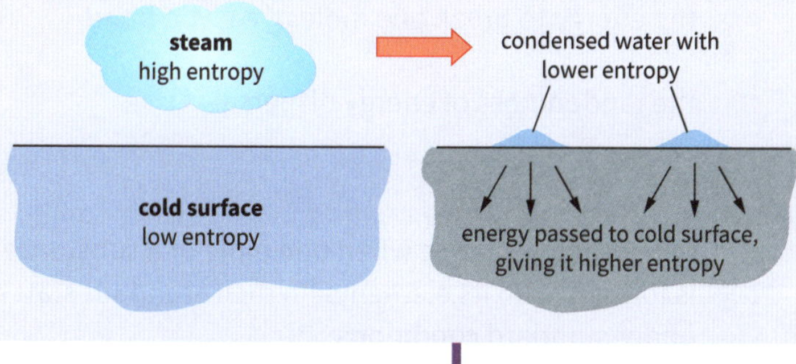


## Free energy

The **Gibbs free energy** change ($\Delta G$) tells you if a reaction is going to be feasible: it is dependent on both enthalpy change ($\Delta H$) and entropy change ($\Delta S$) for a given temperature.

$$\Delta G = \Delta H - T\Delta S$$

- If $\Delta G$ is negative the reaction is feasible at that temperature.
- If $\Delta G$ is positive then the reaction will *not* be feasible at that temperature.

## Entropy and temperature

$\Delta S$ will vary with temperature; there will be significant change when the substance changes state and randomness increases.

$\Delta G = \Delta H - T\Delta S$ can be used to determine the temperature a reaction becomes feasible. At this temperature $\Delta G = 0$.

Just because a reaction is thermodynamically feasible does not mean it **will** happen: the reactants might be very stable (i.e., have a very high activation energy).

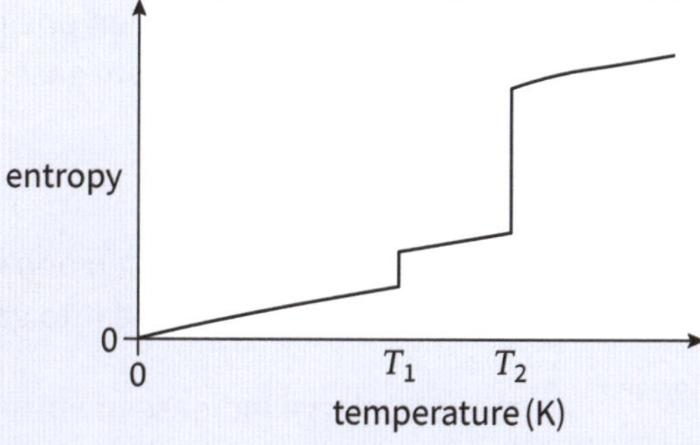

# Retrieval

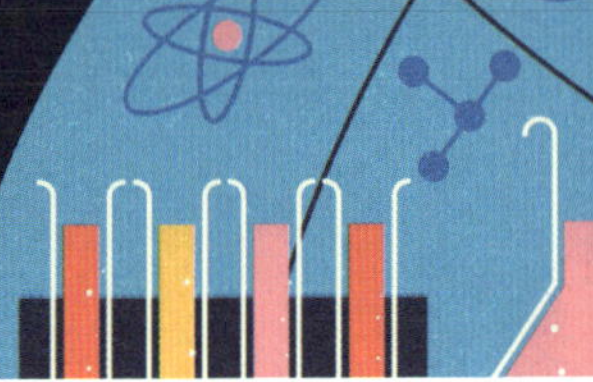

Learn the answers to the questions below, then cover the answers column with a piece of paper and write as many as you can. Check and repeat.

| | Questions | Answers |
|---|---|---|
| 1 | What are standard conditions? | 100 kPa and a stated temperature |
| 2 | What is $\Delta G$? | Gibbs free energy change |
| 3 | What is bond enthalpy? | the energy to break one mole of a given bond |
| 4 | What does entropy measure? | the randomness of energy dispersal |
| 5 | Will a reaction be feasible if $\Delta G$ is positive? | no |
| 6 | What is standard enthalpy of formation? | the enthalpy change when one mole of a substance is formed from its elements (in their standard states) under standard conditions |
| 7 | What is $\Delta S$? | the change in entropy |
| 8 | What is standard enthalpy of combustion? | the enthalpy change when one mole of a substance is completely burnt in oxygen (in their standard states) under standard conditions |
| 9 | What is $\Delta H$? | enthalpy change |
| 10 | What is standard enthalpy of atomisation? | the enthalpy change when one mole of gaseous atoms forms from the element in the standard state (in their standard states) under standard conditions |
| 11 | What is first ionisation energy? | the enthalpy change when one mole of atoms in the gaseous state is converted to one mole of gaseous ions with a positive (+1) charge |
| 12 | Will a reaction be feasible if $\Delta G$ is negative? | yes |
| 13 | What is first electron affinity? | the enthalpy change when one mole of gaseous atoms forms one mole of gaseous ions with a negative (−1) charge |
| 14 | What is the link between feasibility of a reaction and the rate of reaction? | there is no link |
| 15 | What is lattice enthalpy of formation? | the enthalpy change when one mole of a solid ionic compound is formed from gaseous ions |
| 16 | What is the equation used to measure Gibbs free energy? | $\Delta G = \Delta H - T\Delta S$ |
| 17 | What is lattice enthalpy of dissociation? | the enthalpy change when one mole of a solid ionic compound is dissociated into gaseous ions |
| 18 | What happens to entropy as a liquid changes to a gas? | there is a large increase in entropy |

Put paper here

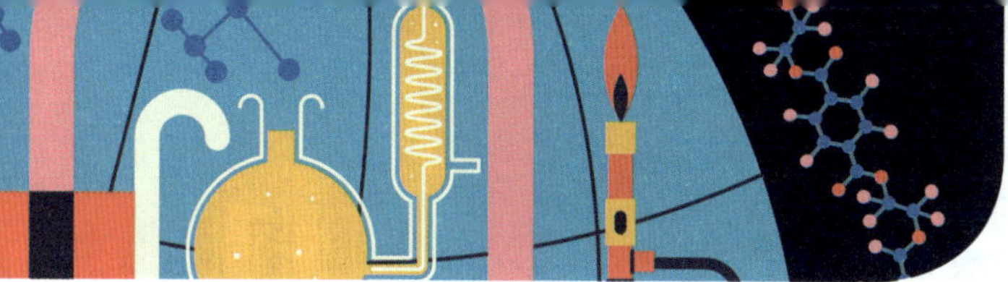



| | | |
|---|---|---|
| 19 | What is standard enthalpy of hydration? | the enthalpy change when one mole of gaseous ions forms aqueous ions |
| 20 | What do Born–Haber cycles show? | the energy/enthalpy changes for an ionic compound |

Put paper here

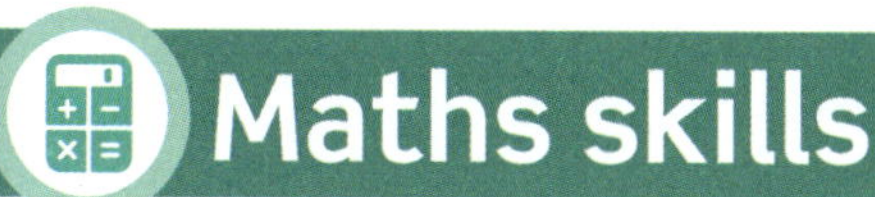

## Maths skills

Practise your maths skills using the worked example and practice questions below.

### Determining units

Almost every quantity in science has a unit with it. A quantity without units is meaningless.

Many units you will be expected to recall, but some you will have to work out based on the situation. Rate constants and equilibrium constants are examples of these.

At A level, compound units, for example, mol/dm$^3$, should be written as a negative power, in the form mol dm$^{-3}$. By convention, positive powers go first.

The reason dm is to the power of negative 3 in mol dm$^{-3}$ is due to power laws. The particular law that is relevant is:

$$x^{-a} = \frac{1}{x^a}$$

for example: $2^{-4} = \frac{1}{2^4}$

### Worked example

**Question**

Determine the units of $k$ in the rate equation: rate = $k$ [A]$^3$

**Answer**

The units of rate are mol dm$^{-3}$ s$^{-1}$. You need to learn this.

The units of [A] are mol dm$^{-3}$. You need to learn this.

Rearrange the equation to make $k$ the subject:

$$k = \frac{\text{rate}}{[\text{A}]^3}$$

and then write out the units:

$$\frac{\text{mol dm}^{-3}\,\text{s}^{-1}}{(\text{mol dm}^{-3})^3}$$

then simplify:

$$\frac{\cancel{\text{mol dm}^{-3}}\,\text{s}^{-1}}{(\text{mol dm}^{-3})^{\cancel{3}\,2}}$$

$$= \frac{\text{s}^{-1}}{\text{mol}^2\,\text{dm}^{-6}}$$

$$= \text{s}^{-1}\,\text{mol}^{-2}\,\text{dm}^6.$$

Units for $k$ are: dm$^6$ mol$^{-2}$ s$^{-1}$.

### Practice

Determine the units of the equilibrium constant or rate constant in each equation.

Remember [A] means concentration of A in mol dm$^{-3}$ and p(A) means partial pressure of A in Pa.

1. rate = $k$[A]$^2$
2. $K_c = \frac{[\text{A}]\,[\text{B}]^2}{[\text{C}]^2}$
3. $K_p = \frac{\text{p(A)}^2}{\text{p(B)}}$
4. rate = $k$[A][B]$^2$
5. $K_c = \frac{[\text{A}]^2\,[\text{B}]^2}{[\text{C}]}$
6. $K_p = \frac{\text{p(A)}^4\,\text{p(B)}^2}{\text{p(C)}\,\text{p(D)}^2}$

# Practice

## Exam-style questions

**1** Magnesium is a reactive Group 2 metal. In a Bunsen flame it readily reacts with oxygen, forming a white solid that is mainly magnesium oxide, and producing a bright white light

**(a)** Define electron affinity.

.......................................................................................................

.......................................................................................................

.......................................................................................................

....................................................................................................... **[3]**

**(b)** Use the data in **Table 1.1** to draw a Born–Haber cycle and calculate the second electron affinity of oxygen.

Give your answer to an appropriate number of significant figures.

| Reaction | Enthalpy (kJ mol$^{-1}$) |
|---|---|
| enthalpy of atomisation Mg | +148 |
| first ionisation energy of Mg | +738 |
| second ionisation energy of Mg | +1450 |
| enthalpy of atomisation of O | +249 |
| first electron affinity of O | −141 |
| lattice formation enthalpy of MgO | −3890 |
| enthalpy of formation of MgO(s) | −602 |

**Table 1.1**

> **Exam tip**
>
> Remember, endothermic processes go up and exothermic processes go down.

second electron affinity of oxygen = ..................... kJ mol$^{-1}$ **[8]**

**(c)** Explain why the first ionisation energy of oxygen is an exothermic process.

.......................................................................................................

.......................................................................................................

....................................................................................................... **[2]**

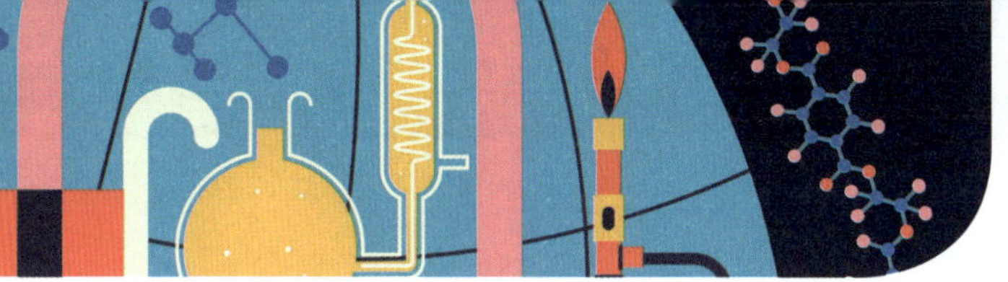



**2** Strontium chloride is an ionic compound that produces the red colour in fireworks.

**Table 2.1** outlines some thermochemical data for strontium chloride.

| Reaction | Enthalpy ($kJ\,mol^{-1}$) |
|---|---|
| enthalpy of atomisation of Sr | +164 |
| first ionisation energy of Sr | +549 |
| second ionisation energy of Sr | +1064 |
| enthalpy of atomisation of Cl | +243 |
| first electron affinity of Cl | −349 |
| lattice formation enthalpy of $SrCl_2$ | −2150 |
| enthalpy of formation of $SrCl_2(s)$ | to be calculated |

**Table 2.1**

**(a)** Define atomisation enthalpy.

..........................................................................................

..........................................................................................

..........................................................................................

.......................................................................................... **[3]**

**(b)** Use the data in **Table 2.1** to calculate the enthalpy of formation of strontium chloride.

enthalpy of formation = .................... $kJ\,mol^{-1}$ **[2]**

**(c)** Explain why the second ionisation energy of strontium is larger than the first ionisation energy.

..........................................................................................

..........................................................................................

.......................................................................................... **[2]**

**(d)** Explain why the bond dissociation enthalpy of the Cl—Cl bond is exactly double the enthalpy of atomisation of chlorine.

..........................................................................................

..........................................................................................

..........................................................................................

.......................................................................................... **[2]**

**Exam tip**

Think carefully about the definition of each term.

3 Lithium fluoride is a white ionic solid that is used in radiation detectors.

**(a)** Define lattice enthalpy.

..........................................................................................

..........................................................................................

..........................................................................................

.......................................................................................... **[3]**

**(b) (i)** **Figure 3.1** shows an incomplete Born–Haber cycle of the formation of lithium fluoride.

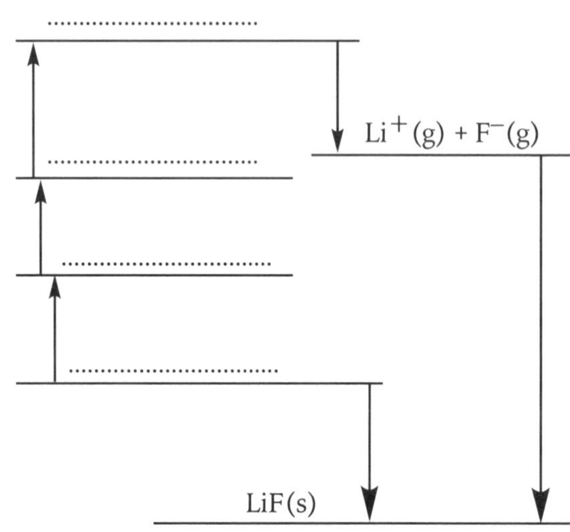


**Figure 3.1**

Complete the Born–Haber cycle by writing the correct species on each line. Include state symbols. **[4]**

**(ii)** Using the data in **Table 3.1**, calculate the value of the lattice formation enthalpy.

| Enthalpy | Value / kJ $mol^{-1}$ |
|---|---|
| formation of LiF | −616 |
| atomisation of Li | +159 |
| ionisation of Li | +520 |
| atomisation of F | +79 |
| electron affinity of F | − 328 |

**Table 3.1**

lattice formation enthalpy = .................... kJ $mol^{-1}$ **[2]**

4 Copper(II) sulfate exists as both hydrated $CuSO_4 \cdot 5H_2O$ and anhydrous $CuSO_4$. Both are ionic solids that are soluble in water.

**(a)** Define enthalpy of hydration. **[3]**

**(b)** Explain why the hydration enthalpy of copper(II) sulfate cannot be measured directly. **[2]**

**(c)** **(i)** Draw a diagram outlining a simple calorimeter that could be used to measure the enthalpy of solution of copper sulfate.

Include labels. **[2]**

**(ii)** Outline a practical procedure that could be used to determine the enthalpy of hydration for copper(II) sulfate. **[6]**

**Exam tip**

Make sure you include measurements you would take, equipment you would use, and the calculations you would have to perform.

**(d)** The lattice dissociation energy is the energy required to completely split up the lattice to form gaseous ions, and has the same magnitude as the lattice enthalpy but it is endothermic.

Use the data in **Table 4.1** to calculate the lattice enthalpy of anhydrous copper(II) sulfate. **[2]**

| **Enthalpy** | $\Delta_{hyd}H^{\ominus}$ $Cu^{2+}(g)$ | $\Delta_{hyd}H^{\ominus}$ $SO_4^{2-}(g)$ | $\Delta_{sol}H^{\ominus}$ $CuSO_4(s)$ |
|---|---|---|---|
| **Value (kJ mol⁻¹)** | −2099 | −1080 | −67 |

**Table 4.1**

5 Ethanoic acid and sodium hydrogen carbonate are two reactants commonly used by children to make artificial volcanoes. They neutralise each other in the following reaction.

$$CH_3COOH(aq) + NaHCO_3(s) \rightarrow CH_3COONa(aq) + CO_2(g) + H_2O(l)$$

**(a)** Define entropy. **[1]**

**(b)** **(i)** Use the data **Table 5.1** to calculate the temperature at which the reaction is feasible.

Start by calculating the enthalpy change first, then calculate entropy.

State the unit of temperature when you give your answer. **[8]**

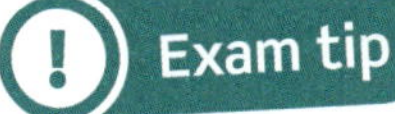

**Exam tip**

Make sure you convert entropy into kJ mol$^{-1}$.

| **Compound** | $CH_3COOH(aq)$ | $NaHCO_3(s)$ | $CH_3COONa(aq)$ | $CO_2(g)$ | $H_2O(l)$ |
|---|---|---|---|---|---|
| **Enthalpy of formation (kJ mol$^{-1}$)** | −483 | −951 | −709 | −394 | −286 |
| **Entropy (J K$^{-1}$ mol$^{-1}$)** | 158 | 102 | 175 | 214 | 70.0 |

**Table 5.1**

**(ii)** Suggest a practical reason why it is unlikely that the reaction will proceed at this temperature. **[1]**

6 **Table 6.1** shows some thermochemical data for the reaction between potassium and oxygen.

| Reaction | Enthalpy ($kJ\,mol^{-1}$) |
|---|---|
| $K(s) \rightarrow K(g)$ | +90 |
| $K(g) \rightarrow K^+(g) + e^-$ | +418 |
| $\frac{1}{2}O_2(g) \rightarrow O(g)$ | +248 |
| $O(g) + e^- \rightarrow O^-(g)$ | −142 |
| $O^-(g) + e^- \rightarrow O^{2-}(g)$ | +844 |
| $2K(g) + \frac{1}{2}O(g) \rightarrow K_2O(s)$ | −362 |

**Table 6.1**

**(a)** Use the data in **Table 6.1** to construct a Born–Haber cycle and determine the lattice formation enthalpy for potassium oxide. **[6]**

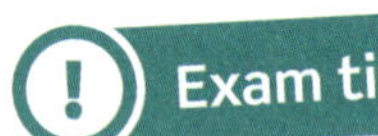

Exam tip

The fully formed ionic solid is the base of the diagram.

**(b) (i)** **Table 6.2** contains some entropy data for the substances involved in the reaction.

| Substance | $K(s)$ | $O_2(g)$ | $K_2O(s)$ |
|---|---|---|---|
| Entropy ($JK^{-1}\,mol^{-1}$) | 67 | 205 | 94 |

**Table 6.2**

Explain why the standard entropy of oxygen is higher than the standard entropy of potassium. **[1]**

**(ii)** Use data in **Table 6.1** and **Table 6.2** to determine the Gibbs free energy of the reaction between potassium and oxygen forming potassium oxide at 300 °C.

Comment on the spontaneity of the reaction. **[6]**

7 Calcium carbonate is a compound found in sedimentary rocks. It is used in many processes including the manufacture of cement. The first stage is the thermal decomposition to form calcium oxide and carbon dioxide.

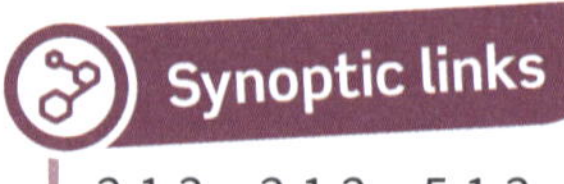

Synoptic links

2.1.3 2.1.2 5.1.3

$$CaCO_3(s) \rightarrow CaO(s) + CO_2(g)$$

The data in **Table 7.1** show how the Gibbs free energy of the reaction varies with temperature.

| Temperature (K) | Gibbs free energy ($kJ\,mol^{-1}$) |
|---|---|
| 200 | 146 |
| 400 | 114 |
| 600 | 82 |
| 800 | 50 |
| 1000 | 18 |
| 1200 | −14 |
| 1400 | −46 |
| 1600 | −78 |

**Table 7.1**

**(a)** Plot a graph of the data and use your graph to determine the temperature at which the Gibbs free energy is 0. **[5]**

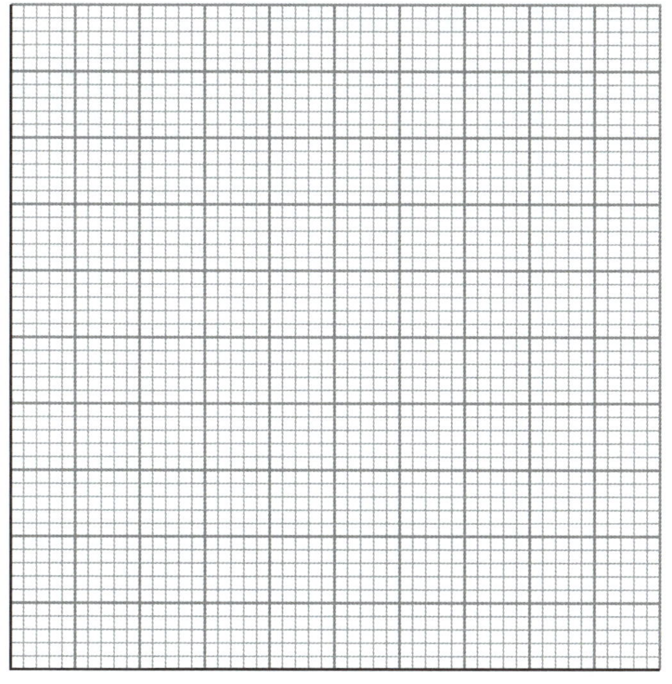

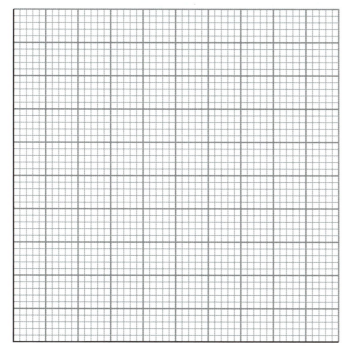

**(b)** Calculate the mass of calcium carbonate that would need to decompose to release 500 $dm^3$ of $CO_2$ gas at standard conditions. **[4]**

**(c)** Give a balanced symbol equation for the reaction of calcium oxide with water to form solid calcium hydroxide. **[1]**

**(d)** Calculate the concentration of a 500 $cm^3$ solution of sulfuric acid that would neutralise 2.28 g of calcium hydroxide. **[4]**

**(e)** Determine the pH of the sulfuric acid used in **7(d)**. **[1]**

**Exam tip**

Remember, all pH values are 2 decimal places.

**8** Hydrogen is made from a reaction between methane and steam:

$$CH_4(g) + H_2O(g) \rightleftharpoons CO(g) + 3H_2(g)$$

**Synoptic links**

3.2.1 3.2.3

**Table 8.1** outlines some thermochemical data about this process.

| Compound | $CH_4(g)$ | $H_2O(g)$ | $CO(g)$ | $H_2(g)$ |
|---|---|---|---|---|
| Enthalpy of formation ( kJ mol$^{-1}$) | −75 | −242 | −111 | 0 |
| Entropy (J K$^{-1}$ mol$^{-1}$) | 186 | 189 | 198 | 131 |

**Table 8.1**

**(a)** Explain why the $\Delta_f H^{\ominus}$ of hydrogen is 0. **[1]**

**(b)** Use the data in **Table 8.1** to calculate the Gibbs free energy of this reaction at 700 °C.

Comment on the feasibility of the reaction at this temperature. **[7]**

**Exam tip**

Set out a table clearly to keep track of each value before calculating $K_p$.

**(c)** 2.00 moles of methane was mixed with 4.00 moles of steam and allowed to reach equilibrium. At equilibrium, 0.70 moles of methane remained. The total pressure was 300 kPa. Calculate the value of $K_p$ for this equilibrium. **[5]**

# Knowledge

## 18 Redox and electrode potentials

### Electrode potentials

When electrodes of different metals are placed in a salt solution and connected, electrons will flow from the more reactive metal to the less reactive metal.

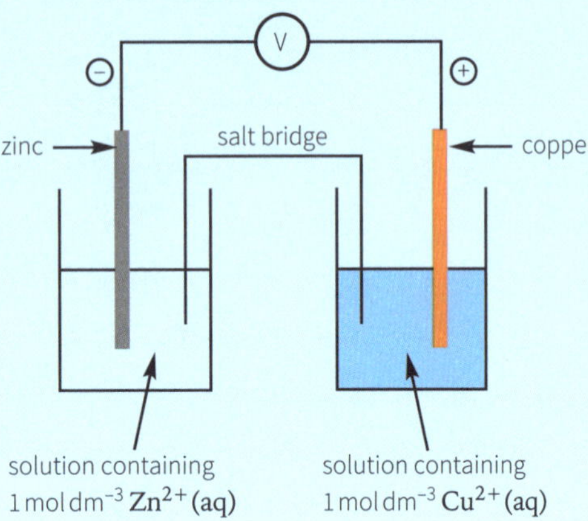


This movement of electrons from one cell to another is the basis for batteries. The half-equations from two half-cells can be used to predict the flow of electrons, the overall reaction, and the electrode potential of the cell.

### Measurement of cell potentials

A **half-cell** is a metal electrode in a solution of its own ions. There is no net transfer of electrons, as an equilibrium is reached between metal atoms and metal ions.

An **electrochemical cell** is two half-cells connected.

- The electrodes are connected by a wire (allows electrons to flow).
- The solutions are connected with a **salt bridge** (allows ions to flow).

Electrons flow between the two cells, the potential difference (voltage, $V$) between the two cells is measured with a voltmeter.

An ion/ion half-cell can be a redox reaction between an element in two different oxidation states, such as a mixture of aqueous $Fe^{2+}$ and $Fe^{3+}$.

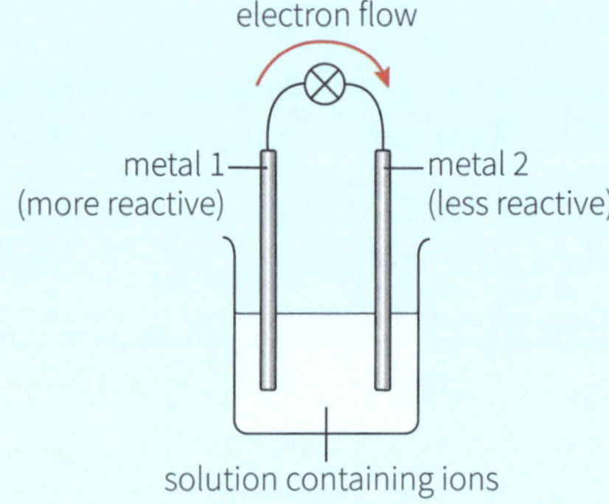


### At the negative electrode

The *negative metal electrode* will lose electrons more readily; this means it is a better **reducing agent**. This metal is *oxidised*. Electrons will flow *from* this electrode.

Metal X atoms form metal ions, increasing the concentration of the metal ions in the solution. The electrons flow through the wire to the positive electrode.

$$X(s) \rightarrow X^{2+}(aq) + 2e^-$$

### At the positive electrode

The *positive metal electrode* will gain electrons, via the wire from the negative half-cell.

Metal Y ions in the solution form metal atoms and metal deposits will form on the electrode. The concentration of the metal ions on the solution will decrease. This metal has been *reduced*.

$$Y^{2+}(aq) + 2e^- \rightarrow Y(s)$$

These two half equations can be added together to produce the overall equation for the reaction.

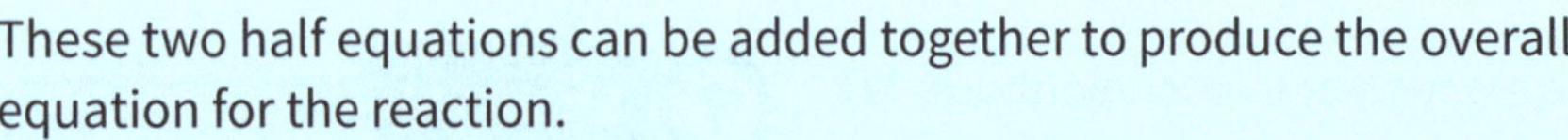

$$X(s) \rightarrow X^{2+}(aq) + 2e^-$$

$$Y^{2+}(aq) + 2e^- \rightarrow Y(s)$$

**Overall:** $X(s) + Y^{2+}(aq) \rightarrow Y(s) + X^{2+}(aq)$

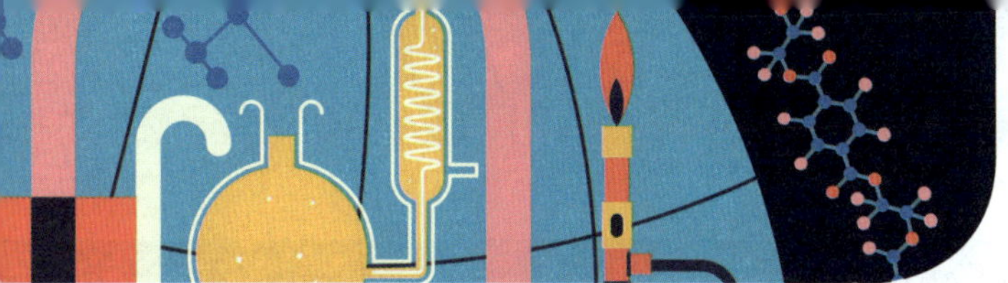



## Predictions from electrode potentials

To predict the movement of electrons and to compare the electrode potential of different half-cells, each half-cell is compared to a **standard hydrogen half-cell**, which is given a potential of 0.00 V. This is done under standard conditions, 100 kPa, 298 K, and $[H^+(aq)] = 1.00\ mol\ dm^{-3}$.

Each half-cell is given a value for the electromotive force (EMF) when measured under standard conditions, the **standard electrode potential** $E^⊖$. The *more negative* the value of $E^⊖$, the better the half-cell can act as a reducing agent; it will give up electrons more easily.

From these values of $E^⊖$, an **electrochemical series** has been produced. The most negative half-cells are placed at the top of the series.

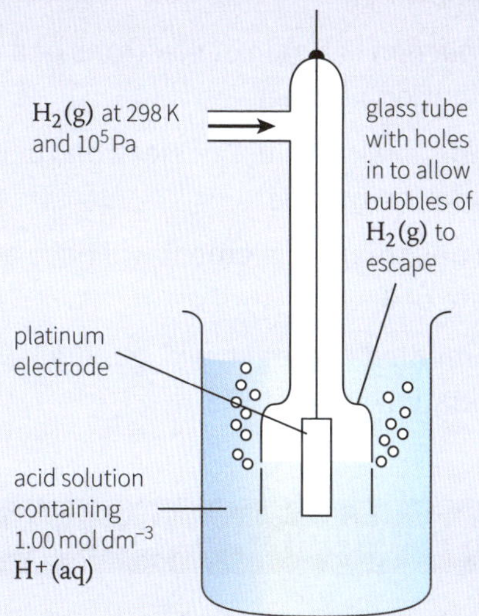

The $E^⊖_{cell}$ is calculated as the difference between the two $E^⊖$ for each half-cell. Redox equations and the electrochemical series are used to predict the direction of redox reactions. Electrons flow from the more negative electrode to the more positive one, so to determine if a reaction will occur or not, the values for $E^⊖$ for each half-cell are compared.

Although these reactions may be feasible there are limitations to these predictions:

- The activation energy needed to start the reaction may be very high.
- The concentrations of the solutions used may not be standard conditions.

## Modern storage cells

The first batteries were made from zinc/copper cells. These days a **non-rechargeable battery** is made from zinc/carbon cells. As the battery is used, the zinc is used up.

Small **rechargeable batteries** are made from **nickel/cadmium cells**. These can be recharged by reversing the direction of the reactions via a large external current to move the electrons back in the reverse direction.

**Lithium batteries** in mobile phones have a positive electrode made from lithium cobalt and a negative electrode made from carbon. The electrodes are arranged in layers with a solid polymer electrolyte between them.

## Fuel cells

**Alkaline hydrogen–oxygen fuel cells** are made from two electrodes separated by a partially permeable membrane. These are an eco-friendly alternative as the only product is water. However, the hydrogen used can be from crude oil or from the electrolysis of water (which uses large quantities of electricity). The energy from the reaction of the hydrogen fuel with oxygen is used to create a voltage.

A **rechargeable battery** in a car's engine is made from lead plates dipped into a solution of sulfuric acid.

Within the battery there are a number of connected cells; each cell is made up of two plates. The negative plate is lead and the positive plate is lead coated with lead(IV) oxide.

## Oxidising and reducing agents

To understand redox titrations you need to recall the following:

- *Reduction* is a gain of electrons or a decrease in oxidation number.
- *Oxidation* is the loss of electrons or an increase in oxidation number.
- An uncombined element will have an oxidation state of 0.
- Combined hydrogen will, generally, have an oxidation state of +1.
- Combined oxygen will, generally, have an oxidation state of −2.
- An *oxidising agent* removes electrons from the species that is being oxidised.
- A *reducing agent* gives electrons to the species that is being reduced.
- Part of the oxidising agent will be reduced and part of the reducing agent will be oxidised.

## Construction of redox equations

For complicated reactions:

- look at the ions that have reacted
- construct half equations and then build these up to get the overall balanced equation.

For example, acidified potassium manganate(VII) can be used as an oxidising agent: the $MnO_4^-(aq)$ ions are reduced to $Mn^{2+}(aq)$ ions.

When balancing half equations, only add certain things:

- electrons to balance the charges
- hydrogen ions to balance the hydrogens atoms
- water to balance the oxygen atoms.

**Step 1** Write the half equation for the reduction reaction:

$$MnO_4^-(aq) + H^+ \rightarrow Mn^{2+}(aq)$$

**Step 2** Balance the electrons: *the oxidation state of manganese has decreased by 5*:

$$MnO_4^-(aq) + H^+ \mathbf{+ 5e^-} \rightarrow Mn^{2+}(aq)$$

**Step 3** Balance the charges:

$$MnO_4^-(aq) + 5e^- \mathbf{+ 8H^+(aq)} \rightarrow Mn^{2+}(aq)$$

**Step 4** Balance the hydrogen and oxygen ions without changing the charge:

$$MnO_4^-(aq) + 5e^- + 8H^+(aq) \rightarrow Mn^{2+}(aq) \mathbf{+ 4H_2O(l)}$$

## Redox titrations

**Redox titrations** are used to find the unknown concentration of oxidising and reducing agents.

- Potassium manganate(VII) under acidified conditions is used to determine *concentrations of reducing agents*.
- Sodium thiosulfate is used to determine the *concentration of oxidising agents* such as iodine.

The first step is using half equations to determine the overall equation of the reaction. Be careful to note the ratio between what you are interested in and the species being measured. Be aware of any hydrated ions, alloys, or additives in tablets, when doing titration calculations.

## Manganate(VII) redox titration

Potassium manganate(VII) can be used to determine the concentration of acidified solutions of reducing agents; common ones are iron and ethanedioic acid. The manganate(VII) ions, $MnO_4^-(aq)$, are reduced and the ions in the unknown solution are oxidised.

The deep purple colour of potassium manganate(VII) makes it difficult to see the bottom of the meniscus, so burette readings are taken from the top of the meniscus. The value used in calculations is still the difference between these two values.

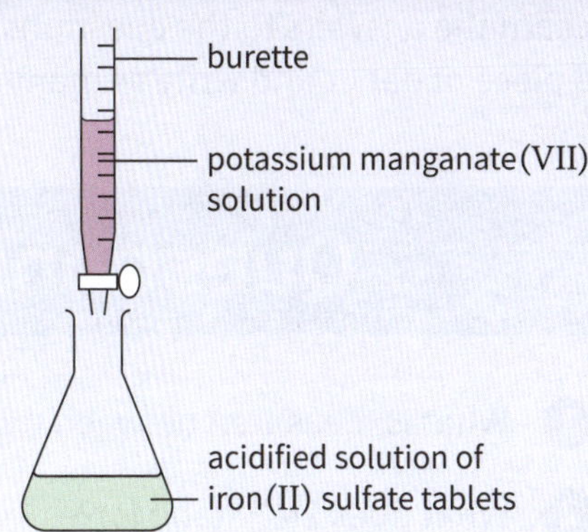

**For iron(III) titrations**

The reduction reaction is:

$$MnO_4^-(aq) + 8H^+(aq) + 5e^- \rightarrow Mn^{2+}(aq) + 4H_2O(l)$$

The oxidation reaction is:

$$Fe^{2+}(aq) \rightarrow Fe^{3+}(aq) + e^-$$

*The overall reaction is:*

$$MnO_4^-(aq) + 8H^+(aq) + 5Fe^{2+}(aq) \rightarrow Mn^{2+}(aq) + 4H_2O(l) + 5Fe^{3+}(aq)$$

**For ethanedioic acid titrations**

The reduction reaction is:

$$MnO_4^-(aq) + 8H^+(aq) + 5e^- \rightarrow Mn^{2+}(aq) + 4H_2O(l)$$

The oxidation reaction is:

$$(COOH)_2(aq) \rightarrow 2CO_2(g) + 2H^+(aq) + 2e^-$$

*The overall reaction is:*

$$2MnO_4^-(aq) + 6H^+(aq) + 5(COOH)_2\,(aq) \rightarrow 2Mn^{2+}(aq) + 8H_2O(l) + 10CO_2(g)$$

## Iodine-thiosulfate redox titration

An iodine–thiosulfate titration can be used to determine the concentration of *oxidising agents* in a solution.

The oxidation reaction is:

$$2S_2O_3^{2-}(aq) \rightarrow S_4O_6^{2-}(aq) + 2e^-$$

The reduction reaction is:

$$I_2(aq) + 2e^- \rightarrow 2I^-(aq)$$

*The overall reaction is:*

$$2S_2O_3^{2-}(aq) + I_2(aq) \rightarrow 2I^-(aq) + S_4O_6^{2-}(aq)$$

1 The $Na_2S_2O_3$ is added to the burette.
2 A solution of oxidising agent and an excess of potassium iodide is added to the conical flask. The iodide ions from the potassium iodide are oxidised to give iodine; this solution is yellow-brown.
3 During the titration the iodine is reduced back to iodine ions and the colour fades, leaving a pale yellow solution. At this point starch is used as an indicator, turning blue-black if iodine is present, then disappearing when the iodine is no longer present. This colour change is easier to monitor.

# Retrieval

Learn the answers to the questions below, then cover the answers column with a piece of paper and write as many as you can. Check and repeat.

| | Questions | Answers |
|---|---|---|
| 1 | What is measured between two half-cells? | potential difference (*V*) |
| 2 | What does EMF stand for? | electromotive force |
| 3 | What are standard conditions for a cell? | 100 kPa, 298 K, and [ions(aq)] = 1.00 mol $dm^{-3}$ |
| 4 | Is the negative metal electrode reduced or oxidised in a cell? | oxidised |
| 5 | Why is a salt bridge used in an electrochemical cell? | to allow ions to flow |
| 6 | Which direction do electrons flow? | from the more negative electrode to the more positive one |
| 7 | Will the concentration of ions in the solution increase or decrease at the negative electrode in a cell? | increase |
| 8 | What is the overall reaction equation in a alkaline hydrogen/oxygen fuel cell? | $2H_2(g) + O_2(g) \rightarrow 2H_2O(l)$ |
| 9 | What is the main difference between a rechargeable battery and a non-rechargeable battery? | rechargeable batteries rely on reversible reactions |
| 10 | Is the positive metal electrode reduced or oxidised in a cell? | reduced |
| 11 | What is the EMF for a standard hydrogen half-cell? | 0.00 V |
| 12 | What are the electrodes connected with in a cell? | a wire |
| 13 | Will the concentration of ions in the solution increase or decrease at the positive electrode in a cell? | decrease |
| 14 | What state is the electrolyte in lithium batteries? | solid polymer |
| 15 | Where are lithium batteries found? | in mobile phones |
| 16 | What are the electrodes made from in a lead battery? | the negative plate is lead and the positive plate is lead coated with lead(IV) oxide |
| 17 | What type of battery is a zinc/carbon battery? | non-rechargeable |
| 18 | How can you recharge a battery? | apply a large external current |
| 19 | Why might hydrogen/oxygen fuel cells be seen as a green alternative? | water is the only product |

Put paper here

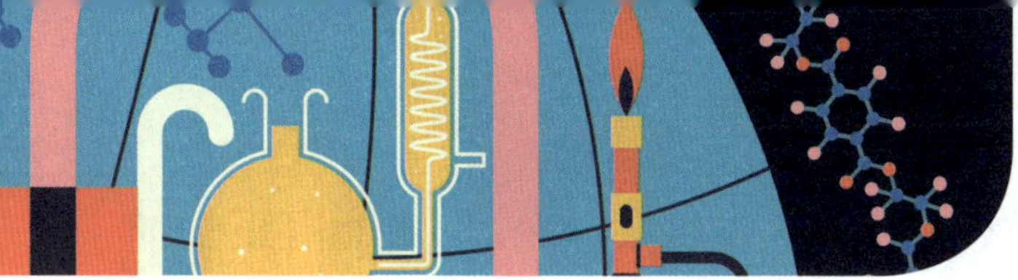



# Practical skills

Practise your practical skills using the worked example and practice questions below.

| Electrochemical cells | Worked example | Practice |
|---|---|---|
| Electrochemical cells convert chemical energy to electrical energy. Measuring the EMF of such a cell can tell you important information about the oxidising and reducing powers of the species involved.<br><br>When measuring $E^\ominus$ it is essential that the system is under standard conditions. This means that:<br>• concentrations must be 1 mol dm$^{-3}$<br>• any gases must be at a pressure of 100 kPa<br>• the temperature is 298 K (20 °C).<br><br>When there is a mixture of ions (e.g. $ClO^-$ and $Cl^-$) then a platinum electrode is used. Often in a school laboratory, a carbon electrode is often used instead.<br><br>As time passes, the concentrations of the ions change. Therefore, it is important to read and record the voltage quickly. | **Questions**<br>Draw and label a diagram of an electrochemical cell made of the following two half-cells:<br>**1** $Cl^-/Cl_2$ half-cells<br>**2** $Ce^{4+}/Ce^{3+}$ half-cells.<br><br>**Answers**<br>It is essential that these diagrams are labelled correctly. Exam questions may provide incorrect diagrams and you need to identify errors and/or correct them. Your diagram should include:<br>**1** For the $Cl^-/Cl_2$ half-cell:<br>• $Cl_2$ at a pressure of 100 kPa<br>• 1 mol dm$^{-3}$ solution of $Cl^-$ (aq)<br>• platinum electrode.<br>**2** For the $Ce^{4+}$ / $Ce^{3+}$ half-cell:<br>• 1 mol dm$^{-3}$ solution of $Ce^{3+}$ (aq)<br>• 1 mol dm$^{-3}$ solution of $Ce^{4+}$ (aq)<br>• platinum electrode.<br><br>Your diagram should also include:<br>• a high resistance voltmeter<br>• a salt bridge<br>• temperature of 298 K. | Explain why the following cannot be used to measure $E^\ominus$.<br>**1** Only a platinum wire is used to connect the two half-cells.<br>**2** An aluminium half-cell is made with 1 mol dm$^{-3}$ aluminium sulfate solution.<br>**3** An $Fe^{3+}(aq)/Fe^{2+}(aq)$ half-cell is made with an iron electrode.<br>**4** The temperature is 0 °C.<br>**5** The temperature is 298 °C.<br>**6** Waiting before reading from the voltmeter. |

# Practice

## Exam-style questions

1 **Table 1.1** contains some standard electrode potential, $E^\ominus$, data.

| Reaction | $E^\ominus$ (V) |
|---|---|
| $Zn^{2+}(aq) + 2e^- \rightleftharpoons Zn(s)$ | −0.76 |
| $Fe^{2+}(aq) + 2e^- \rightleftharpoons Fe(s)$ | −0.44 |
| $H^+(aq) + e^- \rightleftharpoons \frac{1}{2}H_2(g)$ | 0.00 |
| $Cu^{2+}(aq) + 2e^- \rightleftharpoons Cu(s)$ | +0.34 |
| $Fe^{3+}(aq) + 2e^- \rightleftharpoons Fe^{2+}(aq)$ | +0.77 |

**Table 1.1**

**(a) (i)** State standard conditions for determining electrode potentials.

..................................................................................

..................................................................................

.................................................................................. **[2]**

**(ii)** Draw a diagram of the standard hydrogen electrode.

Include labels of any chemicals used.

**Exam tip**

Make sure you can draw all parts of this, including the electrode.

Make sure you include labels explaining the standard conditions that must be met.

**[4]**

**(b)** A student sets up an electrochemical cell with a 1.00 mol dm$^{-3}$ solution of iron(III) sulfate and a platinum electrode on one side and an iron electrode in a 1.00 mol dm$^{-3}$ solution of iron(II) sulfate on the other. The half-cells are connected by a salt bridge and a voltmeter.

**(i)** Give the conventional cell representation for the student's electrochemical cell.

..................................................................................

..................................................................................

.................................................................................. **[2]**

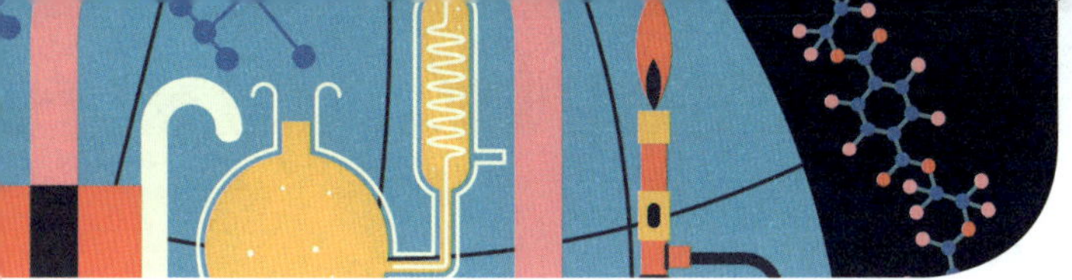



(ii) Calculate the EMF of the cell in **1(b)(i)**.

[1]

(c) Name the species listed in **Table 1.1** that is the strongest reducing agent.

........................................................................................

........................................................................................ [1]

(d) Iron filings are added to 1.00 mol dm$^{-3}$ solutions of $Cu^{2+}$, $Fe^{3+}$, $H^{+}$, $Fe^{2+}$, and $Zn^{2+}$.

Based on electrode potentials, write balanced equations for the three possible reactions that occur.

........................................................................................

........................................................................................

........................................................................................

........................................................................................ [3]

**! Exam tip**

All equations need state symbols, all the time.

**2** **Table 2.1** shows some electrode half-equations and their standard electrode potentials.

| Electrode half-equation | $E^{\ominus}$ (V) |
|---|---|
| $Cl_2(g) + 2e^- \rightleftharpoons 2Cl^-(aq)$ | +1.36 |
| $NO_3^-(aq) + 4H^+(aq) + 3e^- \rightleftharpoons NO(aq) + 2H_2O(l)$ | +0.96 |
| $Fe^{2+}(aq) + e^- \rightleftharpoons Fe^{3+}(aq)$ | +0.77 |
| $Cu^2+(aq) + 2e^- \rightleftharpoons Cu(s)$ | +0.34 |
| $SO_4^{2-}(aq) + 4H^+(aq) + 2e^- \rightleftharpoons SO_2(g) + 2H_2O(aq)$ | +0.17 |
| $2H^+(aq) + 2e^- \rightleftharpoons H_2(g)$ | 0.00 |
| $Fe^{2+}(aq) + 2e^- \rightleftharpoons Fe(s)$ | −0.44 |

**Table 2.1**

(a) Deduce the oxidation state of sulfur in $SO_4^{2-}(aq)$ and $SO_2(g)$.

........................................................................................

........................................................................................

........................................................................................ [2]

(b) State the weakest oxidising agent in **Table 2.1**.

........................................................................................

........................................................................................ [1]

**! Exam tip**

Take care! This is the exam equivalent of a double negative.

**(c)** Write the overall ionic equation for the reaction between iron and any strong acid.

Calculate the EMF of the cell that has the same overall reaction.

ionic equation = ..........................................................................

EMF of the cell = ..................... volts **[2]**

**(d)** Although copper is unreactive with acids it can be oxidised by some acids. Which of the following strong acids could oxidise copper to copper(II)?

- HCl
- $HNO_3$
- $H_2SO_4$. **[4]**

**3** **Table 3.1** shows some half-equations and their standard electrode potentials.

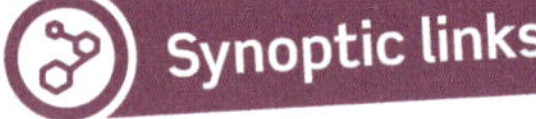


2.1.5 5.2.3

| Equation | $E^\ominus$ (V) |
|---|---|
| $MnO_4^-(aq) + 8H^+(aq) + 5e^- \rightleftharpoons Mn^{2+}(aq) + 4H_2O(l)$ | +1.51 |
| $Cl_2(g) + 2e^- \rightleftharpoons 2Cl^-(aq)$ | +1.36 |
| $2H^+(aq) + 2e^- \rightleftharpoons H_2(g)$ | 0.00 |
| $Fe^{2+}(aq) + 2e^- \rightleftharpoons Fe(s)$ | −0.44 |
| $Fe^{3+}(aq) + e^- \rightleftharpoons Fe^{2+}(aq)$ | +0.77 |

**Table 3.1**

A student was tasked with determining the percentage by mass of iron(II) sulfate, $FeSO_4$, in an iron tablet. 2.00 g of the iron tablet was crushed in a pestle and mortar. The resulting powder was dissolved in 50 $cm^3$ of dilute sulfuric acid. The solution was poured into a volumetric flask and made up to 250 $cm^3$.

25.0 $cm^3$ of the iron sample was titrated with a 0.025 mol $dm^{-3}$ solution of potassium manganate(VII), $KMnO_4$. The average titre was 19.25 $cm^3$.

**(a)** Explain why hydrochloric acid should *not* be used to acidify the reaction between iron(II) sulfate and potassium manganate(VII). Use data from **Table 3.1** to support your answer.

..........................................................................

..........................................................................

..........................................................................

.......................................................................... **[2]**

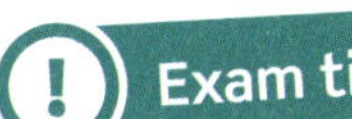


Lay your work out clearly and systematically so the examiner knows what each number represents.

**(b)** Give one practical issue with performing a titration with potassium permanganate in the burette.

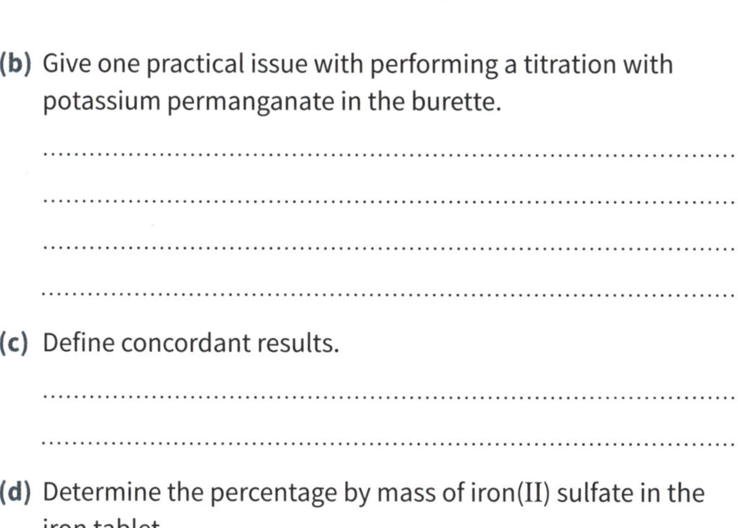

.......................................................................................... **[1]**

**(c)** Define concordant results.

.......................................................................................... **[1]**

**(d)** Determine the percentage by mass of iron(II) sulfate in the iron tablet.

percentage by mass = .................... **[6]**

**4** An electrochemical cell is made of the following half-equations:

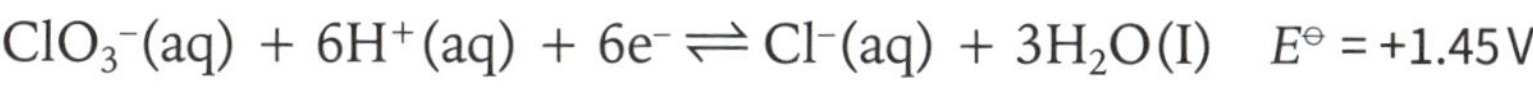

$ClO_3^-(aq) + 6H^+(aq) + 6e^- \rightleftharpoons Cl^-(aq) + 3H_2O(l) \quad E^\ominus = +1.45\,V$

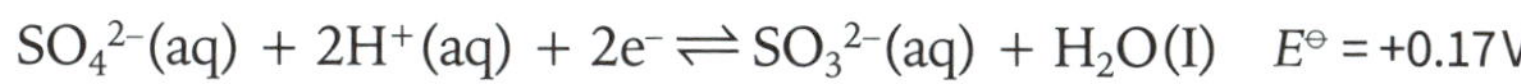

$SO_4^{2-}(aq) + 2H^+(aq) + 2e^- \rightleftharpoons SO_3^{2-}(aq) + H_2O(l) \quad E^\ominus = +0.17\,V$

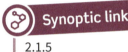

**Synoptic link**

2.1.5

**(a) (i)** Give the oxidation state of chlorine in both $ClO_3^-(aq)$ and $Cl^-(aq)$.

.......................................................................................... **[2]**

**(ii)** Give the overall reaction.

Identify the oxidising and reducing agents.

**[3]**

**Exam tip**

Take care to balance the equation; you are allowed to use halves.

**(b) (i)** Draw a labelled diagram of the practical set-up required to measure the EMF of this electrochemical cell.

**[4]**

**(ii)** Calculate the EMF of the cell.

EMF = .................... V **[1]**

5 **Table 5.1** shows some electrochemical data.

| Electrode half-equation | $E^{\ominus}$ (V) |
|---|---|
| $Au^{+}(aq) + e^{-} \rightleftharpoons Au(s)$ | +1.68 |
| $\frac{1}{2}O_2(g) + 2H^{+}(aq) + 2e^{-} \rightleftharpoons H_2O(l)$ | +1.23 |
| $Ag^{+}(aq) + e^{-} \rightleftharpoons Ag(s)$ | +0.80 |
| $Fe^{3+}(aq) + e^{-} \rightleftharpoons Fe^{2+}(aq)$ | +0.77 |
| $Cu^{2+}(aq) + 2e^{-} \rightleftharpoons Cu(s)$ | +0.34 |
| $Fe^{2+}(aq) + 2e^{-} \rightleftharpoons Fe(s)$ | −0.44 |

**Table 5.1**

**(a)** Explain, in terms of electrons, the role of a reducing agent. **[1]**

**(b)** Deduce the two half-equations that will generate the largest EMF.

Write the overall equation and calculate the EMF. **[2]**

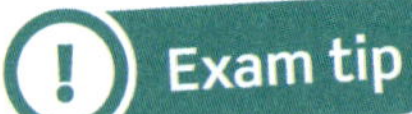

You must give the overall equation to score full marks.

**(c)** Explain why gold is usually not found in aqueous solutions. Use data from **Table 5.1** to support your answer. **[2]**

**(d)** An electrochemical cell was set up between copper and iron.

**(i)** Give the conventional cell representation for the copper and iron cell. **[1]**

**(ii)** Explain the role of the salt bridge in the electrochemical cell. **[2]**

6 **Table 6.1** contains some electrochemical cell data.

2.1.5

| Electrode half-equation | $E^{\ominus}$ (V) |
|---|---|
| $Fe^{3+}(aq) + e^{-} \rightleftharpoons Fe^{2+}(aq)$ | +0.77 |
| $Cl_2(g) + 2e^{-} \rightleftharpoons 2Cl^{-}(aq)$ | +1.36 |
| $2BrO_3^{-}(aq) + 12H^{+}(aq) + 10e^{-} \rightleftharpoons Br_2(aq) + 6H_2O(l)$ | +1.52 |
| $O_3(g) + 2H^{+}(aq) + 2e^{-} \rightleftharpoons O_2(g) + H_2O(l)$ | +2.08 |
| $F_2O(g) + 2H^{+}(aq) + 4e^{-} \rightleftharpoons 2F^{-}(aq) + H_2O(l)$ | +2.15 |

**Table 6.1**

**(a)** State the strongest oxidising agent in **Table 6.1**. **[1]**

**(b)** Give the conventional cell representation of the electrochemical cell made from the data in **Table 6.1** that will produce an EMF of +0.79 V. **[2]**

**(c)** A solution of sodium bromate(V) was added to a beaker containing excess hydrochloric acid.

Using data in **Table 6.1**, write the overall equation for the reaction that occurs.

State two observations that would be made. **[4]**

**Exam tip**

Observations need to be clear. Write what you would see or smell.

**(d)** Using the data from **Table 6.1** suggest a reagent that could be used to prevent the release of chlorine gas into the atmosphere. Write the balanced equation for this reaction. **[2]**

7 Nickel–cadmium cells are rechargeable cells that are used to power many household appliances.

The following two half-equations are involved:

$$NiO(OH) + H_2O + e^- \rightleftharpoons Ni(OH)_2 + OH^- \qquad E^\ominus = +0.52\,V$$

$$Cd(OH)_2 + 2e^- \rightleftharpoons Cd + 2OH^- \qquad E^\ominus = -0.88\,V$$

**(a) (i)** Give the overall equation that occurs when the cell is used. **[2]**

**(ii)** Calculate the EMF of the cell. **[1]**

**(iii)** State the overall equation that occurs when the cell is being recharged. **[1]**

**(b)** State the effect on the standard electrode potential of cadmium if the pH of the solution was increased.

Give a reason. **[2]**

**Synoptic link**

2.1.5

**Exam tip**

Do not say the EMF gets bigger or smaller.

Clearly state if it becomes more positive or negative.

# Knowledge

## 19 Transition elements

### The d-block elements

The **d-block elements** are found in the middle block of the Periodic Table. This group of elements is called the d-block, as electrons are added to the d-sub-shell across the periods.

The **transition elements** (sometimes known as **transition metals**) in Period 4 are from titanium to copper and all have a *partially filled d-sub-shell*. These transition elements have the usual physical properties of metals. They are good conductors of heat and electricity, have high melting and boiling points, and are malleable and ductile. These metallic properties give them a wide range of uses, for example, in building construction and electrical cables.

They also have some unique properties, for example:

- the formation of complexes
- the formation of coloured ions
- variable oxidation states
- the ability to act as a catalyst.

### Electron configuration

The 4s-sub-shell has a lower energy than the 3d-sub-shell, so it is filled first. The only exceptions are copper and chromium, where the 3d-sub-shell is half filled (copper) or fully filled (chromium) before the 4s-sub-shell is filled. This common pattern in the electron arrangements is responsible for the common properties that these elements share. The 4s electrons will be lost first when bonding.

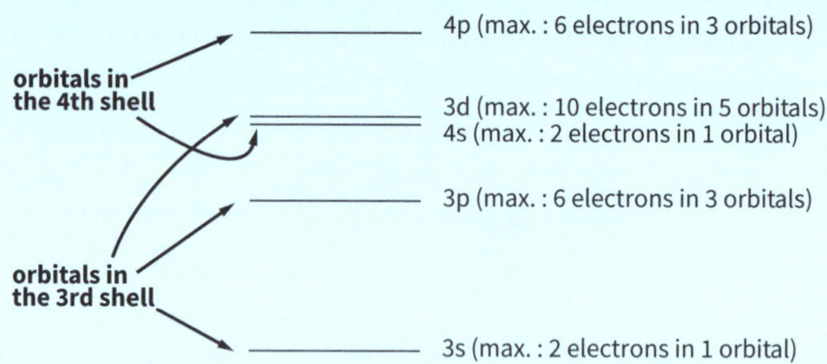


### Electron configuration for Period 4 d-block elements

| Element | Electron configuration |
|---|---|
| **scandium** | **$1s^22s^22p^63s^23p^63d^14s^2$** |
| titanium | $1s^22s^22p^63s^23p^63d^24s^2$ |
| vanadium | $1s^22s^22p^63s^23p^63d^34s^2$ |
| chromium | $1s^22s^22p^63s^23p^63d^54s^1$ |
| manganese | $1s^22s^22p^63s^23p^63d^54s^2$ |
| iron | $1s^22s^22p^63s^23p^63d^64s^2$ |
| cobalt | $1s^22s^22p^63s^23p^63d^74s^2$ |
| nickel | $1s^22s^22p^63s^23p^63d^84s^2$ |
| copper | $1s^22s^22p^63s^23p^63d^{10}4s^1$ |
| **zinc** | **$1s^22s^22p^63s^23p^63d^{10}4s^2$** |

### Scandium and zinc

Scandium and zinc are *not* considered to be transition elements.

Transition elements are elements which form at least one ion with a incomplete d-sub-shell.

- Scandium only forms 3+ ions. The electron configuration is $1s^22s^22p^63s^23p^6$. There are no d-block electrons.
- Zinc only forms 2+ ions. The electron configuration is $1s^22s^22p^63s^23p^63d^{10}$. Here the d-sub-shell is full.

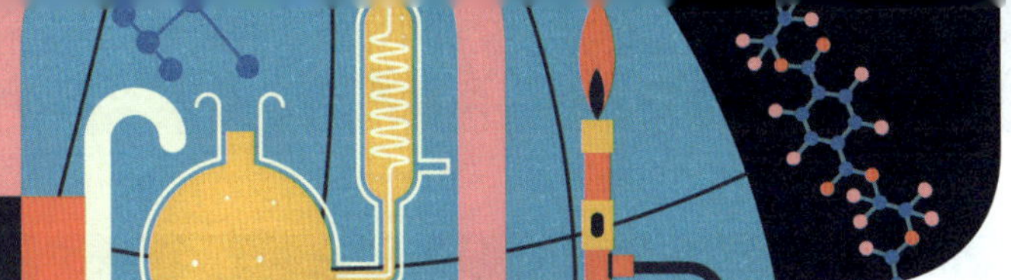



## Oxidation states

Transition metals can show a range of oxidation states and colours. You need to learn the oxidation states and associated colours (here the most common oxidation states are in **bold**).

| Scandium | Titanium | Vanadium | Chromium | Manganese | Iron | Cobalt | Nickel | Copper | Zinc |
|---|---|---|---|---|---|---|---|---|---|
| | +1 | +1 | +1 | +1 | +1 | +1 | +1 | **+1** | |
| | +2 | +2 | **+2** | **+2** | **+2** | **+2** | **+2** | **+2** | **+2** |
| **+3** | +3 | +3 | **+3** | +3 | **+3** | **+3** | +3 | +3 | |
| | **+4** | **+4** | +4 | **+4** | +4 | +4 | +4 | | |
| | | **+5** | +5 | +5 | +5 | +5 | | | |
| | | | **+6** | +6 | +6 | | | | |
| | | | | **+7** | | | | | |

## Catalytic behaviour

Transition elements can be used as **catalysts**. They can be **homogeneous catalysts**, meaning the reactants and catalyst are in the *same phase*. Or, they can be **heterogenous catalysts**, meaning the reactants and the catalyst are in *different phases*.

In the Contact process a vanadium oxide intermediate is produced: the oxidation state of vanadium changes from +5 to +4 and back to +5 again. This ability of transition elements to change oxidation state makes them valuable as catalysts.

The economic benefit of using a catalyst is reduced energy (thus potentially reduced fossil fuel) use. This needs to be balanced with the potential for toxicity from the use of transition metals.

| Process | Product | Transition metal/ compound catalyst | Reaction |
|---|---|---|---|
| Haber process | ammonia | iron (heterogeneous) | $N_2(g) + 3H_2(g) \rightleftharpoons 2NH_3(g)$ |
| hydrogenation of vegetable fats | margarine | nickel (heterogeneous) | $\mathrm{C{=}C}$ (each C bearing H) $+ H_2 \xrightarrow{Ni}$ $\mathrm{-CH-CH-}$ (each C bearing H, H) |
| decomposition of hydrogen peroxide | water and oxygen | manganese dioxide (heterogeneous) | $H_2O_2\ (aq) \xrightarrow{MnO_2} H_2O\ (l) + O_2(g)$ |
| Contact process | sulfuric acid | vanadium(V) oxide (heterogeneous) | $2SO_2(g) + O_2(g) \rightleftharpoons 2SO_3(g)$ |
| iodide ions and peroxodisulfate ions | iodine | iron (homogeneous) | **Step 1:** $S_2O_8^{2-}(aq) + 2Fe^{2+}(aq) \rightarrow 2SO_4^{2-}(aq) + 2Fe^{3+}(aq)$ **Step 2:** $2Fe^{3+}(aq) + 2I^-(aq) \rightarrow 2Fe^{2+}(aq) + I_2(aq)$ |

## Complex ions and coordination number

Transition metals form **complex ions**. This is where a central (transition) metal atom is bonded to **ligands**. These ligands are part of **complexes** as they form a **coordinate bond** with a transition metal by donating a pair of electrons. The number of coordinate bonds around a central metal atom will be the **coordination number** of that complex.

## Ligands

**Monodentate ligands** have *one* pair of electrons that can be donated to form a coordinate bond.

$NH_3$ and $H_2O$ are non-charged ligands that are of a similar size. $Cl^-$ and $CN^-$ are charged ligands.

Non-charged ligands will not affect the overall charge of the complex ion. However, the charge on charged ions will be reflected in the overall charge of the complex.

**Bidentate ligands** will donate *two* pairs of electrons to the central atom. Two common bidentate ligands are $NH_2CH_2CH_2NH_2$ and $C_2O_4^{2-}$.

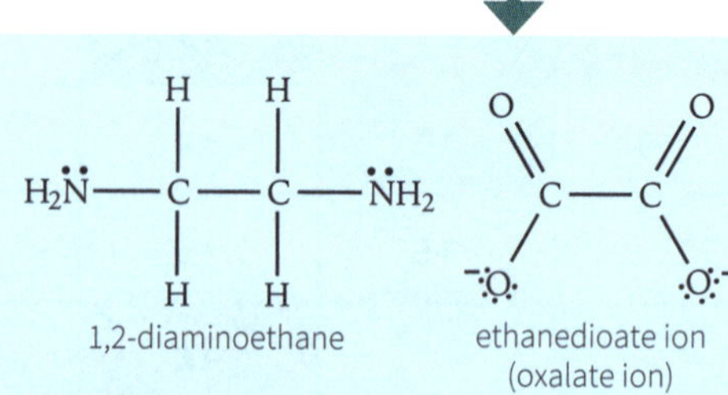


## Shapes of complex ions

There are four main shapes of transition metal complexes:

**(a)** *linear shape*: $[Ag(NH_3)_2]^+$, used as Tollens' reagent

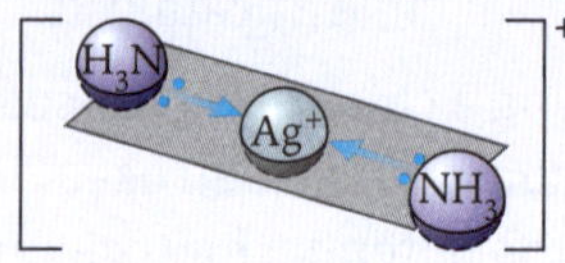

**(b)** *tetrahedral shape*: $Cl^-$ in $[CuCl_4]^{2-}$

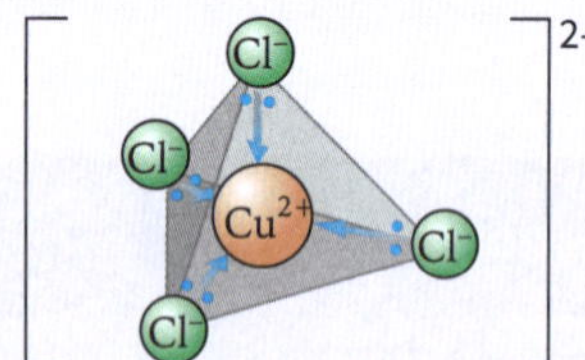

**(c)** *square planar shape*: $[Ni(CN)_4]^{2-}$

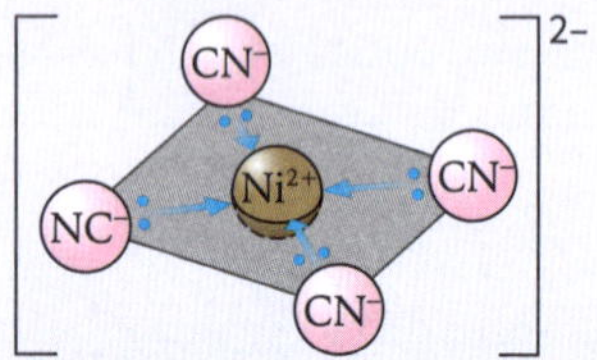

**(d)** *octahedral shape*: $[Cu(H_2O)_6]^{2+}$, has *cis–trans* isomerism.

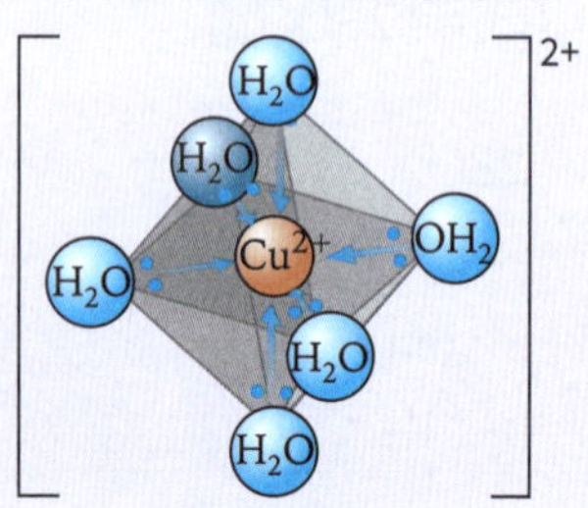

## The importance of iron in haemoglobin

Haemoglobin is found in red blood cells. Haem is the part of haemoglobin that is responsible for carrying oxygen. Haem has a central $Fe^{2+}$ with a coordination number of 6; four of these spaces are bonded to a tetradentate porphyrin ring system.

Oxygen gas, $O_2$, can act as a ligand and binds to the haem molecule. As oxygen is a very poor ligand, the oxygen is easily released from the molecule to the body cells that need it. Carbon monoxide is a poisonous gas; it is a better ligand than oxygen gas so binds strongly to the haem molecule. It can irreversibly bind to the haem and prevent oxygen being carried to cells.

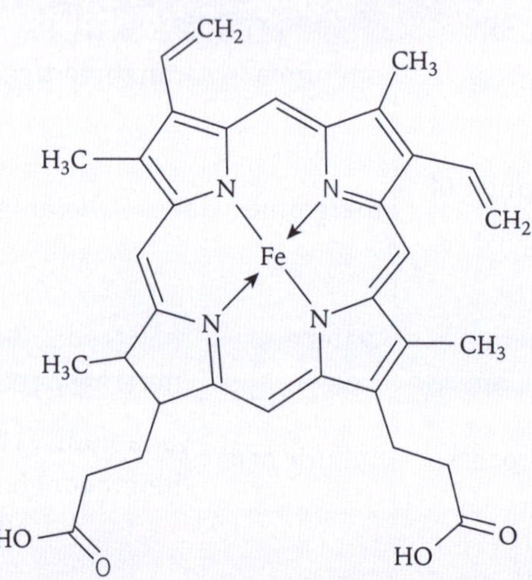

haem (part of haemoglobin)

## Stereoisomerism

Octahedral-shaped complex ions can show *cis–trans* isomerism with only monodentate ligands. Octahedral complexes that involve bidentate ligands can show *cis–trans* isomerism and optical isomerism.

*trans*

*cis* has optical isomers

## *cis*-platin

*cis*-platin: as with many drugs, only one isomer of platin has functional activity. Therefore, only the *cis* form of the complex ion $[Pt(NH_3)_2Cl_2]$ is used as a treatment for cancer, although it has unpleasant side effects. It acts by binding with DNA to prevent cell division.

*cis*-platin *trans*-platin

## Ligand substitution reactions of copper

**Ligand substitution reactions** with non-charged ligands of *ammonia* and *water* take place without a change in coordination number.

Copper(II) sulfate in water gives a pale blue complex ion $[Cu(H_2O)_6]^{2+}$. With excess ammonia, a dark blue complex ion $[Cu(NH_3)_4(H_2O)_2]^{2+}$ forms. Four ammonia ligands are *substituted* for four water ligands. The coordination number of this complex and the overall charge on the ion does not change.

A chloride ion is charged; substitution causes a change in coordination number and overall charge on the ion. When concentrated HCl is added to $[Cu(H_2O)_6]^{2+}$, the pale blue solution will form a yellow solution of $[CuCl_4]^{2-}$.

## Ligand substitution reactions of chromium(III) ions

Violet $[Cr(H_2O)_6]^{3+}$ reacts with excess ammonia to form $[Cr(NH_3)_6]^{3+}$; this reaction happens in two steps:

1 After the addition of a few drops of ammonia, a grey-green precipitate of $Cr(OH)_3$ is formed.

2 This precipitate then dissolves in the excess ammonia to form the purple complex ion $[Cr(NH_3)_6]^{3+}$.

$$[Cr(H_2O)_6]^{3+}(aq) + 6NH_3(aq) \rightarrow [Cr(NH_3)_6]^{3+}(aq) + 6H_2O(l)$$

## Copper(II) ions

$[Cu(H_2O)_6]^{2+}$ is a pale blue solution. With excess aqueous sodium hydroxide a blue precipitate forms.

$$[Cu(H_2O)_6]^{2+}(aq) + 2OH^-(aq) \rightarrow [Cu(H_2O)_4(OH)_2](s) + 2H_2O(l)$$

With excess aqueous ammonia a blue precipitate forms, which further dissolves in ammonia to form a complex ion in a dark blue solution:

$$[Cu(H_2O)_6]^{2+}(aq) + 2NH_3(aq) \rightarrow [Cu(H_2O)_4(OH)_2](s) + 2NH_4^+(aq)$$

Then:

$$[Cu(H_2O)_4(OH)_2](s) + 4NH_3(aq) \rightarrow [Cu(H_2O)_2(NH_3)_4]^{2+}(aq) + 2H_2O(l) + 2OH^-(aq)$$

## Iron(II) ions

$[Fe(H_2O)_6]^{2+}$ is a pale green solution.

With excess aqueous sodium hydroxide, a green precipitate forms. When left standing the precipitate will react with oxygen and the iron(II) will be oxidised to iron(III) and turns brown:

$$[Fe(H_2O)_6]^{2+}(aq) + 2OH^-(aq) \rightarrow [Fe(H_2O)_4(OH)_2](s) + 2H_2O(l)$$

With excess aqueous ammonia the complex reacts in the same way as hydroxide ions. These precipitates do not dissolve:

$$[Fe(H_2O)_6]^{2+}(aq) + 2NH_3(aq) \rightarrow [Fe(H_2O)_4(OH)_2](s) + 2NH_4^+(aq)$$

## Iron(III) ions

$[Fe(H_2O)_6]^{3+}$ is a pale yellow solution.

With excess aqueous sodium hydroxide an orange-brown precipitate forms:

$$[Fe(H_2O)_6]^{3+}(aq) + 3OH^-(aq) \rightarrow [Fe(H_2O)_3(OH)_3](s) + 3H_2O(l)$$

With excess aqueous ammonia the complex reacts in the same way as hydroxide ions. These precipitates do not dissolve:

$$[Fe(H_2O)_6]^{3+}(aq) + 3NH_3(aq) \rightarrow [Fe(H_2O)_3(OH)_3](s) + 3NH_4^+(aq)$$

## Manganese(II) ions

$[Mn(H_2O)_6]^{2+}$ is a pale pink solution.

With excess aqueous sodium hydroxide a light brown precipitate forms: which gets darker as it stands in air:

$$[Mn(H_2O)_6]^{2+}(aq) + 2OH^-(aq) \rightarrow [Mn(H_2O)_4(OH)_2](s) + 2H_2O(l)$$

With excess ammonia the complex reacts in the same way as hydroxide ions. These precipitates do not dissolve:

$$[Mn(H_2O)_6]^{2+}(aq) + 2NH_3(aq) \rightarrow [Mn(H_2O)_4(OH)_2](s) + 2NH_4^+(aq)$$

## Chromium(III) ions

$[Cr(H_2O)_6]^{3+}$ is a violet solution.

With excess aqueous sodium hydroxide, a grey-green precipitate forms, then dissolving in excess aqueous sodium hydroxide forms a dark green solution:

$$[Cr(H_2O)_6]^{3+}(aq) + 3OH^-(aq) \rightarrow [Cr(H_2O)_3(OH)_3](s) + 3H_2O(l)$$

With excess aqueous ammonia this forms a green precipitate, which further dissolves in ammonia to give a complex ion in a dark purple solution:

$$[Cr(H_2O)_6]^{3+}(aq) + 3NH_3\ (aq) \rightarrow [Cr(H_2O)_3(OH)_3](s) + 3NH_4^+(aq)$$

Then:

$$[Cr(H_2O)_3(OH)_3](s) + 6NH_3(aq) \rightarrow [Cr(NH_3)_6]^{3+}(aq) + 3H_2O(l) + 3OH^-(aq)$$

## Iron(II) and iron(III) ions

Iron(II) ions can be oxidised by reacting with acidified manganate(VII) ions. This reaction is used in redox titrations: the purple manganate(VII) ions will be reduced to colourless manganese(II) ions.

Orange-brown iron(III) ions are reduced to pale green iron(II) ions by the reaction with iodide ions. As iodine is formed a brown colour appears.

## Chromium(III) and dichromate(VI) ions

Green chromium(III) ions can be oxidised by hot alkaline hydrogen peroxide to orange dichromate(VI) ions. The reverse reduction reaction, from dichromate(VI) ions back to chromium(III) ions, happens with acidified zinc.

Green chromium(III) ions can be oxidised by hot alkaline hydrogen peroxide to orange dichromate(VI) ions. The reverse reduction reaction, from dichromate(VI) ions back to chromium(III) ions, happens with acidified zinc.

## Copper ions

Copper undergoes **disproportionation**. This means that the same element (species) can be oxidised and reduced in the same reaction.

## Tests for anions

| Anion | Test |
|---|---|
| carbonate ions, $CO_3^{2-}$ | Addition of dilute nitric acid: bubbles of carbon dioxide formed. Carbon dioxide identified by limewater turning cloudy. |
| halide ions, $Cl^-$, $Br^-$, and $I^-$ | Addition of silver nitrate:<br>• Chloride ions form a white precipitate (AgCl) which is soluble in dilute ammonia.<br>• Bromide ions form a cream precipitate (AgBr), soluble in concentrated ammonia.<br>• Iodide ions form a yellow precipitate (AgI), insoluble in concentrated ammonia. |
| sulfate ions, $SO_4^{2-}$ | Addition of barium chloride forms a white precipitate of $BaSO_4$. Acidified conditions are needed to ensure removal of carbonate ions. |

## Tests for cations

Identification of various cations can be done in a test tube.

The identification of ammonium ions, $NH_4^+$, is done by adding hydroxide ions and then warming. A positive result will release ammonia gas, which can be tested for using damp *red* litmus paper, which turns *blue*.

For the transition metals that form coloured complexes, their reactions with dilute sodium hydroxide and ammonia should allow for the identification of $Cu^{2+}$, $Fe^{2+}$, $Fe^{3+}$, $Mn^{2+}$, and $Cr^{3+}$.

# Retrieval

Learn the answers to the questions below, then cover the answers column with a piece of paper and write as many as you can. Check and repeat.

| | Questions | Answers |
|---|---|---|
| 1 | What is the shape of a complex ion with monodentate ligands and a coordination number of 4? | tetrahedral or square planar |
| 2 | How can a transition metal be defined? | an ion with a incomplete d-sub-shell |
| 3 | Which shape of complex ion can show *cis–trans* isomerism and optical isomerism? | octahedral |
| 4 | What colour is Ni(II) in water? | green |
| 5 | Which transition metal is used as a catalyst in the Contact process? | vanadium (as in vanadium(V) oxide) |
| 6 | What is the colour of $[Cr(H_2O)_6]^{3+}$? | violet or green |
| 7 | What is the test to identify ammonium ions? | mix with hydroxide ion and warm; ammonia gas is released, which turns damp red litmus paper blue |
| 8 | What colour precipitate will show a positive test for iodide ions when silver nitrate is added? | yellow |
| 9 | What is the central ion in haemoglobin? | $Fe^{2+}$ |
| 10 | What is the colour change that happens when $[Cu(H_2O)_6]^{2+}$ undergoes a partial ligand substitution with ammonia? | pale blue solution to dark blue precipitate |
| 11 | What is the formula of two bidentate ligands? | $NH_2CH_2CH_2NH_2$ and $C_2O_4^{2-}$ |
| 12 | What colour is Co(II) in water? | pink |
| 13 | What type of isomerism can square planar complex ions show? | *cis–trans* or *E-Z* |
| 14 | What complex ion is formed when copper(II) ions are mixed with excess ammonia? | $[Cu(H_2O)_2(NH_3)_4]^{2+}(aq)$ |
| 15 | What colour is Cr(VI) in water? | orange |
| 16 | When might an intermediate be formed in a reaction? | when the activation energy of individual steps is lower than the overall reaction's activation energy |
| 17 | How are sulfate ions identified? | a white precipitate of $BaSO_4$ forms after addition of barium chloride |
| 18 | Which isomer of $[Pt(NH_3)_2Cl_2]$ can be used as a cancer treatment? | *cis* |

Put paper here

Put paper here

Put paper here

Put paper here

Put paper here

Put paper here

Put paper here

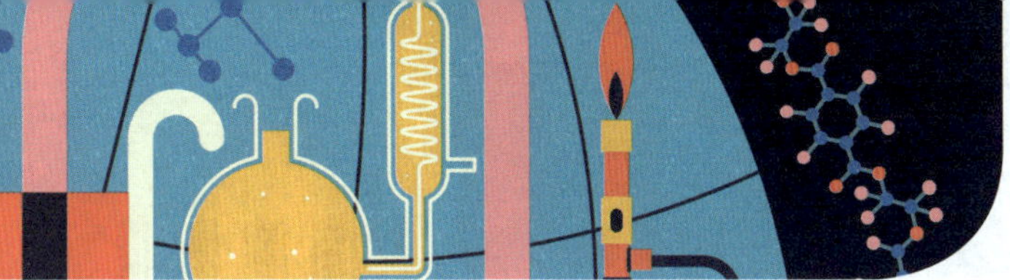



| | | | |
|---|---|---|---|
| 19 | How can chromium(III) be oxidised? | Put paper here | hot alkaline hydrogen peroxide |
| 20 | What is the shape of complex ions with monodentate ligands and a coordination number of 6? | | octahedral |

# Practical skills

Practise your practical skills using the worked example and practice questions below.

## Qualitative analysis

Qualitative analysis is very important to help determine which ions are in a compound. It is important not just to learn the tests and their results, but also why each test produces the results that it does. It is likely that you will be asked for chemical equations to explain the results of experiments and often this may be by giving ionic equations for the reactions. Remember to balance them and include state symbols.

## Worked example

A white solid is tested to see which ions are presents.

**Questions**

1 The student adds $HCl(aq)$ followed by $BaCl_2(aq)$. A white precipitate forms.

a Identify the anion present and write an equation for the formation of the precipitate.

b Explain why $HCl(aq)$ was added first.

2 To identify the cation, $NaOH(aq)$ is added. No precipitate forms. The solution is then heated gently and ammonia is given off.

a How could $NH_3(g)$ be identified?

b Write an equation for the formation of $NH_3(g)$.

**Answers**

1 a anion = $SO_4^{2-}$

$Ba^{2+}(aq) + SO_4^{2-}(aq) \rightarrow BaSO_4(s)$

b To remove any $CO_3^{2-}$ that might be present, which would also give a white ppt.

2 a Moist red litmus paper would turn blue.

b $NH_4^+(aq) + OH^-(aq) \rightarrow NH_3(g) + H_2O(l)$

## Practice

A student is provided with an unlabelled mixture of two ionic compounds. One is soluble in water and the other is not. The student decides to separate them using simple filtration.

1 Explain why separating the two compounds is a sensible idea.

2 The soluble compound forms a green precipitate when $NaOH(aq)$ is added and forms a yellow precipitate when $AgNO_3(aq)$ is added. Identify the soluble compound and write equations for the two precipitation reactions.

3 The insoluble compound is mixed separately with $NaOH(aq)$, $HCl(aq)$, and $AgNO_3(aq)$. A blue precipitate is formed with $NaOH(aq)$, effervescence is seen with $HCl(aq)$, and no visible change happens with $AgNO_3(aq)$. Identify the insoluble compound and write ionic equations for the reactions that happen.

# Practice

## Exam-style questions

1 Iron is a transition metal that can have +2 and +3 oxidation states. It forms complex ions.

(a) Give an equation to show the reaction when an excess of hydrochloric acid is added to an aqueous solution of iron(II) ions.

[1]

(b) State the initial and final shapes of the iron complex ion and explain why the shape had to change.

.................................................................................

.................................................................................

.................................................................................

.................................................................................

.................................................................................

.................................................................................

.................................................................................

.................................................................................

.......................................................................... [4]

**Exam tip**

There is no shortcut to learning the shapes of the complex ions, it just takes practice.

(c) (i) State the change in entropy that occurs during the reaction in **1 (a)**.

.................................................................................

.................................................................................

.......................................................................... [1]

(ii) State the type of isomerism that the complex formed in **1 (b)** will exhibit.

.................................................................................

.................................................................................

.......................................................................... [1]

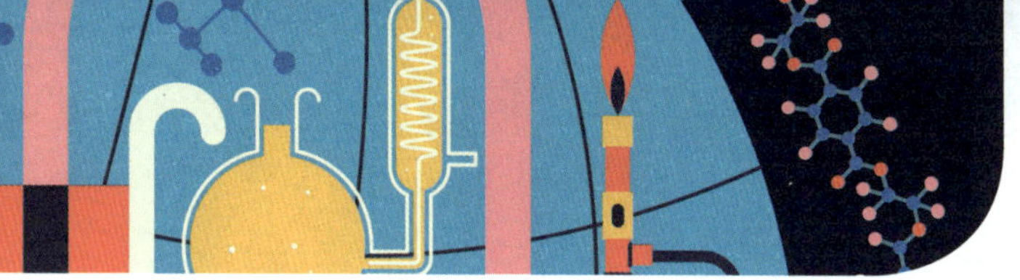



2 Iron is a transition metal that is found inside the haemoglobin of red blood cells.

2.1.3

(a) State the function of haemoglobin in the blood.

.......................................................................................

.......................................................................................

....................................................................................... [1]

(b) Explain why carbon monoxide is toxic.

Refer to haemoglobin in your answer.

.......................................................................................

.......................................................................................

.......................................................................................

.......................................................................................

....................................................................................... [2]

(c) Carbon monoxide becomes toxic to humans when it is above 0.01% of the air being breathed. A boiler is malfunctioning in a holiday apartment that has a volume of 200 $m^3$.

Calculate the mass of methane burnt for the carbon monoxide levels to reach toxic levels at 25 °C.

The gas constant, $R = 8.31\ kJ\ K^{-1}\ mol^{-1}$

mass of methane burnt = ..................... [6]

**3** The platinum complex *cis*-platin is an effective anticancer drug. This is because it binds with the DNA in cancer cells, preventing cell division.

**(a)** State the type of isomerism found in *cis*-platin.

..............................................................................................

..............................................................................................

.............................................................................................. **[1]**

**(b)** State the shape of the *cis*-platin complex ion.

..............................................................................................

..............................................................................................

.............................................................................................. **[1]**

**(c)** Suggest a reason why *trans*-platin is ineffective at treating cancer.

..............................................................................................

..............................................................................................

..............................................................................................

..............................................................................................

.............................................................................................. **[2]**

**(d)** Explain why *cis*-platin is only administered in small doses.

..............................................................................................

..............................................................................................

..............................................................................................

..............................................................................................

.............................................................................................. **[2]**

**Exam tip**

Think carefully about the mechanism with which *cis*-platin disrupts DNA.

**4** Iron is a found in a variety of different minerals. It is mined as haematite, $Fe_2O_3$. Haematite is readily soluble in ethanedioic acid, a bidentate ligand.

**(a)** Define bidentate ligand.

..............................................................................................

..............................................................................................

..............................................................................................

.............................................................................................. **[2]**

**(b)** Give an ionic equation to show the reaction that occurs when iron(III) ions dissolve in a solution of ethanedioic acid to form a complex ion with ethanedioate ions.

..............................................................................................

..............................................................................................

..............................................................................................

.............................................................................................. **[1]**

**Synoptic link**

5.1.3

**Exam tip**

Make sure you learn the formulae of the bidentate ligands.

**(c)** Name the type of isomerism that is demonstrated by the complex ion formed in **4 (b)**.

Give a reason for your answer.

..............................................................................................

..............................................................................................

..............................................................................................

..............................................................................................

.............................................................................................. **[2]**

**(d)** Give an equation to show what would happen when dilute ammonia is added to an aqueous solution of iron(III) ions.

State any observations that would be made.

..............................................................................................

..............................................................................................

..............................................................................................

..............................................................................................

..............................................................................................

..............................................................................................

.............................................................................................. **[4]**

**(e)** To ensure the haematite is fully dissolved, the ethanedioic acid needs to be at a concentration of 1.50 mol $dm^{-3}$.

Calculate the pH of the acid solution used. The $K_a$ of ethanedioic acid is $5.4 \times 10^{-2}$ mol $dm^{-3}$. **[5]**

pH = ....................

**5** A student has a bottle containing a brownish solution labelled 'Iron sulfate'. They are unsure if the solution contains any iron(II) ions.

**(a)** State the formula for the iron(II) species that might be in the bottle.

..........................................................................................

..........................................................................................

.......................................................................................... **[1]**

**(b)** The pH of the solution was found to be 1.62.

Given that the $K_a$ for iron(III) ions is $5.76 \times 10^{-3}$ mol dm$^{-3}$, calculate the concentration of the iron in the bottle.

concentration of iron = ...................... mol dm$^{-3}$ **[4]**

**Exam tip**

Treat anything with a $K_a$ as you would ethanoic acid.

**6** An excess of a given reagent is added to each of the following pairs of aqueous metal ions. For each metal ion, state the initial colour of the solution and the final observation that you would make.

State an overall equation for the formation of the final product from the initial aqueous metal ion for each reaction.

**(a) (i)** An excess of dilute aqueous ammonia is added to separate aqueous solutions containing $[Cr(H_2O)_6]^3$ and $[Cu(H_2O)_6]^{2+}$.

..........................................................................................

..........................................................................................

..........................................................................................

..........................................................................................

..........................................................................................

..........................................................................................

..........................................................................................

..........................................................................................

.......................................................................................... **[4]**

**Exam tip**

You need to ensure you memorise all the colours and formulae of the ions in ligand exchange reactions.

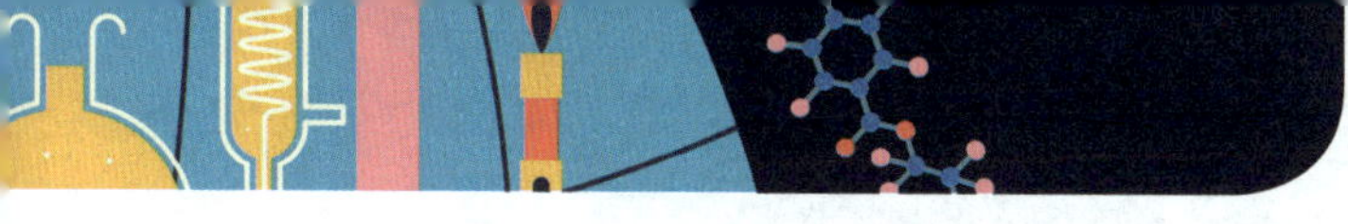



**(ii)** An excess of aqueous sodium hydroxide is added to separate aqueous solutions containing $[Mn(H_2O)_6]^{2+}$ and $[Cr(H_2O)_6]^{3+}$.

..............................................................................................

..............................................................................................

..............................................................................................

..............................................................................................

..............................................................................................

..............................................................................................

..............................................................................................

..............................................................................................

.............................................................................................. **[4]**

**(b)** Write an equation for the reaction that occurs when an aqueous solution of copper(II) ions has iodide ions added.

..............................................................................................

..............................................................................................

..............................................................................................

.............................................................................................. **[1]**

# Knowledge

## 20 Aromatic compounds

### Models for the structure of benzene

Benzene has the formula $C_6H_6$. It took a while to fully determine its structure because the earlier models proposed did not predict the properties of benzene correctly.

The **Kekulé model** suggested that benzene had an alternating structure of double and single bonds. However, this model did not explain all of benzene's properties.

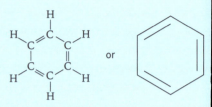

The **delocalised model** supports some of benzene's properties better than the Kekulé model. It shows that benzene has a delocalised ring of electrons which are represented as a ring.

This delocalising of p-orbital electrons makes benzene more stable than the hypothetical cyclohexa-1,3,5-triene.

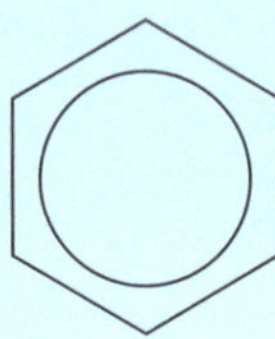

### Experimental evidence for the delocalised model

There are three main pieces of evidence that lead to the structure of benzene:

- its resistance to reactions
- the intermediate bond length
- its enthalpy change of hydrogenation.

Benzene is resistant to reactions and does not display characteristics you would expect from a compound that has double bonds. It does not undergo electrophilic addition reactions and does not decolourise bromine water.

If benzene had an alternating double bond and single bond structure, the double bonds would be 0.134 nm and the single bonds would be 0.153 nm. However, all the bonds within benzene are the same length of 0.139 nm, which is between the length of a double and single bond. This shows that benzene does *not* have an alternating structure.

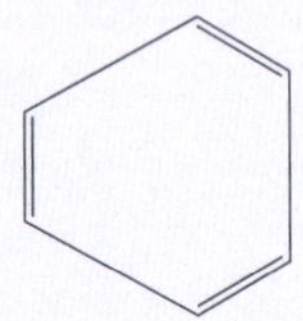

The enthalpy change of hydration for cyclohexane (with one double bond) is $\Delta_{hyd}H^{\ominus} = -120\ \text{kJ mol}^{-1}$, so an estimate for the enthalpy change of hydration for the hypothetical cyclohexa-1,3,5-triene would be $\Delta_{hyd}H^{\ominus} = -360\ \text{kJ mol}^{-1}$, three times larger. However, the enthalpy change of hydration for benzene is $\Delta_{hyd}H^{\ominus} = -208\ \text{kJ mol}^{-1}$, making benzene more stable than an alternating double bond structure.

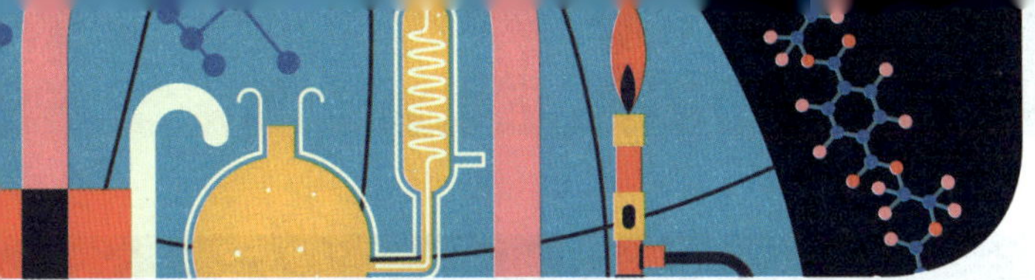



## Naming aromatic compounds

An **aromatic compound** contains a benzene ring. For aromatic compounds that are monosubstituted, the benzene ring is the longest carbon chain, so the smaller group is a prefix to benzene.

If benzene is substituted with a carbon chain longer than 6 carbon atoms or a carbon chain that has another functional group, then the benzene is a **substituent** of the main chain, and the prefix 'phenyl' is used.

If more than one group is substituted into a benzene ring, similar rules to naming a straight chain carbon can be followed. The lowest combination of numbers should be used and substituent groups should be listed in alphabetical order.

$C_2H_5$ — ethylbenzene

Cl — chlorobenzene

$NO_2$ — nitrobenzene

$CH_3$ / H — $-CH_2CH_2CH_2CH_2CH_2CH_3$

2-phenyloctane

$CH_3$, $NO_2$ — 2-nitromethylbenzene

$CH_3$, $O_2N$, $NO_2$, $NO_2$ — 2,4,6-trinitromethylbenzene

## Reactivity of aromatic compounds

Benzene is more resistant to the addition of bromine than other alkenes; it will not decolourise bromine water.

Electrophilic substitution reactions will take precedence over addition reactions in benzene for two main reasons:

1. The ring structure remains intact during most reactions because the aromatic ring is very stable, so a large amount of energy is needed for the ring system to break.
2. The high electron density of the π-system in the centre of the ring means that electrophiles are readily attracted to this area, more than the localised electron density of the π-bond in alkenes.

## Electrophilic substitution

The nitration of benzene is an important reaction. It is the starting point in the manufacture of explosives and amines.

In the process an $NO_2$ group is substituted for a hydrogen on the benzene ring. The first step is the generation of the $NO_2{}^+$ electrophile from concentrated nitric acid and concentrated sulfuric acid.

nitric acid + $H_2SO_4$ (sulfuric acid) $\rightleftharpoons$ … + $HSO_4{}^-$ $\rightleftharpoons$ $H_2O + O{=}N^+{=}O$ (nitronium ion $NO_2{}^+$)

**Step 1:** generation of electrophile

**Step 2**

**Step 3:** hydrogen is released form benzene ring and delocalised ring reforms

Halogenation of benzene can be carried out via a **Halogen carrier**. These could be $AlBr_3$ or $FeBr_3$ for **bromination** or $AlCl_3$ or $FeCl_3$ for **chlorination**. The first step is to generate a Halogen electrophile that will be attracted to the electron-dense ring, for example:

$$Br_2 + FeBr_3 \rightarrow FeBr_4 + Br^+$$

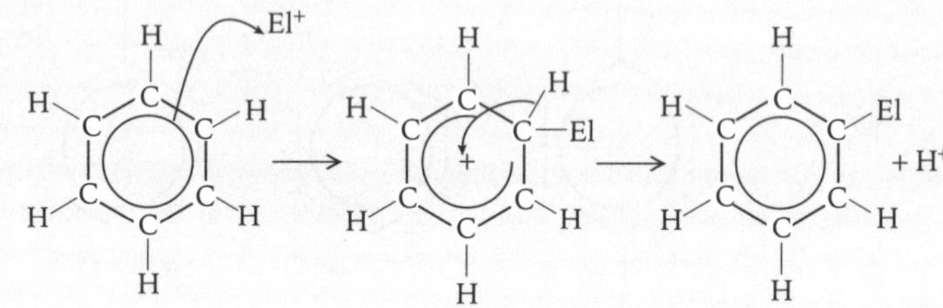

**Friedel–Crafts acylation reactions** add acyl chlorides to benzene rings using an $AlCl_3$ catalyst. **Friedel–Crafts alkylation reactions** use a haloalkane and add an alkyl group to a benzene ring.

**Step 1**

$$CH_3COCl + AlCl_3 \rightarrow CH_3C^+{=}O + AlCl_4{}^-$$

**Step 2**

**Step 3**

$+ H^+$

## Acidity of phenol

Phenols can act as weak acids: the phenol partially dissociates to form a phenoxide ion and a hydrogen ion.

Phenols are less acidic than carboxylic acids but more acidic than alcohols.

- Phenols react with aqueous sodium hydroxide (a strong base) in a neutralisation reaction.
- Phenols do not react with carbonates (a weak base).

## Reactivity of phenols compared to benzene

Phenols are more reactive than benzene due to phenol easily undergoing electrophilic substitution reactions, due to the lone pair of electrons donated by the p-orbital of the oxygen atom in the hydroxyl group. These are added to the delocalised $\pi$-system of electrons making it more susceptible to electrophiles than benzene.

## Nitration of phenol

Phenol reacts with dilute nitric acid at room temperature to form a mixture of 2-nitrophenol and 4-nitrophenol.

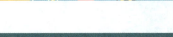

## Bromination of phenol

Phenol will react with bromine water. It undergoes an electrophilic substitution reaction to form a white precipitate of 2,4,6-tribromophenol. This is carried out *without* a Halogen carrier and at room temperature.



## Directing group effects

Organic synthesis routes to build particular compounds can be planned using **directing groups** on benzene rings.

–OH and $–NH_2$ are 2-, 4-directing groups as they are electron-donating groups, whereas $–NO_2$ is a 3-directing group as it is electron withdrawing.



| 2- and 4- directing groups | 3- directing groups |
|---|---|
| $–NH_2$ | RCOR |
| –NHR | –COOR |
| –OH | $–SO_3H$ |
| –OR | –CHO |
| –R | –COOH |
| $–C_6H_6$ | –CN |
| –Halogen | $–NO_2$ |
| | $–NR_3^+$ |

For example, when going from benzene to 1-chloro-4-nitrobenzene the order of the substitution is important.

The $NO_2$ group is 3-directing, so if the nitro group was substituted first, the subsequent substitution of chlorine would be in the 3-position, forming 1-chloro-3-nitrobenzene.

To ensure that both groups end up in the correct position, the Cl group needs to be substituted first as it is 4-directing.

# Retrieval

Learn the answers to the questions below, then cover the answers column with a piece of paper and write as many as you can. Check and repeat.

| | Questions | | Answers |
|---|---|---|---|
| 1 | What was the hypothetical name for benzene? | Put paper here | cyclohexa-1,3,5-triene |
| 2 | What is the name of the product that is produced when phenol reacts with bromine water? | | 2,4,6-tribromophenol |
| 3 | What is the the bond length of the bonds in benzene compared to single and double bonds? | Put paper here | an intermediate length |
| 4 | What are the reactants needed to make 2-nitrophenol and 4-nitrophenol? | | phenol and dilute nitric acid |
| 5 | Where does the $NO_2^+$ electrophile come from in a nitration reaction? | | concentrated nitric and and concentrated sulfuric acid |
| 6 | Which position does –CHO direct to? | Put paper here | 3 |
| 7 | How is phenylamine synthesised? | | nitration of benzene, then reduction of nitrobenzene with tin and hydrochloric acid as a catalyst |
| 8 | Does phenol react with NaOH? | | yes |
| 9 | What reaction can be used to add a carbon chain onto a benzene ring? | Put paper here | Friedel–Crafts alkylation reactions |
| 10 | Which position does –OH direct to on a benzene ring? | | 2 or 4 |
| 11 | Which three pieces of evidence are used as proof of the structure of benzene? | Put paper here | • its enthalpy change of hydrogenation<br>• the intermediate bond length<br>• its resistance to reactions |
| 12 | What does phenol dissociate to? | | a phenoxide ion and a hydrogen ion |
| 13 | What type of reaction is most likely with benzene? | Put paper here | electrophilic substitution |
| 14 | What is the difference between benzene and phenol substitution reaction with Halogens? | | benzene needs a Halogen carrier, phenols do not |
| 15 | What is the catalyst needed when acyl chlorides are added to benzene rings? | | $AlCl_3$ |
| 16 | Why can phenol undergo electrophilic substitution reactions more easily than benzene? | Put paper here | it has one lone pair from oxygen in the –OH bond added to the delocalised electron ring |
| 17 | Which position does the –CN group direct to? | | 3 |

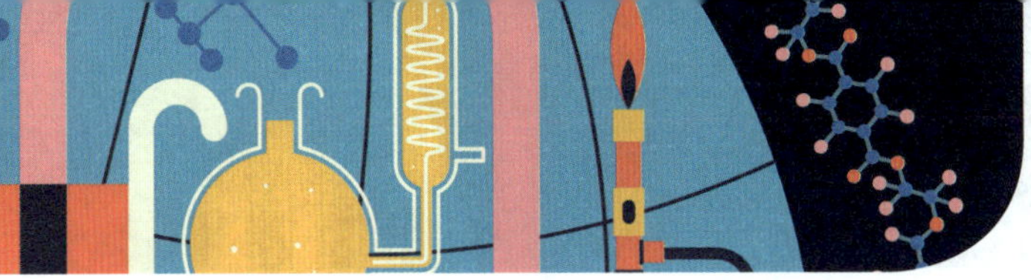



| | | |
|---|---|---|
| 18 | What is the formula for benzene? | $C_6H_6$ |
| 19 | Which are more acidic: phenols or carboxylic acids? | carboxylic acids |
| 20 | Does phenol react with carbonates? | no |

Put paper here

## Maths skills

Practise your maths skills using the worked example and practice questions below.

### Drawing 2-D and 3-D shapes of molecules

Molecules and structures exist in three dimensions and yet as chemists these structures need to be represented in two dimensions on the page. Bonds, wedges, and dashes can show bonds going into the page (away from you) and out of the page (towards you).

Wedges ( ) represent a bond coming towards you.

Dashes ( ) represent a bond going away from you.

### Worked example

**Question**

Draw the skeletal formula of 2-chlorocyclohexanol showing the two functional groups on opposite sides of the carbon ring.

**Answer**

When drawing organic structures, always start with the end of the word. Cyclohexanol looks like this:

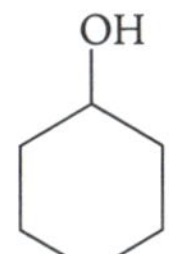


The 2-chloro part needs to be bonded to a neighbouring carbon:

Now add dashes and wedges to show the OH and Cl are facing in opposite directions. There are two ways to do this:

Both of these are correct; in fact, they are optical isomers of each other.

### Practice

1 Draw the following molecules based on the information provided.

a 1,3,5-tribromohexane with all the bromine atoms on the same side of the carbon ring.

b 3,3-dimethylbutan-1-ol showing the tetrahedral arrangement of methyl groups on carbon-3.

c 3,4-dimethylhexane with two chiral centres on the carbon-3 and carbon-4. Use wedges and dashes to show the three optical isomers of 3,4-dimethylhexane.

2 Draw the following inorganic molecules:

a *cis*- and *trans*-isomers of $[Cu(NH_3)_4(H_2O)_2]^{2+}$

b *cis*- and *trans*-isomers of $[Pt(NH_3)_2(Cl)_2]$

c Two optical isomers of $[Co(en)_3]^{2+}$ where 'en' is ethane-1,2-diamine.

## Exam-style questions

1 Benzene, $C_6H_6$, shown in **Figure 1.1**, is an organic compound that is used as the starting point for the synthesis of many useful substances in the chemical industry.

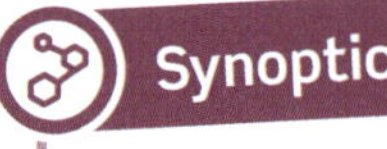


4.1.3

**Figure 1.1**

**(a)** Describe the bonding in a benzene molecule, including its shape and structure.

Compare the bond length between carbon atoms with that of cyclohexane.

.......................................................................................................

.......................................................................................................

.......................................................................................................

....................................................................................................... **[3]**

**(b)** Data given in **Figure 1.2** shows the energy required for the hydrogenation of cyclohexene, and the hydrogenation of benzene. The molecule cyclohexatriene does not exist and is described as being hypothetical.

Exam tip

Do not forget to show your mathematical workings here!

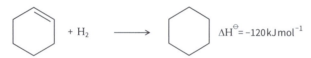

$+ H_2 \longrightarrow$ $\Delta H^{\ominus} = -120\,kJ\,mol^{-1}$

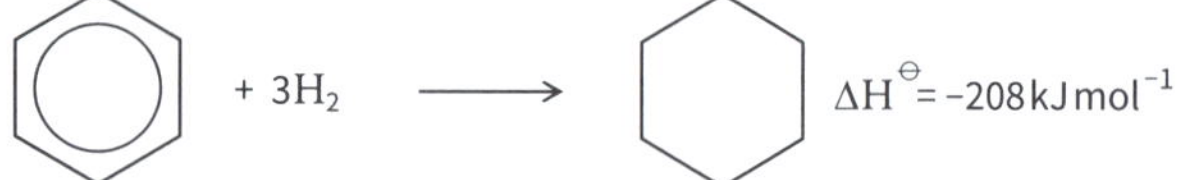


**Figure 1.2**

Explain why the energy required for the hydrogenation of benzene is not three times the energy required for the hydrogenation of cyclohexene.

State what that implies about their relative stabilities.

.......................................................................................................

.......................................................................................................

.......................................................................................................

.......................................................................................................

....................................................................................................... **[4]**

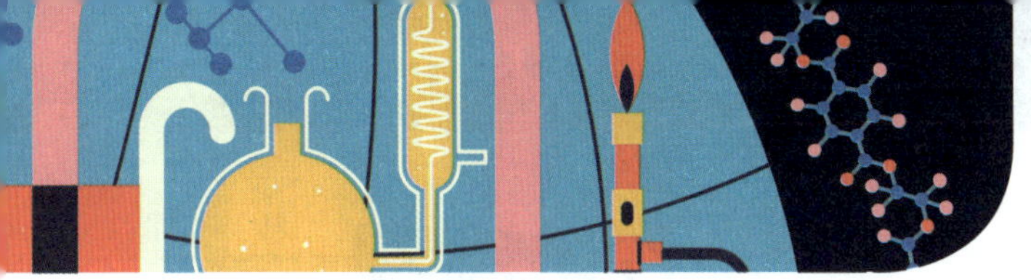



**(c)** A simple test-tube reaction can distinguish between cyclohexene and benzene.

State the reagent needed for this.

Describe the results you would expect to see for both cyclohexene and benzene if this reagent were added, and the test tubes shaken.

Reagent: ..........................................................................................

Observation with cyclohexene: ......................................................................

Observation with benzene: ..................................................................**[3]**

**2** An equation for the formation of an amine from benzene is shown in a two-step synthesis. The first step is shown in **Figure 2.1**.

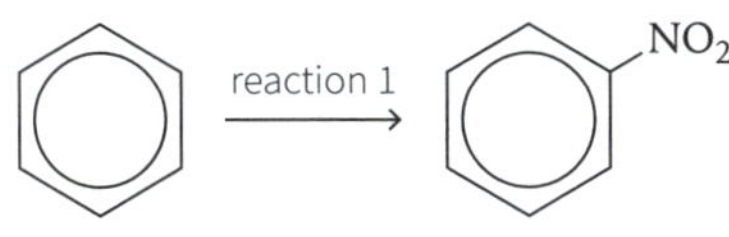


**Figure 2.1**

**(a)** Reaction 1 proceeds by the creation of an electrophile that attacks the benzene.

**(i)** Give the reagent(s) needed to produce this electrophile and write an equation showing its formation.

Reagent: ..........................................................................................

Equation: .................................. ..................................**[2]**

**(ii)** Outline the mechanism showing the attack of this electrophile on benzene.

**[3]**

**(iii)** What type of reaction is reaction 1?

........................................................................................... **[1]**

**(b)** Friedel–Crafts acylation can also be carried out on benzene using the catalyst $AlCl_3$, as shown in the reaction in **Figure 2.2**.

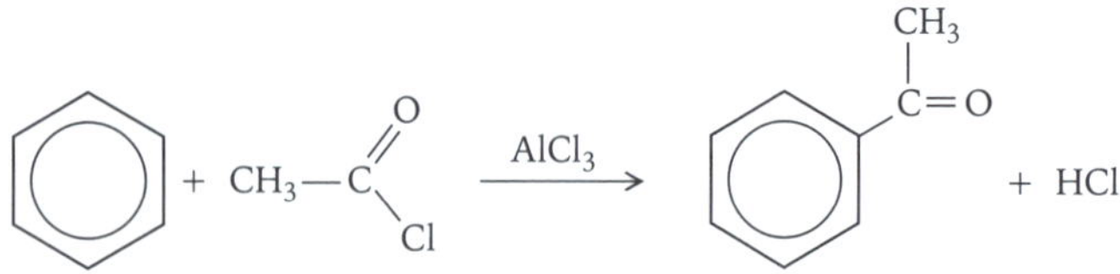


**Figure 2.2**

Draw the shape of an $AlCl_3$ molecule and explain, using VSEPR theory, why it is this shape. Aluminium chloride is a molecule with all its bonds covalent.

> **Exam tip**
>
> Do not forget to state the name of the shape of this molecule in your answer.

Explanation:..............................................................................

..............................................................................

..............................................................................

.............................................................................. **[4]**

**3** Benzene is more likely to undergo substitution reactions than addition reactions.

> **Synoptic links**
>
> 4.1.1 2.1.3

**(a)** Suggest why this is.

..............................................................................

.............................................................................. **[1]**

**(b)** Reagent **B** has the formula $CH_3COCl$. A reaction of benzene with reagent **B** is given in **Figure 3.1**.

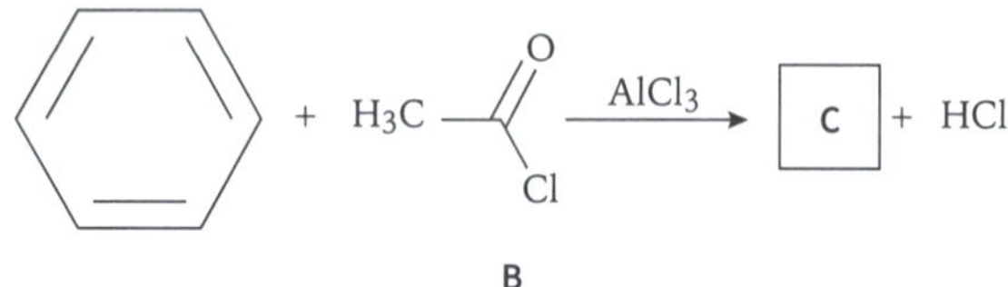


**Figure 3.1**

**(i)** Give the systematic name of reagent **B**.

..............................................................................

.............................................................................. **[1]**

(ii) Draw the structure of the product **C**.

**[1]**

(c) 0.075 mol of benzene is combined with 8 g of reagent **B** to produce 7.8 g of product **C**.

Calculate the percentage yield of the reaction. State which reagent is the limiting reagent.

**Exam tip**

Do not forget the limiting reagent will dictate how much product is produced.

percentage yield = .................... **[5]**

**4** Benzene, shown in **Figure 4.1**, is more resistant to bromination than cyclohexene, also shown in **Figure 4.1**.

**Synoptic link**

4.1.1

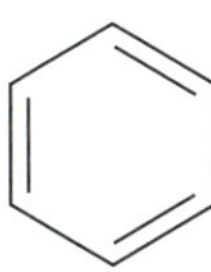

benzene

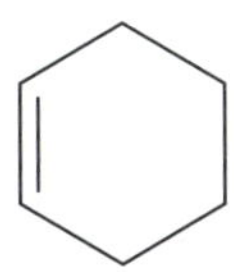

cyclohexene

**Figure 4.1**

(a) Explain why benzene is more resistant to bromination than cyclohexene.

..........................................................................................

..........................................................................................

..........................................................................................

.................................................................................. **[3]**

(b) Give the reagent(s) needed for the bromination of benzene.

..........................................................................................

.................................................................................. **[1]**

**(c)** Outline the mechanism for the bromination of benzene.
Use **X**$^+$ to represent your electrophile.

Students often forget the last step of this mechanism; make sure you finish it off.

**[3]**

**(d)** Give the systematic name of the product formed.

..............................................................................................

.............................................................................................. **[1]**

**5** Phenol, shown in **Figure 5.1**, reacts with bromine, $Br_2$, by electrophilic addition.

6.1.1 2.1.3

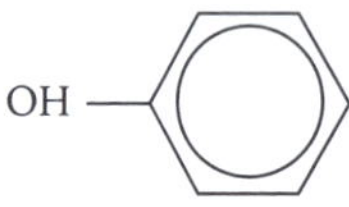


**Figure 5.1**

**(a)** Draw and name the product formed in this reaction.

.............................................................................................. **[2]**

**(b)** Explain why benzene is more resistant to electrophilic addition than phenol.

..............................................................................................

..............................................................................................

.............................................................................................. **[3]**

**(c)** Calculate the volume of bromine gas, in $cm^3$, measured at RTP, needed to react 2350 mg of phenol.

Show your working.

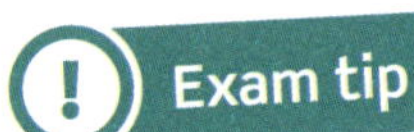


What volume does 1 mole of any gas take up, at RTP?

How many moles of bromine are needed for this amount of phenol?

volume = ..................... $cm^3$ **[5]**

6 A student has a sample of each of three organic compounds, **X**, **Y**, and **Z**, and their structures are shown in **Figure 6.1**.

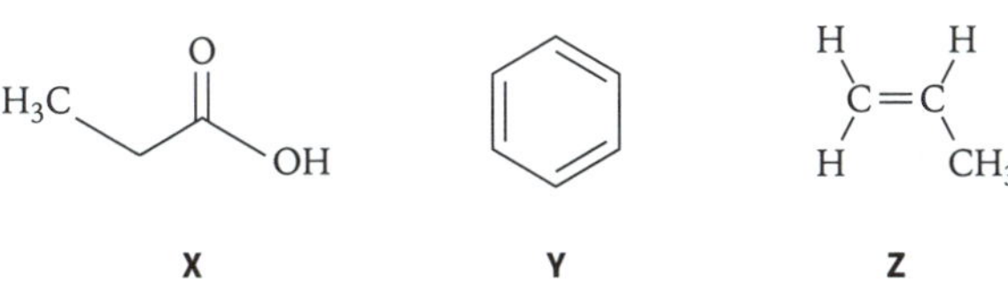


**Figure 6.1**

**Synoptic links**

6.3.1 6.3.2 4.1.1 4.2.4

**(a)** Describe the chemical tests needed to identify the functional groups on compounds **X** and **Z**.

Include reagents and conditions and any observations you would expect.

..............................................................................................

..............................................................................................

..............................................................................................

.............................................................................................. **[3]**

**(b)** A student has NMR spectra from the three compounds.

**(i)** Deduce the number of peaks they would expect to see in the $^{1}H$ NMR spectrum of compound **X**.

..............................................................................................

.............................................................................................. **[1]**

**Exam tip**

How many different H environments are there in this molecule?

**(ii)** Deduce the number of peaks they would expect to see in the $^{1}H$ NMR spectrum of compound **Y**.

..............................................................................................

.............................................................................................. **[1]**

**(iii)** Deduce the number of peaks they would expect to see in the $^{1}H$ NMR spectrum of compound **Z**.

..............................................................................................

.............................................................................................. **[1]**

**(c)** They then collect IR spectra for compounds **X** and **Y**.

Using the *Data Sheet,* explain the differences in the infrared spectra of compounds **X** and **Y**.

..............................................................................................

..............................................................................................

..............................................................................................

.............................................................................................. **[3]**

**(d)** Give the systematic name for compound **X**.

..............................................................................................

.............................................................................................. **[1]**

# Knowledge

## 21 Carbonyl compounds

### Reactions of carbonyl compounds

Aldehydes and ketones are **carbonyl compounds**: they both contain the C=O functional group, although in different places.

The C=O group is located for:

- **aldehydes**, at the end of the carbon chain, for example, propanal
- **ketones**, joined to two carbon atoms in the carbon chain, for example, propanone.

$CH_3CH_2-\underset{\substack{|\\H}}{C}=O$

propanal, an aldehyde

$H-\underset{\substack{|\\H}}{\overset{\substack{H\\|}}{C}}-\overset{\substack{O\\\|}}{C}-\underset{\substack{|\\H}}{\overset{\substack{H\\|}}{C}}-H$

propanone, a ketone

### Oxidation reactions

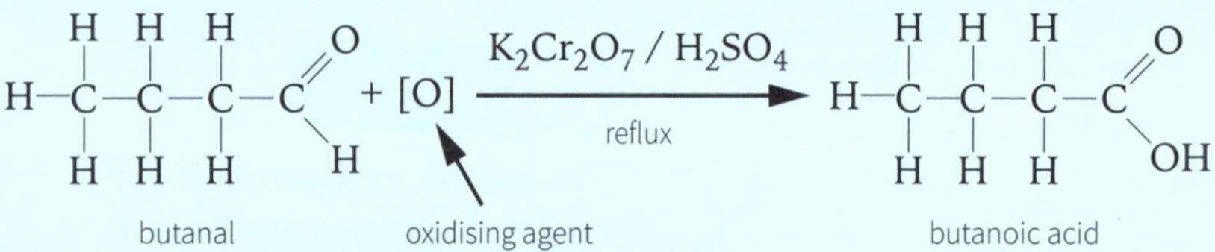


Aldehydes can be oxidised to carboxylic acids when they are refluxed with $Cr_2O_7^{2-}/H^+$; for example, with acidified potassium dichromate.

Ketones cannot undergo oxidation reactions.

### Nucleophilic addition reactions

As the C=O bond is polar, $C^{\delta+}=O^{\delta-}$, carbonyl compounds react with some nucleophiles by **nucleophilic addition**. The nucleophile is attracted to the slightly positive carbon and is added across the double bond.

### Reaction with $NaBH_4$

Aldehydes are reduced by warming with $NaBH_4$ in a nucleophilic addition reaction to a *primary alcohol*.

$CH_3CH_2CH_2CHO + 2[H] \xrightarrow{NaBH_4\ /\ H_2O} CH_3CH_2CH_2CH_2OH$

butanal — reducing agent — butan-1-ol

$CH_3COCH_3 + 2[H] \xrightarrow{NaBH_4\ /\ H_2O} CH_3CH(OH)CH_3$

propanone — reducing agent — propan-2-ol

Ketones are reduced by $NaBH_4$ in a nucleophilic addition reaction to a *secondary alcohol*.

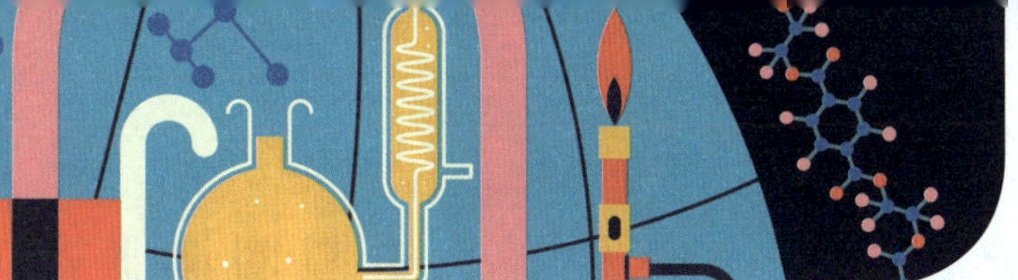



In this reaction the nucleophile is the $OH^-$ ion, which has a lone pair of electrons.

1. The lone pair is attracted to the slightly positive carbon and forms a dative covalent bond.
2. The $\pi$-bond breaks by **heterolytic fission** and forms a negatively charged intermediate.
3. The intermediate is protonated by a hydrogen atom in a water molecule to form an alcohol.

$$H_3C-\overset{\delta+}{C}(=\overset{\delta-}{O})-CH_3 \;+\; :H \longrightarrow H_3C-C(\ddot{O}^-)(H)-CH_3 \xrightarrow{\overset{\delta+}{H}-\overset{\delta-}{O}-\overset{\delta+}{H}} H_3C-C(OH)(H)-CH_3 + OH^-$$

intermediate forms

intermediate is protonated

## Reactions with HCN

$$CH_3CH_2CHO + HCN \xrightarrow{H_2SO_4/NaCN} CH_3CH_2CH(OH)CN$$

hydroxynitrile

Hydrogen cyanide, HCN adds across the C=O bond. As HCN is too dangerous to be used directly, acidified sodium cyanide is used to provide it indirectly.

This is a useful way of increasing the length of a carbon chain.

In this reaction the nucleophile is $CN^-$ ion, which has a lone pair of electrons.

1. The lone pair is attracted to the slightly positive carbon and forms a dative covalent bond.
2. The π-bond breaks by heterolytic fission and forms a negatively charged intermediate.
3. This is then protonated by water or $H^+$ ions to form hydroxynitrile.

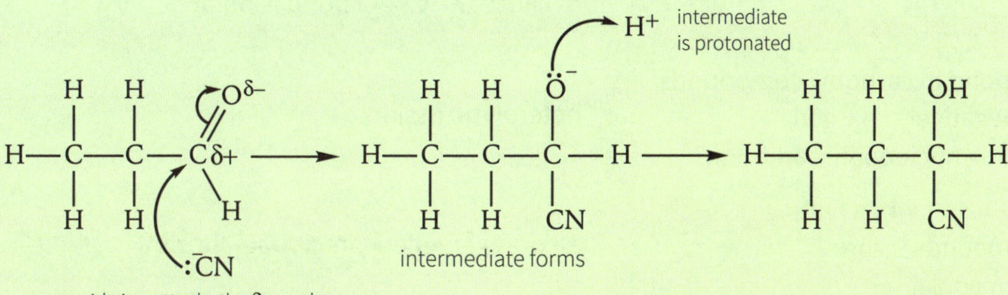


cyanide ion attacks the δ+ carbon atom and forms a covalent bond

## Tests for carbonyl groups

2,4-dinitrophenylhydrazine is used to detect the presence of a carbonyl group in an organic compound.

A positive result gives an orange/yellow precipitate. The melting point of the derivate can then be compared to reference samples to determine the identity.

Ketones and aldehydes can be differentiated by Tollens' reagent, or ammoniacal silver nitrate. Aldehydes oxidise to carboxylic acids and the silver ions reduce to silver; a distinctive silver mirror is produced.

# Retrieval

Learn the answers to the questions below, then cover the answers column with a piece of paper and write as many as you can. Check and repeat.

| | Questions | Answers |
|---|---|---|
| 1 | What are aldehydes oxidised to? | carboxylic acids |
| 2 | Which part of a carbonyl is a nucleophile attracted to? | $C^{\delta+}$ |
| 3 | What does $NaBH_4$ reduce an aldehyde to? | a primary alcohol |
| 4 | What does $NaBH_4$ reduce a ketone to? | a secondary alcohol |
| 5 | Which reagent can be used to distinguish between aldehydes and ketones? | Tollens' reagent |
| 6 | What species acts as the nucleophile in the reaction between carbonyl compounds and $NaBH_4$? | $H^-$ |
| 7 | How is HCN safely added to a reaction? | via acidified sodium cyanide |
| 8 | What reagent can be used to determine the presence of a carbonyl group? | 2,4-dinitrophenylhydrazine |
| 9 | What is the functional group in both aldehydes and ketones? | C=O |
| 10 | What is the product of the reaction between a carbonyl compound and hydrogen cyanide? | hydroxynitrile |
| 11 | Why can carbonyl compounds react by nucleophilic addition? | because the C=O bond is polar |
| 12 | How does the π-bond in carbonyl compounds break to form a negatively charged intermediate during nucleophilic addition? | heterolytic fission |
| 13 | What is a positive result when testing for carbonyl compounds using 2,4-dinitrophenylhydrazine? | an orange/yellow precipitate forms |
| 14 | What is the suffix used when naming aldehydes? | -anal |
| 15 | What is the suffix used when naming ketones? | -anone |
| 16 | Where is the C=O group located in an aldehyde? | at the end of the carbon chain |
| 17 | Where is the C=O group located in a ketone? | joined to two carbon atoms in the carbon chain |

Put paper here

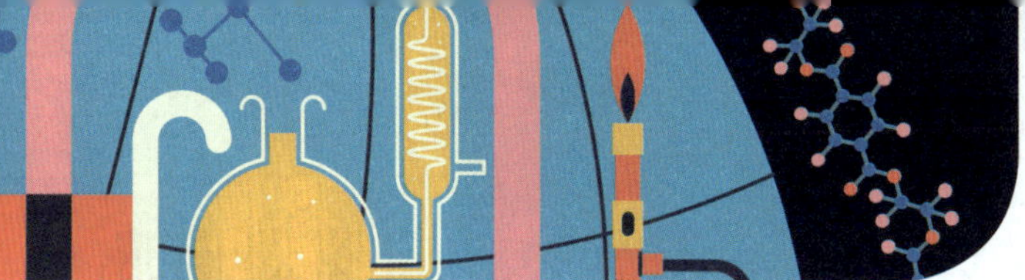



| | | Put paper here | |
|---|---|---|---|
| 18 | What reagent can be used to oxidise aldehydes? | | acidified potassium dichromate. |
| 19 | Can ketones be oxidised? | | no |
| 20 | In the reaction between a carbonyl and HCN, what acts as a nucleophile? | | $CN^-$ ion |

## Maths skills

Practise your maths skills using the worked example and practice questions below.

### Identifying chiral centres

Symmetry is essential to working out if two molecules are isomers and what types of isomerism they show. For A level you need to identify chiral centres in organic molecules to show they have optical isomerism and draw them.

Optical isomers are non-super-imposable mirror images of each other.

This happens when a carbon atom (or any atom) is bonded to four different groups. This way, there is no way of rotating a molecule to make it look like its mirror image.

The simplest way to work out if a molecule has chiral centres is to see if it has four different groups bonded to a carbon atom. When looking at skeletal formulae it is important to remember the missing hydrogen atoms.

### Worked example

**Question**

Identify the chiral centre in the molecule of 1-chloro-2methylbutane below:

**Answer**

The chiral centre is on the second carbon atom (in the middle of the diagram). This is because it has four different groups attached to it: methyl, ethyl, chloromethyl, and a hydrogen. The hydrogen is not shown so you need to remember to imagine it. The two diagrams below show the optical isomers:

### Practice

Find the chiral centres in each molecule below. Some may have more than one. In each case, draw (using wedges and dashes) the optical isomers.

If possible, use a molymod kit to build these.

**1**

**2**

**3**

# Practice

## Exam-style questions

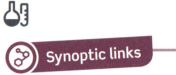

**1** Compound **B**, $CH_3CHO$, can react by nucleophilic addition with KCN, followed by dilute acid, to form a racemic mixture.

**(a) (i)** Draw the displayed formula of compound **B**, clearly showing all bonds.

**[1]**

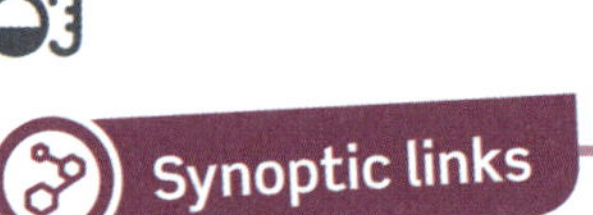

Synoptic links

2.1.3 4.1.1

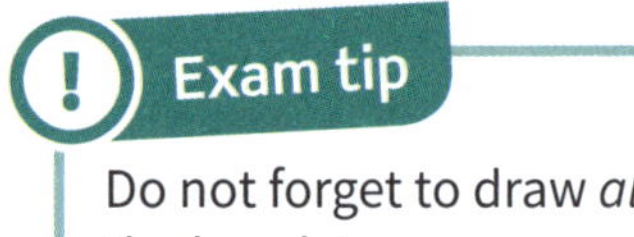

Exam tip

Do not forget to draw *all* the bonds!

**(ii)** Outline the mechanism for this reaction.

**[3]**

**(b)** Compound **C**, shown in **Figure 1.1**, can be distinguished from compound **B** using a simple chemical test.

```
     H   O   H
     |   ‖   |
  H—C — C — C—H
     |       |
     H       H
```

**Figure 1.1**

Outline the experimental steps you would take to determine which compound was which.

Give all reagents and conditions and any observations you would expect.

..........................................................................................

..........................................................................................

..........................................................................................

..........................................................................................

..........................................................................................

..........................................................................................

..........................................................................................

.......................................................................................... **[4]**

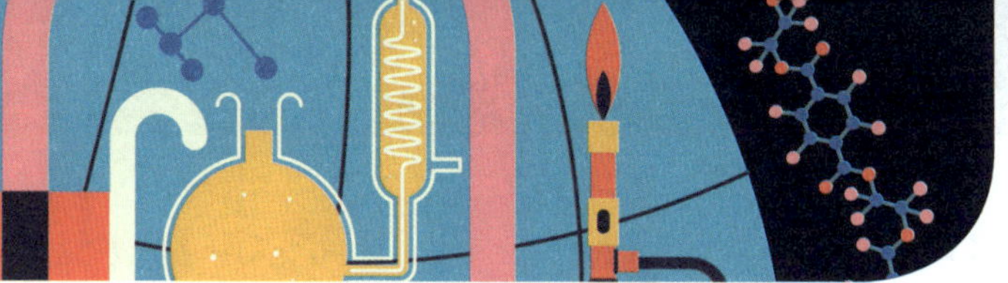



2 Aldehydes and ketones can both be reduced by $NaBH_4$ in aqueous solution. A compound **D** has the structural formula $CH_3C(CH_3)_2CH_2CHO$ and when reacted with $NaBH_4$ in aqueous solution forms a product, **E**.

**(a) (i)** Draw the structure of compound **D**.

Outline the mechanism for the reduction of compound **D**, using $OH^-$ to represent the electrophile.

**[3]**

**(ii)** Give the systematic name of product **E**.

........................................................................................

........................................................................................

........................................................................................ **[1]**

**(iii)** Predict the number of peaks in the $^{13}C$ NMR spectrum of product **E**.

........................................................................................

........................................................................................

........................................................................................ **[1]**

**(b)** A chemical test can be used to distinguish between aldehydes and ketones.

State the reagent(s) needed and give any observations that would be expected.

Reagent: ........................................................................................

........................................................................................

Observation with aldehydes: ........................................................................................

........................................................................................

Observation with ketones: ........................................................................................

........................................................................................ **[3]**

**Synoptic links**

4.1.1 6.3.2

**Exam tip**

Identify the functional group first, and bear this in mind. Then start drawing from the left-hand side of the molecule, reading the formula left to right.

3 A chemist reacts ethanal, $CH_3CHO$, with KCN, followed by a dilute acid, forming a product that exists as optical isomers.

(a) (i) State what the role of the KCN is in this reaction.

..........................................................................................

.......................................................................................... [1]

(ii) Outline a mechanism for this reaction.

[3]

(iii) Name this mechanism.

..........................................................................................

.......................................................................................... [1]

(b) Explain why the product formed is not optically active overall.

..........................................................................................

..........................................................................................

.......................................................................................... [2]

**Exam tip**

Think about the 3D shape of the product(s) formed. What type of mixture of these would be produced?

4 Two chemicals, **P** and **Q**, are shown in **Figure 4.1**. They are isomers of each other.

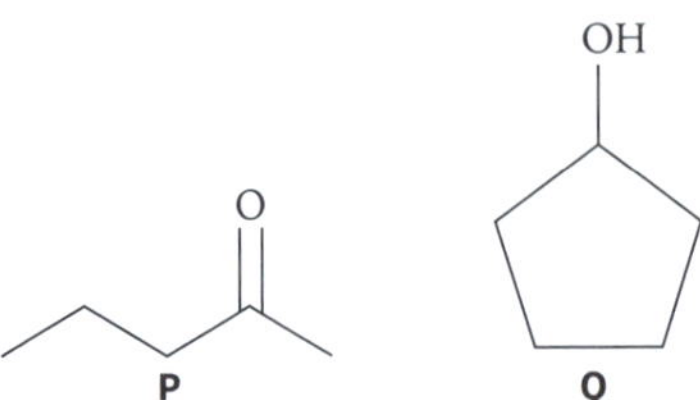


**Figure 4.1**

**Synoptic links**

4.1.1 4.2.4

(a) Define the term isomer.

..........................................................................................

.......................................................................................... [1]

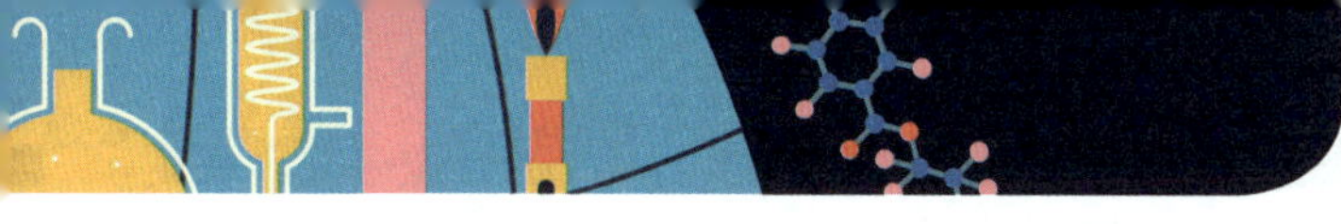



**(b) (i)** Use the *Data Sheet* to help you answer this question. Suggest the wavenumber of an absorption that will appear in the IR spectrum of **P** but not **Q**.

........................................................................................................

........................................................................................................ **[1]**

> **Exam tip**
>
> Name the functional group found in **P**, but not in **Q**, and use the Data Sheet to find its associated wavenumber.

**(ii)** Use the *Data Sheet* to help you answer this question. Suggest the wavenumber of an absorption that will appear in the IR spectrum of **Q** but not **P**.

........................................................................................................

........................................................................................................ **[1]**

**(c) (i)** State which chemical, **P** or **Q**, would produce a racemic mixture, containing a 50 : 50 mix of optical isomers, when reacted with KCN.

........................................................................................................

........................................................................................................ **[1]**

**(ii)** Outline a mechanism for this reaction and give the name of this mechanism.

**[4]**

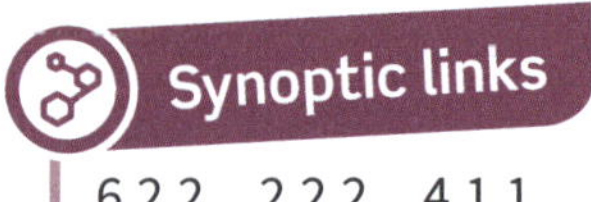

5 Ketones, such as butanone shown in **Figure 5.1**, can be reduced to produce secondary alcohols using $NaBH_4$. The reduced product is sometimes an optical isomer.

**Synoptic links**
6.2.2 2.2.2 4.1.1

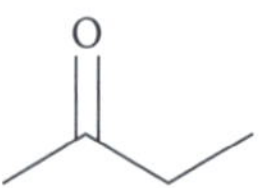


**Figure 5.1**

**(a)** Define the term optical isomer.

..................................................................................................

.................................................................................................. **[1]**

**(b) (i)** Outline a mechanism for the reduction, using $[H^-]$ as the nucleophile.

**[3]**

**(ii)** Draw the 3D representations of the two optical isomers produced, showing how they are related.

**[2]**

**(iii)** Give the IUPAC name for one of the structures you have drawn.

..................................................................................................

.................................................................................................. **[1]**

6 Acetoin, shown in **Figure 6.1**, is a chemical flavour, added to foods to give a buttery taste.

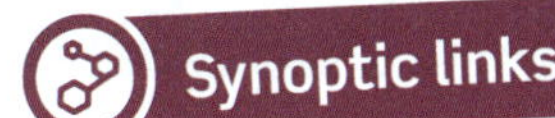

4.1.1 6.3.2 4.2.1 4.2.4

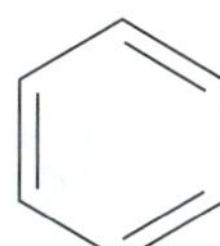

**Figure 6.1**

**(a)** **(i)** Give the IUPAC name of this compound.

.......................................................................................

....................................................................................... **[1]**

**(ii)** The chemist collected some spectral data on acetoin. State the number of peaks the chemist would see in the $^{1}H$ NMR spectrum of acetoin.

.......................................................................................

....................................................................................... **[1]**

**(iii)** State the number of peaks the chemist would see in the $^{13}C$ NMR spectrum of acetoin.

.......................................................................................

....................................................................................... **[1]**

**(b)** A chemist then prepared a solution of acidified potassium dichromate, and added a few drops of acetoin to it. They then refluxed it.

**(i)** State what you would expect the chemist to see.

.......................................................................................

....................................................................................... **[1]**

**(ii)** State the type of reaction that acetoin undergoes.

.......................................................................................

....................................................................................... **[1]**

**(iii)** Draw the organic product formed in this reaction.

**[1]**

**(iv)** Use the *Data Sheet* to explain whether the IR spectrum of acetoin would change after this reaction, and which peaks you would expect to see before and after.

.......................................................................................

.......................................................................................

....................................................................................... **[2]**

**Exam tip**

Do not forget to give the wavenumbers of the peaks mentioned.

# Knowledge

## 22 Carboxylic acids and esters

### Properties of carboxylic acids

Carboxylic acids have a carbonyl C=O functional group and a hydroxyl –OH functional group on the same carbon.

methanoic acid

ethanoic acid

propanoic acid

### Solubility

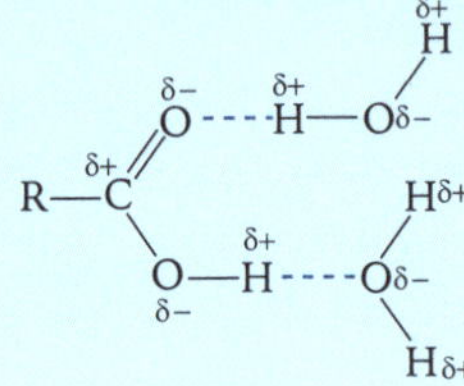

The partial negative charges, and the lone pair of electrons on the oxygen atoms, and the positive charges on the hydrogen atoms of a carboxylic acid mean that it will form hydrogen bonds with water. This makes carboxylic acids soluble in water. The longer the carboxylic acid, the less soluble it is.

### Reactions

Carboxylic acids will partially dissociate the hydrogen from the hydroxyl group, making them weak acids.

Carboxylic acids can react as other acids do in reactions. A **carboxylate ion** is the conjugate base of a carboxylic acid; in carboxylate salts the hydrogen of the –OH group has been replaced by a metal.

$$CH_3COOH \rightleftharpoons CH_3COO^- + H^+$$

### Esters

When a carboxylic acid reacts with an alcohol in the presence of an acid catalyst, a condensation reaction occurs and an ester is formed.

$$CH_3-C(=O)-OH + CH_3CH_2-O-H \xrightarrow{H^+} CH_3-C(=O)-O-CH_2CH_3 + H_2O$$

ethanoic acid; ethanol; ethylethanoate (an ester)

The name of an ester derives from both the carboxylic acid it was made from and the alcohol.

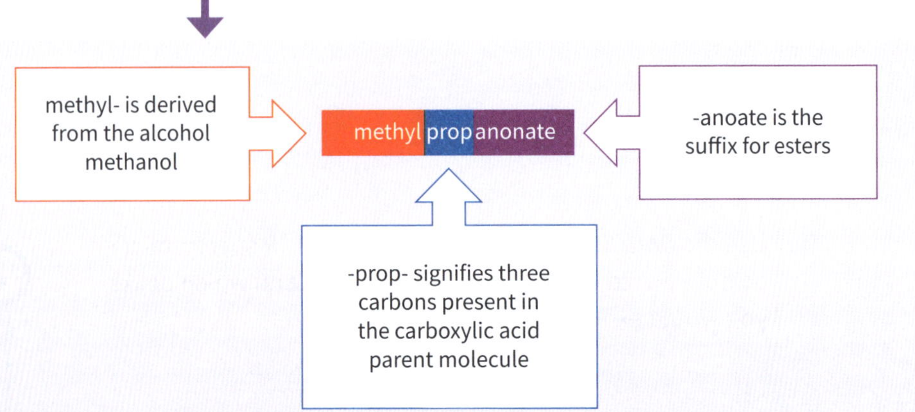
methyl- is derived from the alcohol methanol
methylpropanonate
-anoate is the suffix for esters
-prop- signifies three carbons present in the carboxylic acid parent molecule

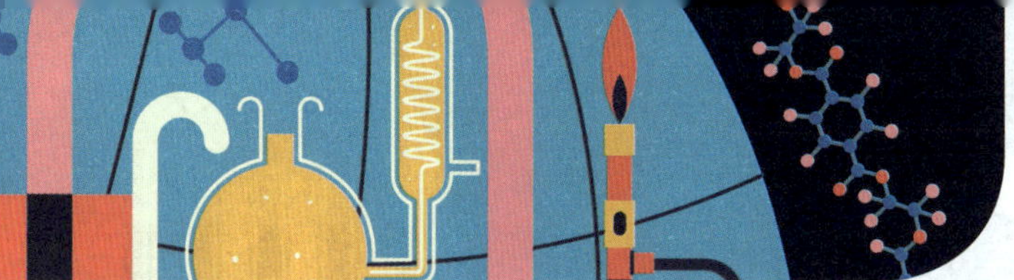



## Acyl chlorides

There are several derivatives of carboxylic acids that all include the acyl group, R–C=O. The difference in electronegativity of elements with these functional groups may lead to polarity within compounds. This means they take part in **nucleophilic addition–elimination reactions**. Acyl chlorides can be prepared from carboxylic acids using $SOCl_2$.

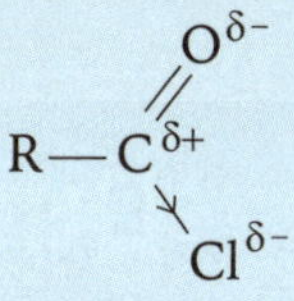

$R—C^{\delta+}(=O^{\delta-})—Cl^{\delta-}$

acyl chlorides

## Reactions

Acyl chlorides and acid anhydrides will react with alcohols in nucleophilic addition–elimination reactions to form esters:

$$CH_3COCl(aq) + C_2H_5OH(aq) \rightarrow CH_3COOC_2H_5(aq) + HCl(aq)$$

This includes the **esterification** of phenol, which does not form an ester from carboxylic acids.

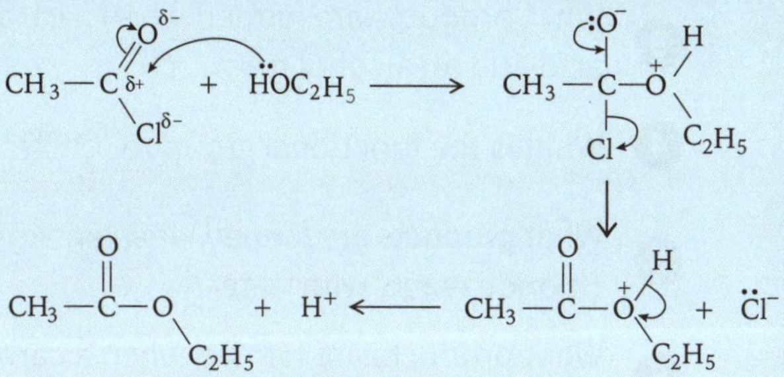

**Esters** can undergo hydrolysis in either acidic or alkaline conditions.

- With *acids* they form carboxylic acids and alcohols.
- With *bases* they form salts of carboxylic acids.

**Acyl chlorides** and **acid anhydrides** react with water in nucleophilic addition–elimination reactions, both reactions will form a carboxylic acid:

$$CH_3COCl(aq) + H_2O(l) \rightarrow CH_3COOH(aq) + HCl(aq)$$

**Acyl chlorides** and **acid anhydrides** react with ammonia in nucleophilic addition–elimination reactions to form amides.

$$CH_3COCl(aq) + 2NH_3(aq) \rightarrow CH_3CONH_2(aq) + HCl(aq)$$

**Acyl chlorides** and **acid anhydrides** react with primary amines in nucleophilic addition–elimination reactions to form an N-substituted amide.

$$CH_3COCl(aq) + CH_3NH_2(aq) \rightarrow CH_3CONHCH_3(aq) + HCl(aq)$$

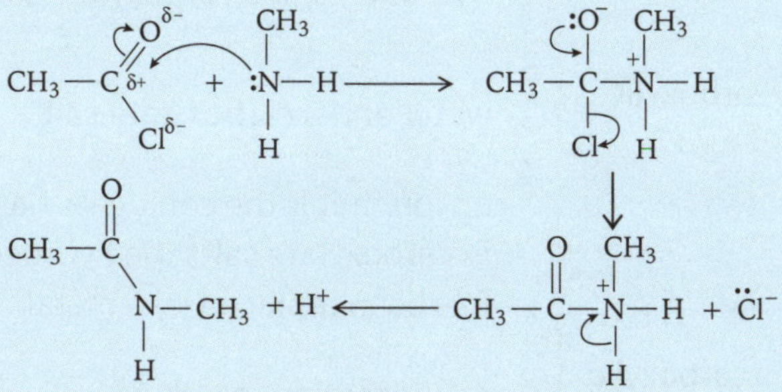

# Retrieval

Learn the answers to the questions below, then cover the answers column with a piece of paper and write as many as you can. Check and repeat.

| | Questions | Answers |
|---|---|---|
| 1 | Which two functional groups do carboxylic acids have on the same carbon? | carbonyl and hydroxyl |
| 2 | Why are carboxylic acids soluble in water? | they form hydrogen bonds with water |
| 3 | What products are formed when a carboxylic acid and an alcohol react? | an ester and water |
| 4 | What is the functional group RCOCl? | acyl chloride |
| 5 | What products are formed when an acyl chloride reacts with water? | a carboxylic acid and hydrochloric acid |
| 6 | What products are formed when a carboxylic acid reacts with a metal carbonate? | water, carbon dioxide, and a carboxylate salt |
| 7 | What is the name of the ester formed from methanol and propanoic acid? | methyl propanoate |
| 8 | What products are formed when an acyl chloride reacts with ammonia? | an amide and hydrochloric acid |
| 9 | What other reactant is needed to form an acyl chloride from a carboxylic acid? | $SOCl_2$ |
| 10 | What products are formed when an acyl chloride reacts with an alcohol? | an ester and hydrochloric acid |
| 11 | Why can carboxylic acids form hydrogen bonds? | the partial negative charges and the lone pair of electrons on the oxygen atoms and the positive charges on the hydrogen atoms can form hydrogen bonds with water |
| 12 | What products form when a carboxylic acid reacts with a metal oxide? | water and a carboxylate salt |
| 13 | What products are formed when a carboxylic acid reacts with a metal? | hydrogen and a carboxylate salt |
| 14 | What products are formed when a carboxylic acid reacts with an alkali? | water and a carboxylate salt |
| 15 | What is a carboxylate ion? | an ion that is the conjugate base of a carboxylic acid: in carboxylate salts the hydrogen of the –OH group has been replaced by a metal |
| 16 | What type of reaction occurs when a carboxylic acid reacts with an alcohol to form an ester? | condensation reaction |
| 17 | How can I form an ester using phenol? | by reacting it with an acyl chloride |

Put paper here

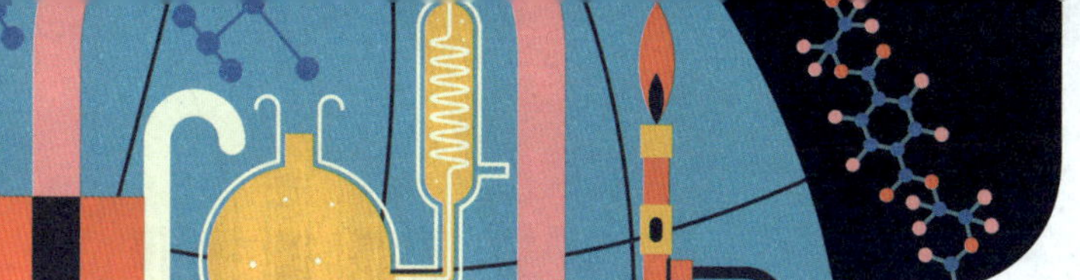



| | Question | Answer |
|---|---|---|
| 18 | What products are formed when esters undergo hydrolysis in acidic conditions? | carboxylic acids and alcohols |
| 19 | What products are formed when esters undergo hydrolysis in alkaline conditions? | salts of carboxylic acids |
| 20 | What are the products formed when an acyl chloride reacts with water in a nucleophilic addition-elimination reaction? | a carboxylic acid and hydrochloric acid |

Put paper here

## Practical skills

Practise your practical skills using the worked example and practice questions below.

### Identifying organic functional groups

Chemical analysis tells you extremely important information about compounds. It is important that chemical tests are memorised.

Any chemical reaction that shows a visible change can be used as a chemical test. These include:

- using $Br_2$ solution to test for alkenes
- using acidified potassium dichromate to test for primary and secondary alcohols and aldehydes
- using Tollens' reagent to test for aldehydes
- using sodium carbonate to test for carboxylic acids
- using silver nitrate to test for haloalkanes.

### Worked example

A student is provided with the following chemicals in unlabelled bottles:

- 1-chloropentane
- 1-bromopentane
- 1-iodopentane
- pentan-1-ol

The student mixes each separately with 1 $cm^3$ of ethanol and 1 $cm^3$ of 0.05 mol $dm^{-3}$ silver nitrate, heats to 50 °C, and leaves for a minute. Some chemicals produce precipitates.

**Questions**

1 Which chemical produces a precipitate the fastest?

2 Pentan-1-ol should be kept away from naked flames. Suggest how the student could safely heat this to 50 °C.

**Answers**

1 1-iodopentane

2 Place the test tubes in a hot water bath *not* a Bunsen burner.

### Practice

A student has samples of the following chemicals:

- cyclohexene
- cyclohexanol
- cyclohexanone
- cyclohexanecarboxylic acid

1 a Which chemical would react with immediately $Br_2$? Write an equation for this reaction.

b Which chemical would react with $Na_2CO_3$? Write an equation for this reaction.

2 The remaining two compounds can be distinguished between by heating with acidified potassium dichromate(VI). A student suggests heating directly with a Bunsen burner.

a Why is this not a sensible way to proceed?

b Suggest how the student could amend the practical procedure.

# Practice

## Exam-style questions

Synoptic links

4.2.1 4.1.1 2.1.3

1 **(a)** Pentan-1-ol, shown in **Figure 1.1**, can be used as feedstock for multiple syntheses, including conversion into a carboxylic acid.

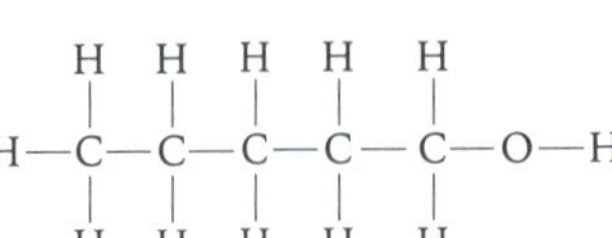

**Figure 1.1**

**(i)** Name the reagents and conditions needed to oxidise pentan-1-ol to a carboxylic acid.

........................................................................................ **[1]**

**(ii)** Deduce the IUPAC name of the carboxylic acid produced.

........................................................................................ **[1]**

**(b)** Esterification with a different carboxylic acid can create ester **A**, shown in **Figure 1.2**.

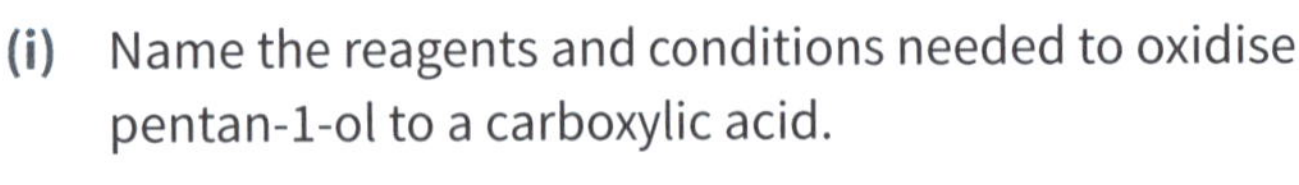

**Figure 1.2**

Deduce and draw the displayed formula of the carboxylic acid used as a reagent to create ester **A**.

**[1]**

**(c)** 151 mg of pentan-1-ol was reacted completely with an excess of carboxylic acid to produce ester **A**.

Calculate the percentage yield of the reaction if 161 mg of ester **A** was produced.

**Exam tip**

Start by working out the number of moles of pentan-1-ol, and use that to work out the theoretical number of moles of ester produced.

percentage yield = .................... % **[4]**

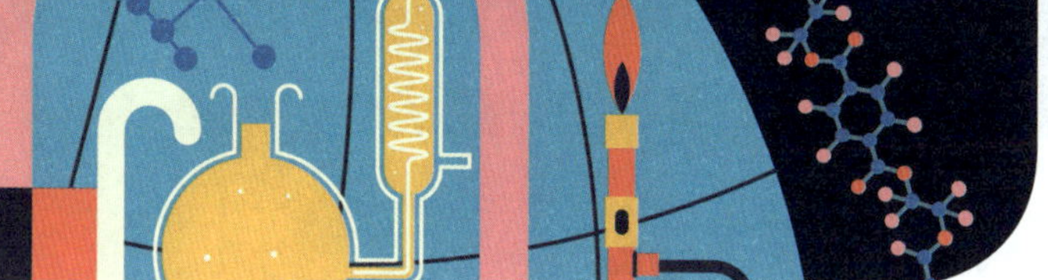



2 Biodiesel is made up of long-chain carboxylic acids reacted with alcohols in a dehydration reaction.

**Synoptic link**

2.2.2

**(a) (i)** Suggest why some car manufacturers might choose to design cars that run on biodiesel instead of petrol.

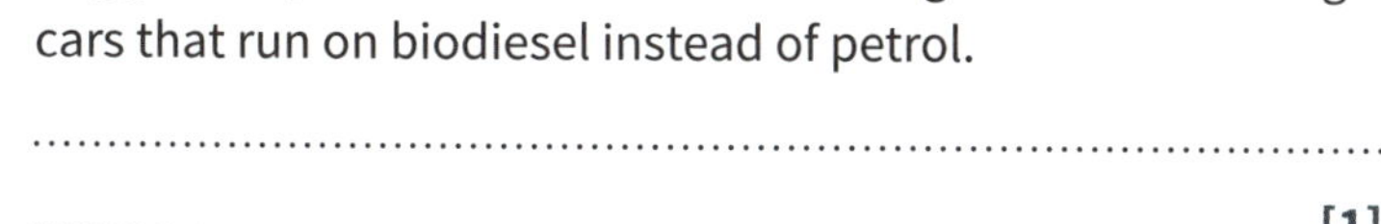

..................................................................................................

.................................................................................................. **[1]**

**(ii)** A general structure of a biodiesel molecule, triglyceride, is given in **Figure 2.1**, with $R_1$, $R_2$, and $R_3$ representing long alkyl chains.

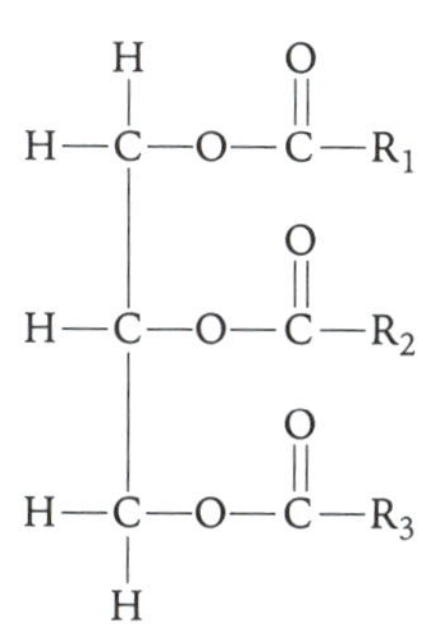


**Figure 2.1**

Circle one ester bond in the molecule in **Figure 2.1**. **[1]**

**(b) (i)** Esters can be hydrolysed to return to their starting products. Give the reagents and conditions needed for this to happen.

.................................................................................................. **[1]**

**(ii)** Draw the displayed formula of the tri-ol produced when triglyceride is hydrolysed.

**[1]**

**(c)** Explain the difference in the average boiling point of biodiesel and a straight-chain ester with a similar molecular mass.

..................................................................................................

..................................................................................................

..................................................................................................

..................................................................................................

.................................................................................................. **[3]**

**Exam tip**

What are the different forces that might be acting between molecules? How might they differ between a straight chain and a branched chain molecule?

3 Carboxylic acids can be used to create reagents, such as compound **F** shown in **Figure 3.1**. This can then be used to create carbon–carbon bonds in a selection of reactions.

2.1.3

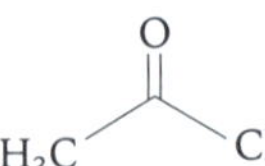


**Figure 3.1**

**(a) (i)** Give the reagent needed to create compound **F** from ethanoic acid.

..........................................................................................

.................................................................................. [1]

**(ii)** Outline the mechanism for the acylation of ethanol with compound **F** to form ester **G**, shown in **Figure 3.2**.

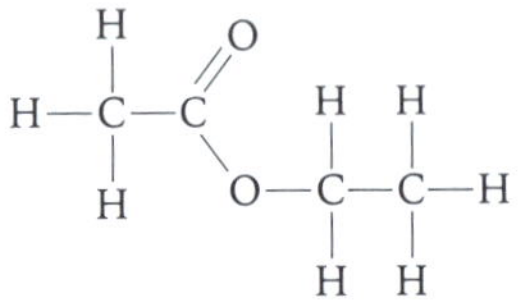

Curly arrows show the movement of a pair of electrons.

**Figure 3.2**

[3]

**(b)** 1.727 g of compound **F** was converted into ester **G**, where compound **F** was the limiting reagent. 1.540 g of ester **G** was formed.

Calculate the percentage yield of ester **G**.

**Exam tip**

When unsure of how to start a question like this, calculate the moles of something and see if that helps to guide you; it may even get you a mark!

percentage yield = ...................... [4]

**4** A student can synthesise aspirin by the reaction of salicylic acid with ethanoic anhydride, as shown in **Figure 4.1**, to produce aspirin and ethanoic acid. Ethanoic acid is extremely soluble in water.

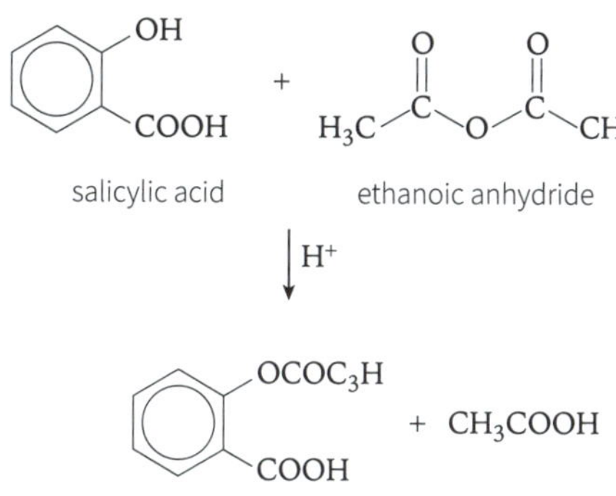


**Figure 4.1**

**Synoptic link**

2.1.3

**(a)** Describe how a student would purify the sample of aspirin at the end of this process.

.......................................................................................................

.......................................................................................................

.......................................................................................................

.......................................................................................................

.......................................................................................................

.......................................................................................................

.......................................................................................................

....................................................................................................... **[4]**

**Exam tip**

This is worth four marks, so make sure to give clear and full steps!

**(b)** Calculate the atom economy of this reaction.

atom economy = ........................ **[2]**

**(c)** Give a reason why industry might choose to use ethanoic anhydride rather than ethanoyl chloride in the large-scale synthesis of aspirin.

.......................................................................................................

....................................................................................................... **[1]**

**5** **(a)** Esters are commonly used in perfumes and fragrances.

**(i)** Draw the displayed formula of the ester, ethyl propanoate.

**[2]**

**(ii)** Name the peaks you would expect to see in the IR spectrum of this compound.

..........................................................................................

.......................................................................................... **[1]**

**(b)** Another ester, ethyl ethanoate, is shown in **Figure 5.1**.

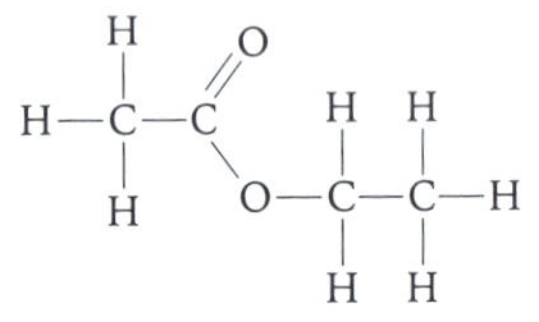

$$H-\underset{\underset{H}{|}}{\overset{\overset{H}{|}}{C}}-C\begin{matrix}\nearrow O\\ \searrow O-\underset{\underset{H}{|}}{\overset{\overset{H}{|}}{C}}-\underset{\underset{H}{|}}{\overset{\overset{H}{|}}{C}}-H\end{matrix}$$

**Figure 5.1**

**(i)** Write an equation for the reaction of ethyl ethanoate with hot sodium hydroxide solution.

Show all functional groups clearly.

..........................................................................................

.......................................................................................... **[1]**

**(ii)** Write an equation for the reaction of ethyl ethanoate with hot dilute hydrochloric acid solution.

Show all functional groups clearly.

..........................................................................................

.......................................................................................... **[1]**

**(iii)** Ethyl ethanoate is less water soluble than ethanoic acid, $CH_3COOH$.

Explain why.

You may use diagrams to help you.

..........................................................................................

..........................................................................................

.......................................................................................... **[2]**

**Synoptic links**

4.1.1 4.2.4

**Exam tip**

What is present in ethanoic acid that is not in the ester?

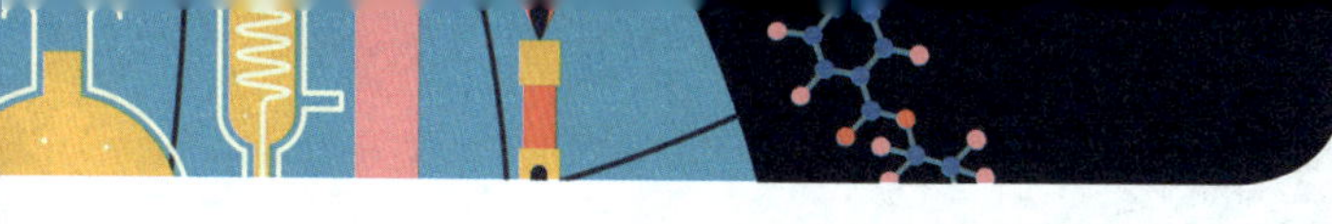



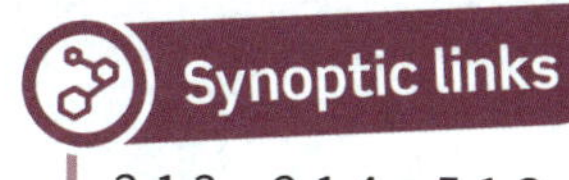


2.1.3 2.1.4 5.1.3

**6** **(a)** A chemist carries out the reaction below using 2.16 g of ethanoic acid and an excess of magnesium hydroxide. The chemist purifies the solid to obtain 2.85 g of magnesium acetate tetrahydrate, $(CH_3COO)_2Mg{\bullet}4H_2O$:

$$2CH_3COOH + Mg(OH)_2 \rightarrow (CH_3COO)_2Mg + 2H_2O$$

**(i)** Describe a method to obtain a pure sample of magnesium acetate tetrahydrate, $(CH_3COO)_2Mg{\bullet}4H_2O$.

..........................................................................................

..........................................................................................

..........................................................................................

..........................................................................................

.......................................................................................... **[3]**

**Exam tip**

This question is worth three marks, so make sure you give three points.

**(ii)** Determine the percentage yield.

percentage yield = ......................% **[5]**

**(iii)** Describe a method you would use to check the purity of the final product.

..........................................................................................

..........................................................................................

..........................................................................................

.......................................................................................... **[2]**

**(b)** Propanoic acid, $CH_3CH_2COOH$, is a weak acid.

**(i)** Explain what is meant by the term weak acid.

..........................................................................................

.......................................................................................... **[1]**

**(ii)** Write an expression for the $K_a$ of propanoic acid.

..........................................................................................

.......................................................................................... **[1]**

**(iii)** Write the full balanced equation for the reaction of propanoic acid with calcium oxide.

You do not need to include state symbols.

**[2]**

# Knowledge

## 23 Nitrogen compounds

### Amines

Amines are organic compounds which have a nitrogen atom in place of a carbon atom. They can be primary, secondary, or tertiary, depending on the number of hydrogen atoms attached to the nitrogen.

1°amine — propan-1-amine

2°amine — N-methylpropan-1-amine

3°amine — N,N-dimethylpropan-1-amine

### Reactivity of amines

$$CH_3CH_2NH_2 + H^+ \longrightarrow [CH_3CH_2NH_3]^+$$

Amines are weak bases, they are proton acceptors. The lone pair of electrons on the nitrogen atom is able to accept a proton .

They react with *dilute acids* to form salts, following the rules for general salt equations.

### Preparation of aliphatic (non-aromatic) amines

Amines can be prepared by:

- the nucleophilic substitution reaction between ammonia and a halogenoalkane
- reduction of nitriles.

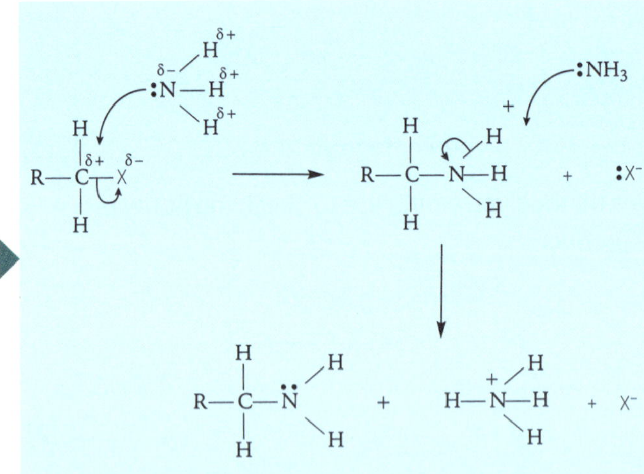

Primary aliphatic amines are produced in the reaction between ammonia and halogenoalkanes. Overa the Halogen is replaced by $NH_2$ in a nucleophili substitution reaction. Th reaction is done in ethar as the solvent to preven the substitution of wate to form an alcohol.

Primary amines can be produced by the reduction of nitriles. Halogenoalkanes react with a cyanide ions in aqueous ethanol to form a nitrile:

$RBr + CN^- \rightarrow R\text{–}C{\equiv}N + Br^-$

This nitrile can then be reduced to a primary amine:

$R\text{–}C{\equiv}N + 2H_2 \rightarrow R\text{–}CH_2NH_2$

### The reaction of amines with halogenoalkanes

When ammonia or amines react with halogenoalkanes they can produce primary, secondary, or tertiary amines.

$$CH_3CH_2Cl + NH_3 \xrightarrow{-H^+} CH_3CH_2NH_2 + HCl$$

Reaction with ammonia

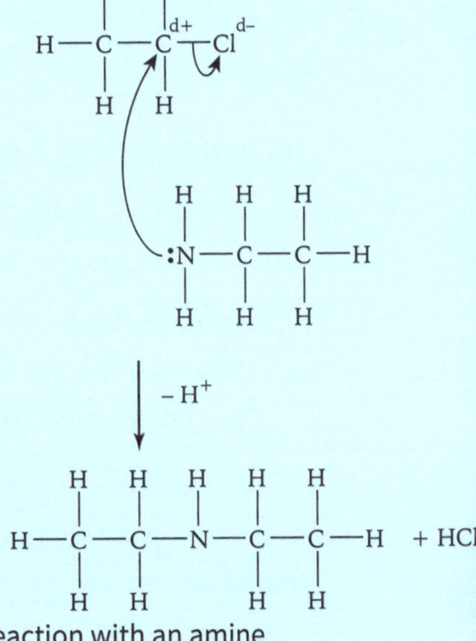

Reaction with an amine

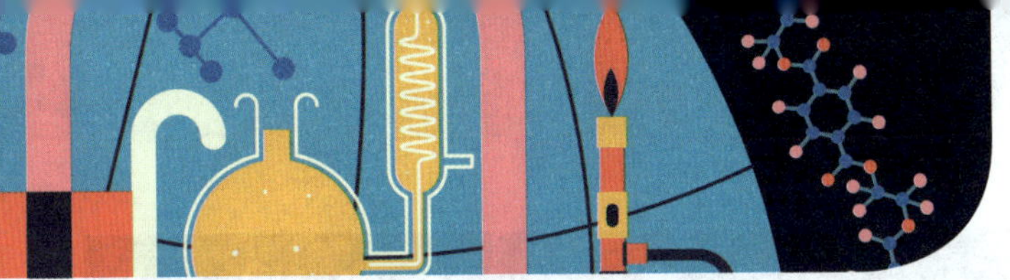



## Preparation of aromatic amines

Phenylamine can be produced from benzene.

**Step 1** Nitrobenzene is produced from reacting benzene with concentrated nitric acid and sulfuric acid.

**Step 2** Nitrobenzene is reduced to phenylamine using a mixture of tin and hydrochloric acid as a catalyst.

$$C_6H_5NO_2 + 6H^+ + 6e^- \longrightarrow C_6H_5NH_2 + 2H_2O$$

nitrobenzene + $6H^+$ + $6e^-$ (from reducing agent) → phenylamine + $2H_2O$

Aromatic amines are important in the manufacture of dyes.

## Reactions of amino acids

$$H_2N{-}CH(R){-}COOH \longrightarrow H_3N^+{-}CH(R){-}COO^-$$

receives $H^+$ from a $-COOH$ group

$H^+$ is donated to an $-NH_2$ group

a zwitterion

As amino acids have both an amine group and a carboxylic acid group, they have both acidic and basic properties. The carboxylic acid group will lose protons whilst the amine group will accept protons. This means amino acids have both a positive charge and a negative charge at the same time. This form of an amino acid is called a **zwitterion**.

- In strong basic conditions the amino acid loses a proton.
- In strongly acidic conditions the amino acid gains a proton.

$$\underset{\text{zwitterion}}{H_3\overset{+}{N}{-}CH(R){-}COO^-} + \underset{\text{strong base}}{OH^-} \rightleftharpoons \underset{\text{anionic form}}{H_2N{-}CH(R){-}COO^-} + H_2O$$

$$\underset{\text{zwitterion}}{H_3\overset{+}{N}{-}CH(R){-}COO^-} + \underset{\text{strong acid}}{H^+} \rightleftharpoons \underset{\text{cationic form}}{H_3\overset{+}{N}{-}CH(R){-}COOH}$$

Amino acids undergo a range of reactions.

- The carboxylic acid group reacts with aqueous alkali to form a salt and water.
- The carboxylic acid group reacts with an alcohol to form an ester.
- The amine group reacts with acids to form a salt.

$$H_2N{-}CH_2{-}COOH + NaOH \longrightarrow H_2N{-}CH_2{-}COO^-Na^+ + H_2O$$

$$H_2N{-}CH(CH_2OH){-}COOH + C_2H_5OH + H^+ \longrightarrow H_3N^+{-}CH(CH_2OH){-}COO{-}C_2H_5 + H_2O$$

$$H_2N{-}CH(CH_3){-}COOH + HCl \longrightarrow H_3\overset{+}{N}{-}CH(CH_3){-}COOH + Cl^-$$

## Chirality

**Stereoisomers** have the same structural formula but the atoms are arranged differently in space. This can be *E-Z* isomerism when there is restricted rotation around a double bond, or optical isomerism when there is a chiral carbon.

A **chiral carbon** has *four different groups* attached to it.

Optical isomers will be mirror images of each other, and will be non-superimposable.

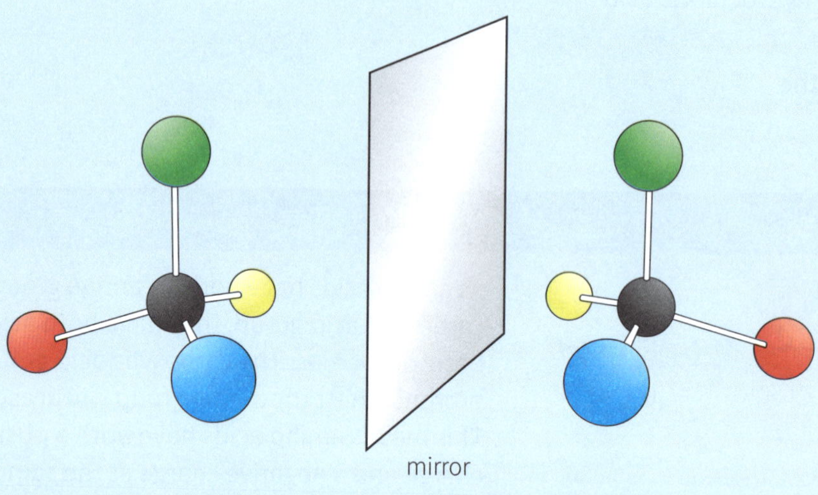


## Optical isomers

Optical isomers, or enantiomers, may have different properties, such as tasting or smelling differently. Both ibuprofen and thalidomide have chiral centres and their different enantiomers produce different therapeutic effects.

Optical isomerism can also be found in complex ions.

Most amino acids have chiral carbons. Generally, only one of the enantiomers is likely to fit into the active sites on enzymes.

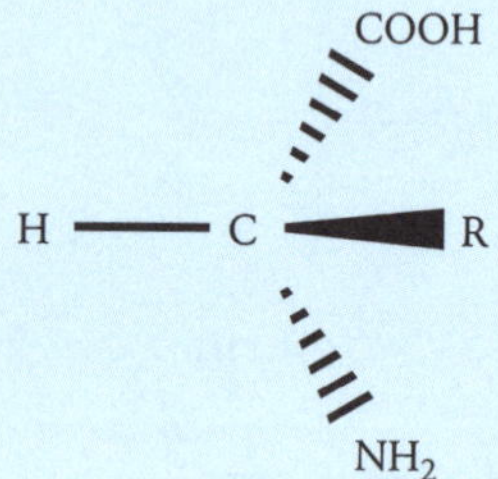


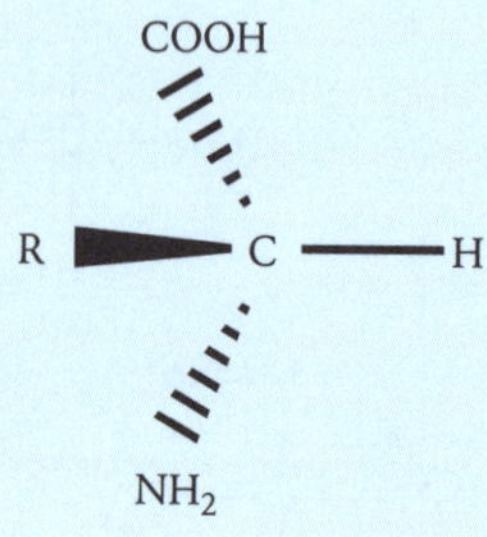

## Chiral centres

The chiral centre can be identified by finding the carbon atom that has four different groups attached to it. This can be difficult to identify when that carbon is part of a ring structure.

(+)-enantiomer(effective isomer)

(−)-enantiomer (teratogenic isomer)

stereocentre

## Amides

Acyl chlorides react with ammonia and amines to form amides. Similarly to amines, these can be primary, secondary, or tertiary.

primary amide: one carbon atom bonded to N

$CH_3CH_2-C(=O)-NH_2$

propanamide

secondary amide: two carbon atoms bonded to N

$H_3C-C(=O)-N(H)-CH_3$

N-methylethanamide

tertiary amide: three carbon atoms bonded to N

$H-C(=O)-N(CH_3)-CH_3$

N,N-dimethylmethanamide

## Amino acids, amides, and chirality

Amino acids have an amine group and a carboxylic acid group. The basic structure of all amino acids is the same but the R-group can be a range of different groups and gives the amino acids individual properties. With the exception of glycine where the R-group is a hydrogen, the central carbon is **chiral**. This means that they show **optical isomerism**.

amine group

R-group

$H_2N-C(R)(H)-COOH$

carboxylic acid group

α-carbon: the first carbon atom attached to the –COOH group

# Retrieval

Learn the answers to the questions below, then cover the answers column with a piece of paper and write as many as you can. Check and repeat.

| | Questions | Answers |
|---|---|---|
| 1 | What type of amine is N,N-dimethylpropan-1-amine? | tertiary |
| 2 | What are the two types of stereoisomerism? | *E-Z* and optical |
| 3 | What solvent is used when ammonia is reacted with a halogenoalkane? | ethanol |
| 4 | What conditions are needed to produce phenylamine? | a mixture of tin and hydrochloric acid as a catalyst |
| 5 | What is a chiral carbon? | a carbon with four different groups attached |
| 6 | What is the group called that is different in each amino acid? | R-group |
| 7 | In strong basic conditions, what is the charge on an amino acid? | negative |
| 8 | In strong acidic conditions, what is the charge on an amino acid? | positive |
| 9 | Name two examples of compounds where the enantiomers behave differently. | ibuprofen and thalidomide |
| 10 | What is an enantiomer? | an optical isomer |
| 11 | What two functional groups can be found on an amino acid? | carboxylic acid and amine |
| 12 | Why are amines weak bases? | they are proton acceptors |
| 13 | What do amines form when reacted with dilute acids? | salts |
| 14 | What is produced when a carboxylic acid reacts with an aqueous alkali? | salt and water |
| 15 | What is produced when a carboxylic acid reacts with an alcohol? | an ester |
| 16 | What type of isomerism do nearly all amino acids show? | optical isomerism |
| 17 | What type of isomers are mirror images? | optical isomers/enantiomers |
| 18 | What reactants are needed to make primary aliphatic amines? | ammonia and halogenoalkanes |
| 19 | Why are amines nucleophiles? | lone pair of electrons on nitrogen |
| 20 | What type of reaction is the reaction between ammonia and halogenoalkanes? | nucleophilic substitution reaction |

Put paper here

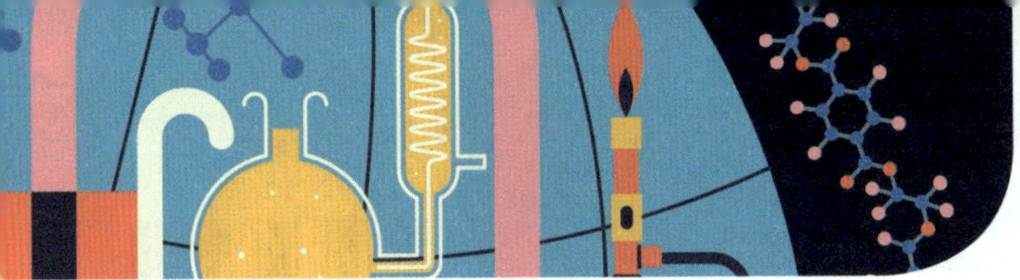



## Maths skills

Practise your maths skills using the worked example and practice questions below.

### Estimating results

You need to be able to estimate what will happen to quantities in calculations if one is changed.

For example, in the formula: $a = \frac{b}{c^2}$, if $c$ is increased then $a$ will decrease as you would be dividing by a bigger number.

Similarly, in the equation:

$d = e - 2f$, decreasing $f$ will increase $d$ as there is less being subtracted.

In general, making the denominator (the bottom part of a fraction) bigger will decrease the subject of the equation. Making the numerator (the top part of the equation) bigger will increase the subject of the equation.

The reverse is true in each case.

### Worked example

**Questions**

What would be the effect on $y$ of increasing $x$ in each equation? See if you can answer these without a calculator first. If you are struggling trying to calculate them, try $x = 10$ then $x = 100$ and see how $y$ changes.

1 $y = \frac{100}{x}$

2 $y = 10^{-x}$

3 $y = \frac{50}{-\log_{10}x}$

**Answers**

1 As $x$ is in the denominator (the bottom), increasing $x$ will decrease $y$. You would be dividing 100 by a bigger number, so $x$ would be a smaller number.

2 You can rearrange this to be $y = \frac{1}{10^x}$. Increasing $x$ increases $10x$ which increases the denominator, so $y$ decreases

3 Increasing $x$ increases $\log_{10}x$, which makes $-\log_{10}x$ more negative. The denominator is smaller, so $y$ is bigger (less negative).

### Practice

What would be the effect of increasing $[H^+]$ or $p(H_2)$ (i.e. the hydrogen containing species) on the subject of each equation?

1 $pH = -\log_{10}[H^+]$

2 $K_p = \frac{p(NH_3)^2}{p(N_2) \times p(H_2)^3}$

3 $K_p = \frac{p(I_2)\, p(H_2)}{p(HI)^2}$

4 $\text{rate} = K[H^+][(CH_3)_2O]$

5 $K_a = \frac{[H^+]^2}{0.5 - [H^+]}$

assume $[H^+]$ can only be between 0 and 0.5

# Practice

## Exam-style questions

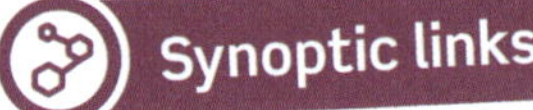


4.1.3 4.1.1

1 Amino acids are incredibly important substances used to create proteins in the body. The basic structure of the amino acid alanine is shown in **Figure 1.1**.

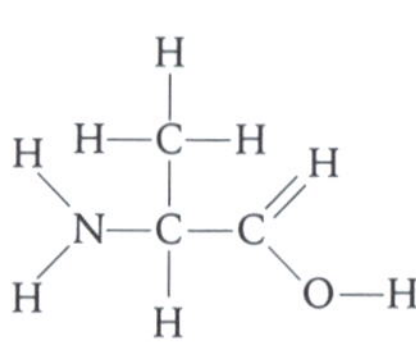


**Figure 1.1**

**(a)** Most amino acids exist as stereoisomers and alanine is an optical isomer.

**(i)** Define the term stereoisomerism

..................................................................................................

..................................................................................................

..................................................................................................

..................................................................................................

..................................................................................................

.......................................................................................... **[2]**

**(ii)** Define the term optical isomerism.

..................................................................................................

..................................................................................................

..................................................................................................

.......................................................................................... **[1]**

**(iii)** Give the 3D structure of one isomer and identify the chiral carbon(s) in the molecule by putting an asterisk * next to the carbon.

**[2]**

**(b)** Alanine is an α-amino acid.

**(i)** Give the IUPAC name for alanine.

..................................................................................................

.......................................................................................... **[1]**

**(ii)** Give the general formula for an α-amino acid.

..................................................................................................

.......................................................................................... **[1]**

**Exam tip**

Start by naming the most 'important' part(s) of the compound: this goes at the end of the name. Then work back from there.

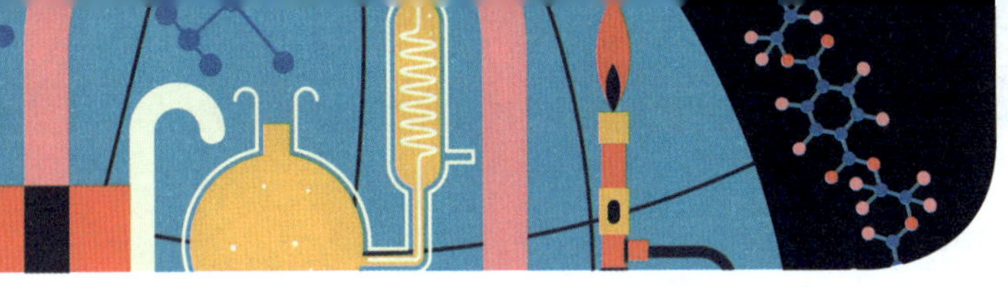



(c) The amino acid can act as a base and react with acids.

(i) State how the amino acid acts as a base.

.................................................................................................

................................................................................................. [1]

(ii) Give the equation for the reaction of alanine, $CH_3CH(NH_2)COOH$, with hydrochloric acid, HCl. You do *not* need to include state symbols.

.................................................................................................

................................................................................................. [1]

**2** A company is creating some feedstock for the eventual creation of a new azo dye.

The synthesis of this feedstock chemical, dye **A**, has three stages, as shown in **Figure 2.1**.

6.1.1

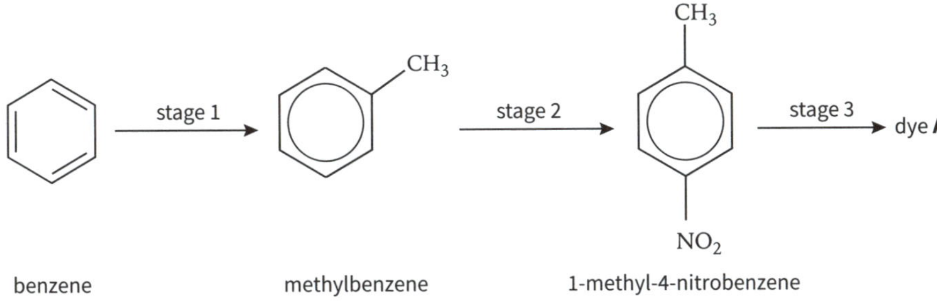


**Figure 2.1**

(a) Name the reagents needed for stage 1.

.................................................................................................

................................................................................................. [1]

There are two reagents needed for this mark.

(b) Name the type of reaction in stage 1.

.................................................................................................

................................................................................................. [1]

(c) State the reagent(s) needed for stage 2 and outline the mechanism for this reaction.

Reagent(s): .................................................................................

Mechanism: .................................................................................

[4]

**(d)** Name the reagent(s) needed for stage 3 and name the product, dye **A**.

Reagent(s): ..........................................................................

..........................................................................................

Name of dye **A**: .....................................................................

.......................................................................................... **[2]**

**(e)** Name the type of reaction in stage 3.

..........................................................................................

.......................................................................................... **[1]**

**3** This question is about the structure of amines.

**(a)** There are three secondary amines that contain four carbon atoms per molecule.

Give the skeletal formulae of these three secondary amines.

6.3.2 6.2.3 4.1.1

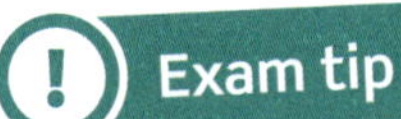

Double-check you have not drawn the same chemical twice!

**[3]**

**(b)** Trimethylamine, shown in **Figure 3.1**, is the only tertiary amine that contains three carbon atoms.

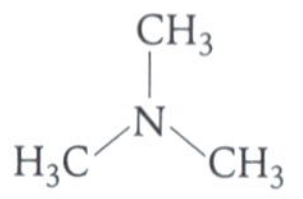

$$H_3C-N(CH_3)-CH_3$$

**Figure 3.1**

**(i)** State the number of peaks you would expect to see in the $^{1}H$ NMR spectrum of trimethylamine.

..........................................................................................

.......................................................................................... **[1]**

**(ii)** Suggest why you would see this number of peaks in the $^{1}H$ NMR spectrum of this amine.

..........................................................................................

.......................................................................................... **[1]**

**(c)** The compound N-ethylacetamide, shown in **Figure 3.2**, can be synthesised from a primary amine when it is reacted with an acyl chloride, ethanoyl chloride.

O
$H_3C$ N $CH_3$
H

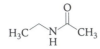

**Figure 3.2**

**(i)** Give the structure of this primary amine.

**[1]**

**(ii)** State the name of this primary amine.

.................................................................................................

................................................................................................. **[1]**

**4** The human body relies on 21 essential amino acids to function.

The structure and common names of one of these amino acids, alanine, is shown in **Figure 4.1**.

**Synoptic link**

6.1.3

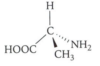

H
C
HOOC $NH_2$
$CH_3$

**Figure 4.1**

**(a)** Demonstrate the chiral centre in alanine by circling it. **[1]**

**(b)** Alanine is one of the amino acids that shows optical isomerism. One of these optical isomers is shown in **Figure 4.1**.

**(i)** Give the structure of the other isomer of alanine.

**Exam tip**

Check you have drawn the optical isomer, and not just the same structure but facing a different direction.

**[1]**

**(ii)** Define the term optical isomers.

.................................................................................................

.................................................................................................

................................................................................................. **[2]**

**(c)** Amino acids can act as acids in some reactions. Serine is another amino acid, shown in **Figure 4.2**.

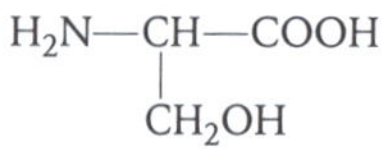

$$H_2N{-}\underset{\displaystyle CH_2OH}{\underset{|}{CH}}{-}COOH$$

**Figure 4.2**

**(i)** Give an equation to show the reaction of serine with sodium hydroxide, NaOH.

..............................................................................................

.............................................................................................. **[1]**

**(ii)** Give an equation to show the reaction of serine with methanol, $CH_3OH$, to produce an ester.

**[1]**

**(iii)** Give the structure of the ester product, clearly showing the functional group.

**[1]**

**5** There are four stereoisomers of the amino acid, isoleucine.

**(a)** One stereoisomer of isoleucine is shown in **Figure 5.1**.

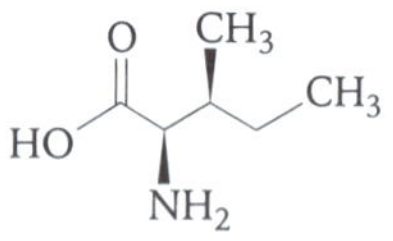

O CH$_3$ CH$_3$ HO NH$_2$

**Figure 5.1**

**(i)** Define the term amino acid.

..............................................................................................

.............................................................................................. **[1]**

**(ii)** Suggest a chiral centre in this amino acid by circling it, on **Figure 5.1**.

**[1]**

**(iii)** Draw the 3D representations of the three remaining stereoisomers of isoleucine.

**[3]**

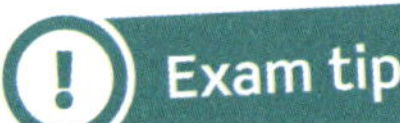


Check that all three of your stereoisomers show different optical isomerism.

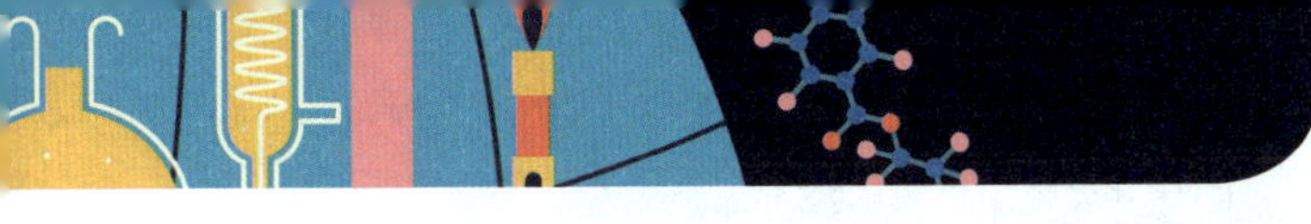



**(b)** The general structure of amino acids can be written as $RCH(NH_2)COOH$.

Give an equation for the general reaction of an amino acid with an acid, HA.

**[1]**

**6** A student was investigating the reactions and uses of organic amines.

**(a)** The student reacted an excess of $C_2H_5NH_2$ with two different acids.

Give the formula of the salts formed when an excess of $C_2H_5NH_2$ reacts with:

**(i)** sulfuric acid

.......................................................................................

....................................................................................... **[1]**

**(ii)** ethanoic acid

.......................................................................................

....................................................................................... **[1]**

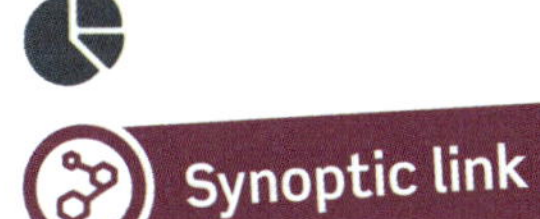

Synoptic link

6.2.4

**(b)** The student wanted to create the compound propylamine, $CH_3CH_3CH_3NH_2$.

**(i)** Give the reagents and conditions the student needs to create this from the starting material 1-bromopropane, $CH_3CH_3CH_3Br$, in one step.

.......................................................................................

.......................................................................................

....................................................................................... **[1]**

**(ii)** State what type of reaction this is.

.......................................................................................

.......................................................................................

....................................................................................... **[1]**

**(iii)** State the reagents and conditions the student needs to make a nitrile, $CH_3CH_3CN$, from 1-bromoethane, $CH_3CH_3Br$.

.......................................................................................

.......................................................................................

....................................................................................... **[1]**

**(iv)** The student then reduces the nitrile to create ethylamine. State the reagents and conditions needed for this.

.......................................................................................

.......................................................................................

....................................................................................... **[1]**

**Exam tip**

Do not forget to include charges.

# Knowledge

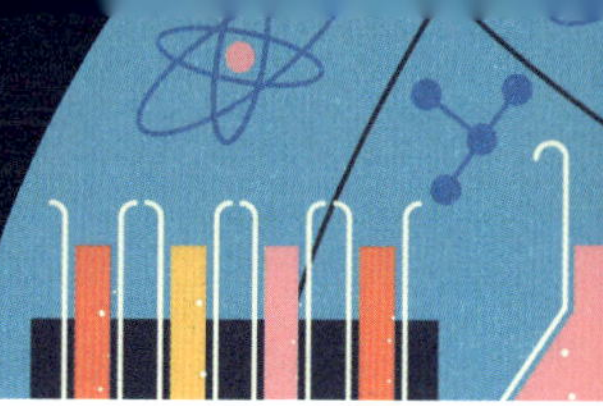

## 24 Polymers

### Condensation polymers

Condensation polymers form polyesters and polyamides.

**Condensation polymerisation** is the joining of **monomers** with the loss of a small molecule, usually water or hydrogen chloride.

### Polyesters

The condensation reaction between dicarboxylic acids and diols forms a polyester where the small molecule water is lost.

Terylene is a polyester made from benzene-1,4-dicarboxylic acid and ethane-1,2-diol. It is used in textiles.

$n\,HO{-}C(=O){-}C_6H_4{-}C(=O){-}OH + n\,H{-}O{-}CH_2{-}CH_2{-}OH$

benzene-1, 4-dicarboxylic acid ethane-1, 2-diol

$\downarrow$

$HO{+}C(=O){-}C_6H_4{-}C(=O){-}O{-}CH_2{-}CH_2{-}O{+}_n H + (2n-1)\,H_2O$

repeating unit 'Terylene' or 'Dacron'

### Polyamides

Dicarboxylic acids (or acyl chlorides) and diamines form polyamides.
An amide link forms and a water molecule (or HCl) is lost.

**Kevlar** is a polyamide combining 1,4-diaminobenzene and benzene-1,4-dicarbonyl chloride in a condensation reaction.

$n\,H_2N{-}C_6H_4{-}NH_2 + n\,ClOC{-}C_6H_4{-}COCl \rightarrow H{-}({-}NH{-}C_6H_4{-}NH{-}C(=O){-}C_6H_4{-}C(=O){-})_n{-}Cl + (2n-1)HCl$

1, 4-diaminobenzene benzene-1, 4-dicarbonyl chloride 'Kevlar'

**Nylon-6,6** is also a polyamide made from a condensation polymerisation reaction, that is used in textiles.

$$\left({-}NH{-}(CH_2)_6{-}NH{-}C(=O){-}(CH_2)_4{-}C(=O){-}\right)_n$$

The repeating unit of nylon-6,6

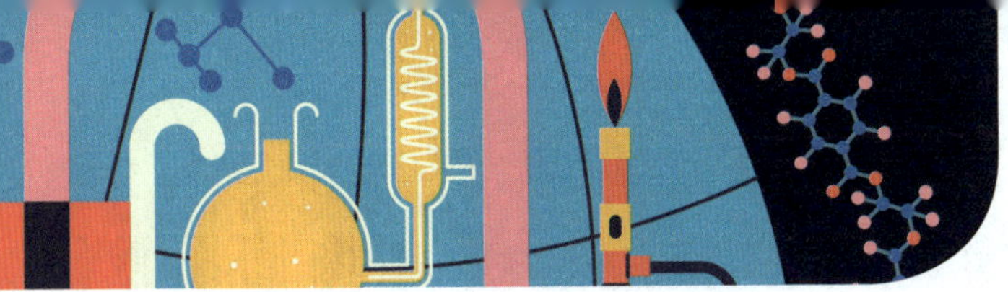



## Identifying monomers and repeating units

You need to be able to draw the polymer chain from given monomers resulting from a condensation or addition polymerisation reaction and identify the monomer from a polymer.

### Addition polymerisation

monomers $H_2N-(CH_2)_6-NH_2$ $HOOC-(CH_2)_4-COOH$ $H_2N-(CH_2)_6-NH_2$ $HOOC-(CH_2)_4-COOH$

$\downarrow$

$\cdots-NH-(CH_2)_6-NH-CO-(CH_2)_4-CO-NH-(CH_2)_6-NH-CO-(CH_2)_4-CO-\cdots$

{repeating unit}

$+ H_2O$ $+ H_2O$ $+ H_2O$ $+ H_2O$

### Condensation polymerisation

monomers $CH_2{=}CHCl$ $CH_2{=}CHCl$ $CH_2{=}CHCl$

$\downarrow$

$-CH_2{=}CHCl-CH_2{=}CHCl-CH_2{=}CHCl-$

repeating unit

## Hydrolysis of polyesters

$[-CO-C_6H_4-CO-O-CH_2-CH_2-CH_2-O-]_n$

base hydrolysis NaOH/$H_2O$ ⋮ $H^+/H_2O$ acid hydrolysis

$n^+Na^-OOC-C_6H_4-COO^-Na^+$ $n HOCH_2CH_2CH_2OH$ ⋮ $n HOCH_2CH_2CH_2OH$ $n HOOC-C_6H_4-COOH$

## Hydrolysis of amides

$[-CO-C_6H_4-CO-NH-C_6H_4-NH-]_n$

base hydrolysis NaOH/$H_2O$ ⋮ $H^+/H_2O$ acid hydrolysis

$n\ {}^+Na^-OOC-C_6H_4-COO^-Na^+$ + $n\ H_2N-C_6H_4-NH_2$ ⋮ $n\ H_3\overset{+}{N}-C_6H_4-\overset{+}{N}H_3$ + $n\ HOOC-C_6H_4-COOH$

# Retrieval

Learn the answers to the questions below, then cover the answers column with a piece of paper and write as many as you can. Check and repeat.

| | Questions | | Answers |
|---|---|---|---|
| 1 | How can polyesters and polyamides be broken down? | Put paper here | hydrolysis |
| 2 | What is the name of the link formed between dicarboxylic acids and diamines? | | amide link |
| 3 | What is lost in a condensation polymerisation? | | a small molecule such as water |
| 4 | What type of polymer is Terylene? | Put paper here | polyester/condensation |
| 5 | Which functional groups link to form a polyester? | | an alcohol and a carboxylic acid |
| 6 | What functional groups link to form a polyamide? | | amines and carboxylic acids or amides and acyl chlorides |
| 7 | What small molecule is lost in the reaction of amides with acyl chlorides? | Put paper here | HCl |
| 8 | What is needed for hydrolysis under basic conditions? | | $NaOH/H_2O$ |
| 9 | What are polyesters used for? | | textiles |
| 10 | What type of polymer is nylon-6,6? | Put paper here | polyamide/condensation |
| 11 | What type of polymer is Kevlar? | | polyamide/condensation |
| 12 | What is condensation polymerisation? | | the joining of monomers with the loss of a small molecule, usually water or hydrogen chloride |
| 13 | What conditions can the hydrolysis of polyesters be carried out in? | Put paper here | acidic or basic conditions |
| 14 | What conditions can the hydrolysis of polyamides be carried out in? | | acidic or basic conditions |
| 15 | How do we show the structure of a polymer chain? | Put paper here | using repeat units |
| 16 | What is a diol? | | a compound containing two hydroxyl groups (–OH) |
| 17 | What is a dicarboxylic acid? | | a compound containing two carboxyl groups (–COOH) |
| 18 | What is a diamine? | Put paper here | a compound containing two amine groups ($–NH_2$) |
| 19 | What is addition polymerisation? | | the joining of monomers with no by-products |
| 20 | Define hydrolysis. | | chemical reaction with water which causes the breakdown of a compound |

# Maths skills

Practise your maths skills using the worked example and practice questions below.

| Ratios, fractions, and percentages | Worked example | Practice |
|---|---|---|
| Understanding of percentages, fractions, and ratios is extremely important in percentage composition questions. Avoid making mistakes, especially in unusual questions, by understanding the mathematical concepts of ratios, fractions, and percentages.<br><br>For example, it is useful to recognise fractions in their decimal form. For the ratio:<br>1:2.66666<br>0.6666 is $\frac{2}{3}$.<br>Multiply both sides by 3 and you get a ratio of 3:8 which is much more helpful. | **Question**<br>Hydrocarbon **W** contains by mass 90% carbon and 10% hydrogen and has an $M_r$ of 80 g mol$^{-1}$. Determine the molecular formula of **W**.<br><br>**Answer**<br>Divide the percentage mass by the $A_r$:<br>$\frac{90}{12} = 7.5; \frac{10}{1} = 10$<br>Divide by the smallest ratio:<br>$\frac{7.5}{7.5} = 1; \frac{10}{7.5} = 1.333$<br>Clearly $CH_{1.3333}$ is not a formula. You should recognise that 1.3333 is a number a whole number divided by 3. So, multiply both ratios by 3.<br>$1 \times 3 = 3 ; 1.333 \times 3 = 4$<br>Therefore, the empirical formula is $C_3H_4$ with an $M_r$ of 40.<br>The $M_r$ of **W** is 80.<br>To determine the molecular formula of **W**, double the atoms of each element in the empirical formula, so the $M_r$ is 80.<br>The molecular formula of **W** is $C_6H_8$. This could be cyclohexa-1,3-diene. | Determine the molecular formula of the following hydrocarbons. (They may not be hydrocarbons covered in your specification. Do not worry about this, the point is to give you tricky or unusual examples.)<br><br>**1** 85.7% C and $M_r$ = 84 g mol$^{-1}$<br>**2** 96.0% C and $M_r$ = 50 g mol$^{-1}$<br>**3** 87.3% C and $M_r$ = 110 g mol$^{-1}$<br>**4** 93.8% C and $M_r$ = 128 g mol$^{-1}$ |

For answers and more practice questions visit www.oxfordrevise.com/scienceanswers

Even more practice and interactive revision quizzes are available on kerboodle

# Practice

## Exam-style questions

1 A huge number of polymers is used in industry. PVC and Kevlar are two of them.

**(a)** Complete **Table 1.1** showing the monomers, repeating units, and type of polymerisation of these polymers.

Use $(C_6H_4)$ to represent the benzene groups in Kevlar.

| Polymer | PVC | Kevlar |
|---|---|---|
| Repeating unit | $\left[ -\underset{H}{\overset{H}{C}}-\underset{H}{\overset{Cl}{C}}- \right]_n$ | |
| Monomer | | $H_2N-C_6H_4-NH_2$ 1,4-diaminobenzene; $ClOC-C_6H_4-COCl$ 1,4-benzenedicarbonyl chloridett |
| Type of polymerisation | | |

**Table 1.1** **[4]**

**(b)** Explain if there is a difference in atom economy in the synthesis of these polymers.

........................................................................................

........................................................................................

........................................................................................

........................................................................................

........................................................................................

........................................................................................

........................................................................................

........................................................................................

........................................................................................ **[3]**

**Exam tip**

Check that all of your atoms have the correct number of bonds and that you have not gained atoms from anywhere.

**(c)** Discuss the advantages and disadvantages of the disposal and of the recycling of PVC.

........................................................................................

........................................................................................

........................................................................................

........................................................................................

........................................................................................

........................................................................................

........................................................................................

........................................................................................

........................................................................................ **[3]**

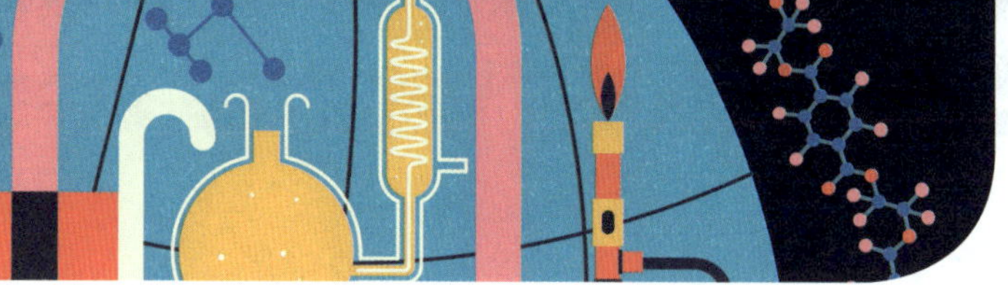



**2** Nylon-6,6 is synthesised from two monomers using condensation polymerisation, as shown in **Figure 2.1**.

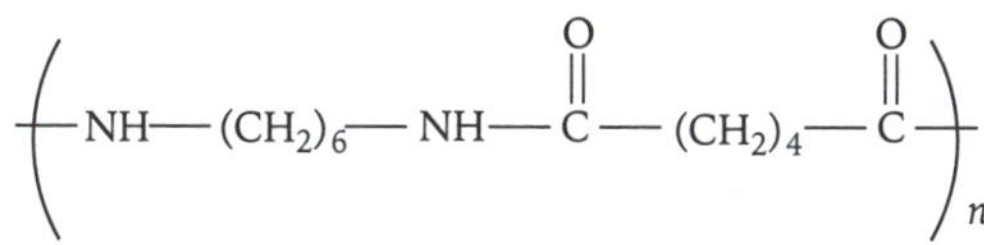


the repeating unit of nylon-6,6

**Figure 2.1**

**(a)** Suggest the structures of these two monomers.

**[2]**

**(b)** Explain why the melting point of nylon-6,6 is higher than that of its monomers.

..........................................................................................

..........................................................................................

..........................................................................................

..........................................................................................

.......................................................................................... **[2]**

**(c)** Nylon 6 is not made using condensation polymerisation. It is formed by ring-opening polymerisation using azepan-2-one, also known as caprolactam, shown in **Figure 2.2**.

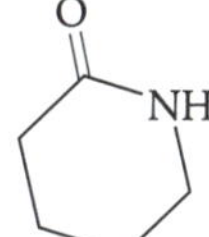


**Figure 2.2**

Give the IUPAC name for nylon 6.

..........................................................................................

.......................................................................................... **[1]**

**(d)** Give the repeating unit of nylon 6.

**[1]**

**Synoptic links**

2.2.2 2.1.3 1.2.1

**Exam tip**

Start by drawing in/imagining where the polymer breaks to form two monomers and draw the two separate parts. What could you use to fill in the gap?

**Exam tip**

What are the rules you know for naming polymers?

**(e)** The synthesis of nylon 6 in a school laboratory had a percentage yield of 60%. The students separated out the nylon 6 and measured its density as 1.06 g ml$^{-1}$.

Calculate the number of moles of azepan-2-one the students must have started with if 4 cm$^3$ of nylon 6 was produced. Azepan-2-one has an $M_r$ of 113 g mol$^{-1}$.

moles of azepan-2-one = .................... **[4]**

**(f)** Suggest two reasons why the students' percentage yield was so low.

..........................................................................................

..........................................................................................

.................................................................................... **[2]**

**3** Proteins are sequences of amino acids joined by amine links. A polymer made of two amino acids is shown in **Figure 3.1**.

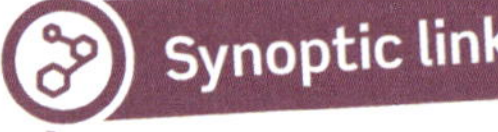

**Synoptic link**

6.3.1

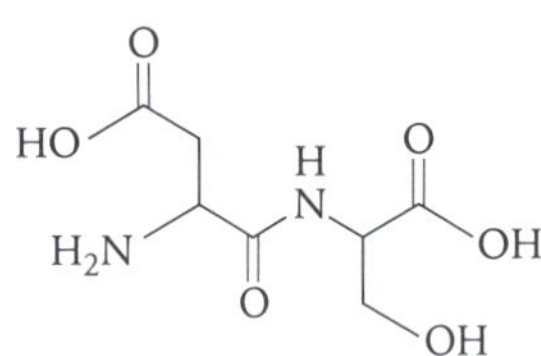


**Figure 3.1**

**(a)** Identify the amide link by circling it in **Figure 3.1**. **[1]**

**Exam tip**

Make sure you are clear with your circle.

**(b)** State the type of polymerisation that occurs here.

..........................................................................................

.................................................................................... **[1]**

**(c)** Give the structures of the amino acids made when the polymer shown in **Figure 3.1** is hydrolysed.

**[2]**

**(d)** Explain what laboratory technique could be used to separate the amino acids after hydrolysis and how this could be used to identify them.

..........................................................................................

..........................................................................................

..........................................................................................

.................................................................................... **[3]**

**Exam tip**

What process is used to make chemicals show up?

How do we then compare them to others?

4 This question is about 2-aminobutanoic acid.

**(a)** An amino acid, 2-aminobutanoic acid, was found to have the composition by mass of 46.6% C, 8.7% H, 13.6% N, and 31.1% O.

Determine the molecular formula of the amino acid if its molecular mass is 103.0 g $mol^{-1}$. Show your working.

.......................................................................................

.......................................................................................

.......................................................................................

.......................................................................................

.......................................................................................

....................................................................................... **[4]**

**Synoptic link**

2.1.3

**Exam tip**

Do not forget to work out the molecular formula using the molecular mass given too, for the last mark.

**(b)** 2-aminobutanoic acid has the chemical formula $C_4H_9O_2N$.

Give the structures of the two optical isomers of 2-aminobutanoic acid and show how they are related.

**[2]**

**(c)** 2-aminobutanoic acid can join with another molecule of itself to form a polymer.

**(i)** Give the structure of the product after the polymerisation of two molecules of 2-aminobutanoic acid.

**[1]**

**(ii)** State what type of polymerisation this is.

.......................................................................................

....................................................................................... **[1]**

5 This question is about hydrolysis.

**(a)** The equation below shows the acid hydrolysis of the ester ethyl ethanoate by sulfuric acid:

$$CH_3COOCH_2CH_2CH_3 + H_2O \rightarrow CH_3COOH + CH_3CH_2OH$$

**(i)** State why this equation does not show the sulfuric acid.

..............................................................................

.............................................................................. [1]

**(ii)** Give the displayed formulae of the two products formed. Show the functional groups clearly.

[2]

**(iii)** Give the formulae of the two products formed if you hydrolysed propyl ethanoate the same way.

..............................................................................

..............................................................................

..............................................................................

..............................................................................

.............................................................................. [2]

**(b)** Esters can be hydrolysed using alkali too.

**(i)** Give the equation for the hydrolysis of propyl propanoate using sodium hydroxide.

..............................................................................

..............................................................................

.............................................................................. [2]

**(ii)** Give the IUPAC name of one of the products.

..............................................................................

.............................................................................. [1]

**Synoptic link**

4.1.1

**Exam tip**

Include the charges on the salt formed.

6 Proteins are polymer structures made up of repeating units of monomers.

(a) **Figure 6.1** shows a short section of a protein chain, **A**.

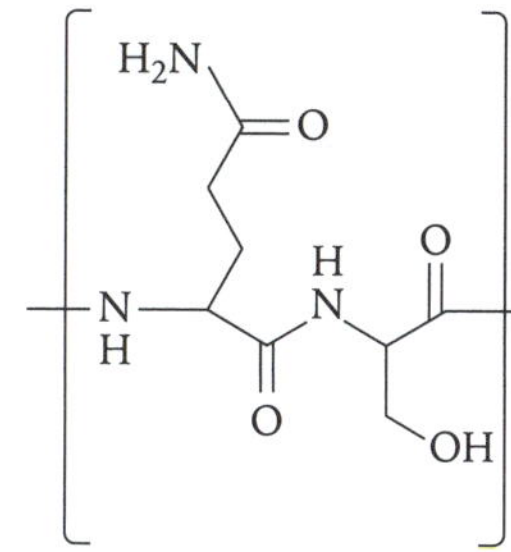
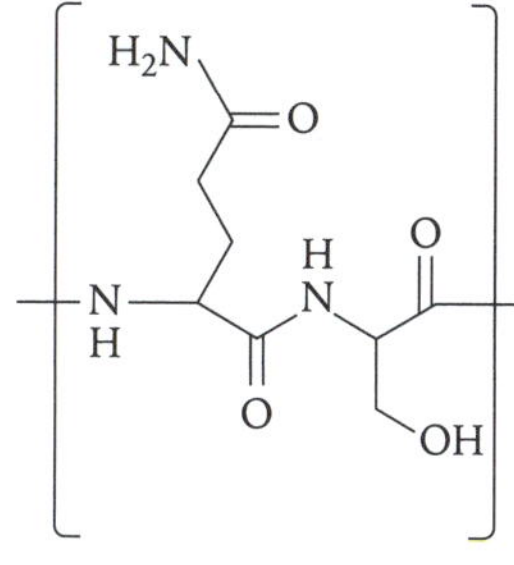


**Figure 6.1**

Identify the amide link in the **Figure 6.1**. **[1]**

(b) A student reacts the protein with hot aqueous sodium hydroxide, NaOH.

(i) State the name for this type of reaction.

..............................................................................................

.......................................................................................... **[1]**

(ii) Give the structures of the products formed from this section of the protein.

**[2]**

(c) The monomers and the polymers will have different melting points.

Explain why the melting points will be different.

Describe any forces involved.

..............................................................................................

..............................................................................................

..............................................................................................

..............................................................................................

..............................................................................................

.......................................................................................... **[3]**

**Synoptic link**

2.2.2

**Exam tip**

Ensure your circle is clear and does not go through any atoms.

Knowledge

# 25 Organic synthesis

## Formation of –C–C≡N

A nitrile, –CN, can be used to increase the chain length of a carbon compound.

The formation of the –C–C≡N bond can be by:

- *nucleophilic substitution* for haloalkanes and alcohols,
- *nucleophilic addition* for carbonyl compounds using HCN.

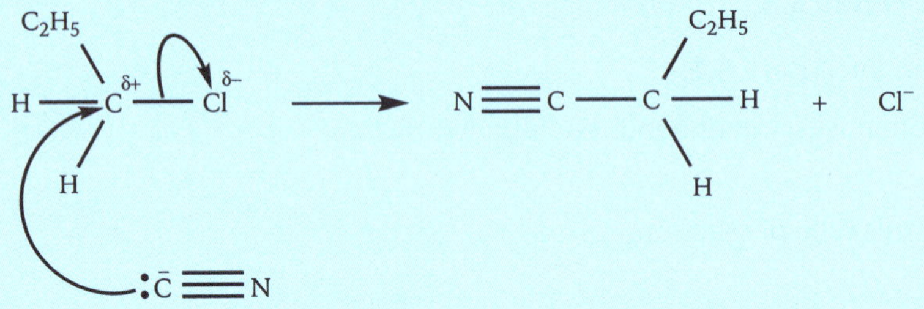

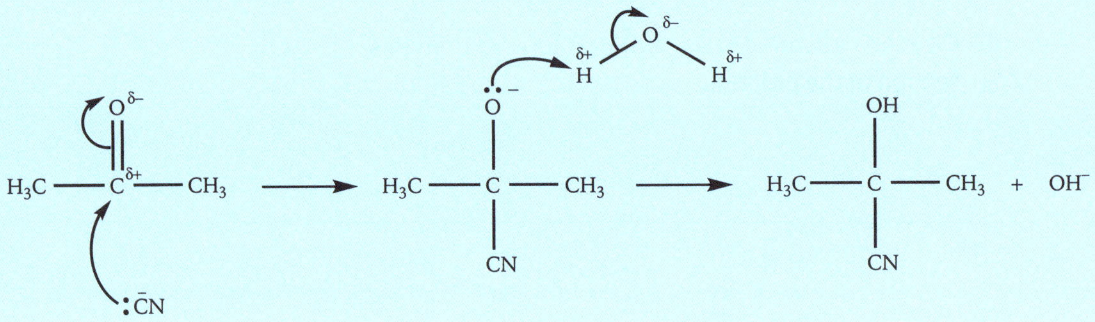

## Reaction of nitriles

Nitriles are reduced to amines using hydrogen and a nickel catalyst.

$$CH_3CH_2C{\equiv}N + 2H_2 \rightarrow CH_3CH_2CH_2NH_2$$

Nitriles undergo acid hydrolysis to form carboxylic acids. This is done by heating with water and hydrogen chloride.

$$CH_3CH_3CH_2C{\equiv}N + 2H_2O + HCl \rightarrow CH_3CH_2CH_2COOH + NH_4Cl$$

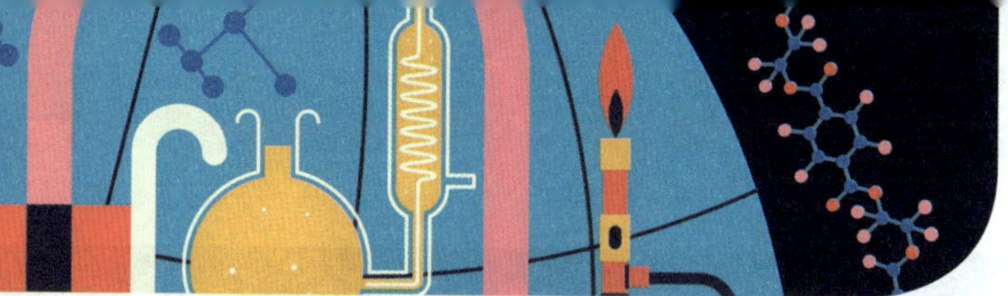



## Forming C–C bonds with benzene

The Friedel–Crafts reaction can be used in the formation of a substituted aromatic C—C bond in the presence of a halogen carrier.

- An *alkylation* reactions adds a haloalkane.
- An *acylation* reaction adds an an acyl chloride.

$$+\ C_2H_5Cl \xrightarrow{AlCl_3} (C_2H_5) + HCl$$

ethylbenzene

$$+\ C_2H_5COCl \xrightarrow{AlCl_3} C(=O)C_2H_5 + HCl$$

phenylpropanone

## Practical skills

Organic solids and solutions can be prepared by a range of techniques. Quickfit apparatus can be used to set up distillation and to heat reactions under reflux.

Solids are separated from organic solutions using a range of techniques such as using separating funnels and filtration under reduced pressure, followed by recrystallisation.

The purity of a solid can be determined by measuring its melting point; this will show if a compound is pure or a mixture.

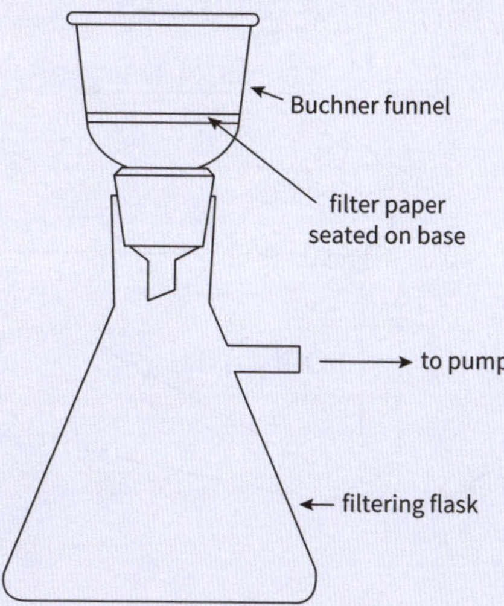

## Synthetic routes

An important part of modern chemistry is being able to synthesise one compound from another. This might be a compound with a number of functional groups, such as the hydroxyl group, carboxylic acid groups, ketones, and aldehydes.

From the functional groups shown in a compound you need to be able to predict the reactions of that compound with common tests, for example with a carbonate.

A chemist will aim to design a pathway that uses non-hazardous materials and reduces waste. You can use the following maps to plan organic synthesis pathways.

alcohol
alkene
carboxylic acid

haloalkane
ketone

## Common reactions

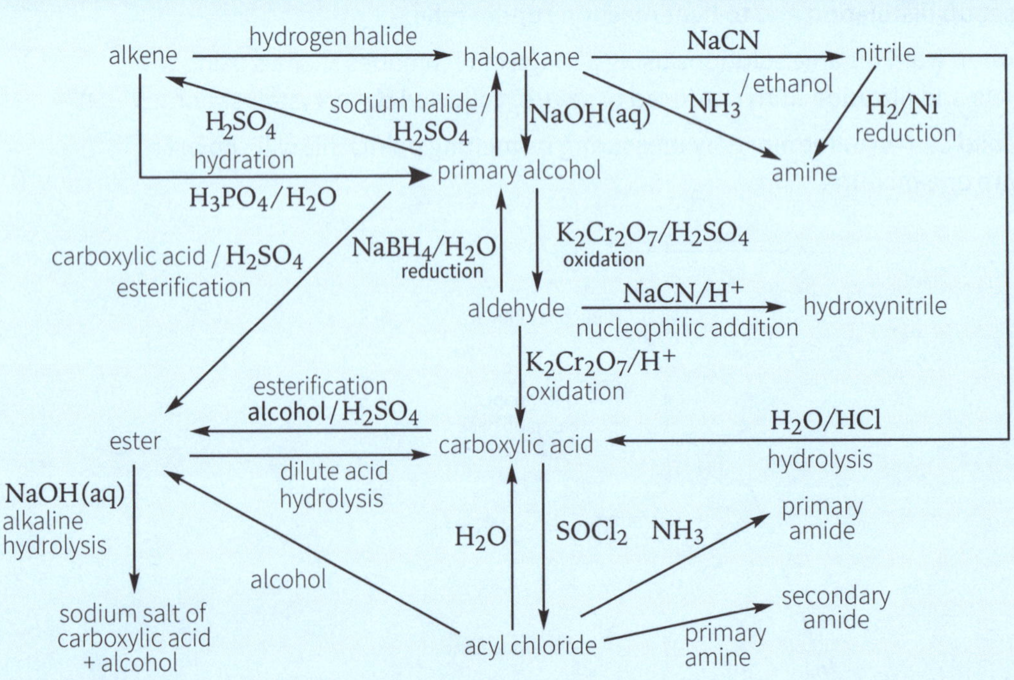

## Reactions of benzene

$CH_3$

$NO_2$

$NH_2$

$NH_2$

Br

Br

Sn/HCl

reduction

$Br_2$

Br

Br

$CH_3Cl/AlCl_3$

alkylation

$HNO_3/H_2SO_4$

nitration

H

HO—C—$CH_3$

O

$CH_3$

C

$CH_3COCl/AlCl_3$

acylation

$Br_2/FeBr_3$

bromination

Cl

$NaBH_4$

reduction

$Cl_2/AlCl_3$

chlorination

## Reactions of phenol

OH

Br

Br

OH

$NO_2$

and

OH

NaOH(aq)

$HNO_3$

nitration

$NO_2$

OH

$Br_2$

bromination

neutralisation

Br

$O^-Na^+$

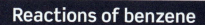

# Retrieval

Learn the answers to the questions below, then cover the answers column with a piece of paper and write as many as you can. Check and repeat.

| | Questions | Answers |
|---|---|---|
| 1 | What reagent is needed for the nitration of phenol? | nitric acid |
| 2 | What does a primary alcohol fully oxidise to? | carboxylic acid |
| 3 | What is formed when an aldehyde undergoes nucleophilic addition with hydrogen cyanide? | hydroxynitrile |
| 4 | What reagents are needed for the acylation of benzene? | $CH_3COCl$ and $AlCl_3$ |
| 5 | What reagents are needed for the reduction of an aldehyde to a primary alcohol? | $NaBH_4$ and $H_2O$ |
| 6 | What oxidising agent is used for alcohols? | acidified potassium dichromate |
| 7 | What is the intermediate product formed when converting a carboxylic acid to an amide? | acyl chloride |
| 8 | What does a primary alcohol partially oxidise to? | aldehyde |
| 9 | What is the intermediate product formed when converting a primary alcohol to a nitrile? | haloalkane |
| 10 | What type of reaction happens when converting a nitrile to a carboxylic acid? | hydrolysis |
| 11 | How can a carbon chain length be increased? | addition of cyanide to form a nitrile |
| 12 | What two functional groups are needed to form an ester? | alcohol and carboxylic acid |
| 13 | What catalyses the conversion of a primary alcohol to a haloalkane? | sulfuric acid |
| 14 | What is *Quickfit* apparatus used for? | distillation and heating reactions under reflux |
| 15 | What type of reaction happens when bromine is added to phenol? | bromination |
| 16 | What is the intermediate product when converting an acyl chloride to an aldehyde? | carboxylic acid |
| 17 | What reagents are needed for the reduction of nitrobenzene? | tin and HCl, then NaOH |
| 18 | What intermediate species is produced when converting a haloalkane to a carboxylic acid? | nitrile/alcohol then carbonyl |

Put paper here

| | | Put paper here | |
|---|---|---|---|
| 19 | Which catalyst is needed for the chlorination of benzene? | | $AlCl_3$ |
| 20 | Which catalyst is needed for the production of an alcohol from an alkene? | | phosphoric acid |

# Practical skills

Practise your practical skills using the worked example and practice questions below.

## Preparing an organic solid

When preparing and purifying an organic solid there are two additional steps compared to purifying an organic liquid:

- recrystallisation
- melting point determination to confirm purity.

Recrystallisation is done by the following method:

1. Dissolve the impure product in the minimum amount of hot solvent.
2. Leave the filtrate to cool and crystallise.
3. Filter under reduced pressure.

Measuring the melting point can confirm if something is pure, as pure substances have well-defined melting points. A melting point range of less than 2 °C would suggest a fairly pure substance.

## Worked example

Recrystallisation is used to purify a solid organic product.

### Questions

1. Initially the impure solid is dissolved in the minimum amount of hot solvent. Explain why the minimum amount of hot solvent is used.
2. When filtering the hot solution, it is often advisable to keep the filter funnel and filter paper hot or warm. Suggest why this is.
3. The warm filtrate is allowed to cool and the product precipitates. It is then filtered under reduced pressure while cold. Explain why:
   - **a** it is filtered under reduced pressure
   - **b** it is cold.

### Answers

1. This means a **saturated** solution is formed. This means when it cools the product will precipitate.
2. To prevent the product precipitating out in the filter paper and being lost.
3. **a** It is faster and leaves a drier residue.
   **b** So the product does not dissolve in solution.

## Practice

A student prepares a clean dry sample of aspirin (2-ethanoyloxybenzoic acid) from salicylic acid (2-hydroxybenzoic acid) and ethanoic anhydride.

1. Draw the organic structures above.
2. The reactants were heated with an acid catalyst under reflux. Explain why reflux should be used.
3. The student needs a precise measurement of the mass of salicylic acid. Suggest how the student should measure the mass of the acid (it is a solid).
4. The product was filtered under pressure and left to dry. Outline how the student could determine if the product was pure.

# Practice

## Exam-style questions

1 This question is about different synthetic routes for making esters.

**Synoptic links**

4.1.1 6.13

**(a)** Methyl propanoate can be made from an alcohol and a carboxylic acid.

Give an equation for the formation of methyl propanoate, $CH_2CH_3COOCH_3$, from methanol and propanoic acid.

**[2]**

**(b)** Methyl propanoate can also be made from the reaction between an alcohol and an acyl chloride.

Suggest a method to test the purity of the sample of methyl propanoate.

..............................................................................................

..............................................................................................

..............................................................................................

.............................................................................................. **[2]**

**Exam tip**

This is for two marks so include how to check the purity as well as your choice of method.

**(c)** Propanoic anhydride can be used instead of propanoyl chloride in the preparation of methyl propanoate from methanol.

**(i)** Give the displayed structure of propanoic anhydride.

**[1]**

**(ii)** Give one advantage of using propanoic anhydride instead of propanoyl chloride in the industrial manufacture of methyl propanoate from methanol.

..............................................................................................

..............................................................................................

.............................................................................................. **[1]**

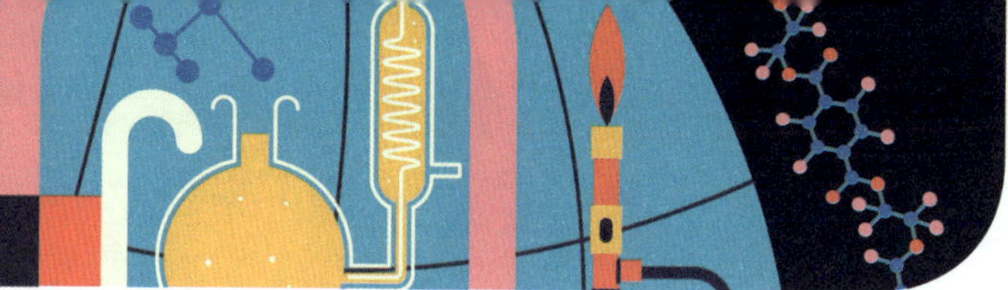



**2** The reaction scheme in **Figure 2** shows how to create compound **C** from compounds **A** and **B**.

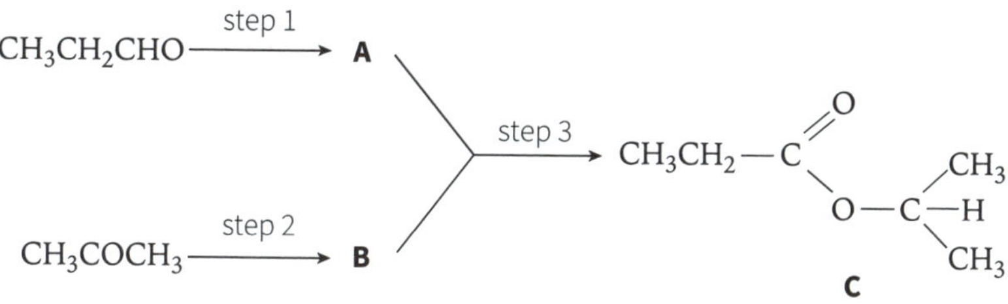


**Figure 2.1**

Synoptic link

4.1.1

**(a)** Step 1 is an oxidation reaction.

**(i)** Give the structure of product **A**.

**[1]**

**(ii)** Suggest the reagents and conditions needed for this step.

..............................................................................................

..............................................................................................

.............................................................................................. **[1]**

**(iii)** State the name of compound **A.**

..............................................................................................

..............................................................................................

.............................................................................................. **[1]**

**(b)** Step 2 produces an alcohol, **B**.

**(i)** Suggest the reagents and conditions needed for this step.

..............................................................................................

..............................................................................................

.............................................................................................. **[1]**

**(ii)** State the name for this type of reaction.

..............................................................................................

.............................................................................................. **[1]**

(iii) Give the structure of compound **B**.

**[1]**

**(c)** Step 3 is an esterification reaction, also known as a condensation reaction.

**(i)** State the reagents and conditions needed for step 3.

.......................................................................................

.......................................................................................

....................................................................................... **[1]**

**(ii)** State the reagents and conditions needed to hydrolyse the ester back to its starting reagents.

.......................................................................................

....................................................................................... **[1]**

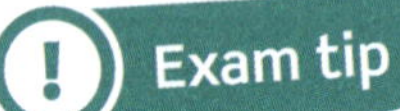

**Exam tip**

If this is hydrolysis, there must be water in there somewhere.

**(iii)** Identify the functional group(s) in compound **C** by circling it/them. **[1]**

**3** This question is about the synthesis of aminoethane.

**(a)** Chloroethane can be used to create aminoethane when reacted with an excess of ammonia.

Give an equation for this reaction. You do not need to include state symbols.

.......................................................................................

.......................................................................................

....................................................................................... **[2]**

**Synoptic link**

6.2.1

**(b)** Aminoethane can also be prepared by a reduction reaction.

Suggest a starting reagent and the conditions that can be used to prepare aminoethane by reduction.

.......................................................................................

.......................................................................................

....................................................................................... **[2]**

**(c)** These two methods have different atom economies.

Explain which of the methods used in **3(a)** and **3(b)** has the higher atom economy.

.......................................................................................

.......................................................................................

....................................................................................... **[2]**

**Exam tip**

You are not supposed to know this off the top of your head, so compare the atom economy of these two reactions to make your decision.

4 The reaction scheme in **Figure 4.1** shows a selection of reactions starting from butanone.

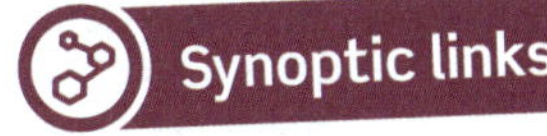

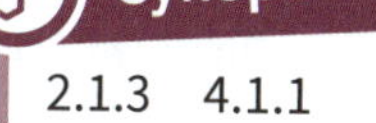

**Synoptic links**

2.1.3 4.1.1

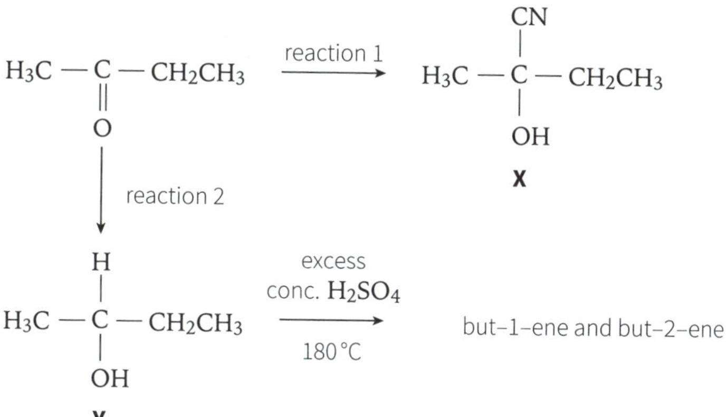


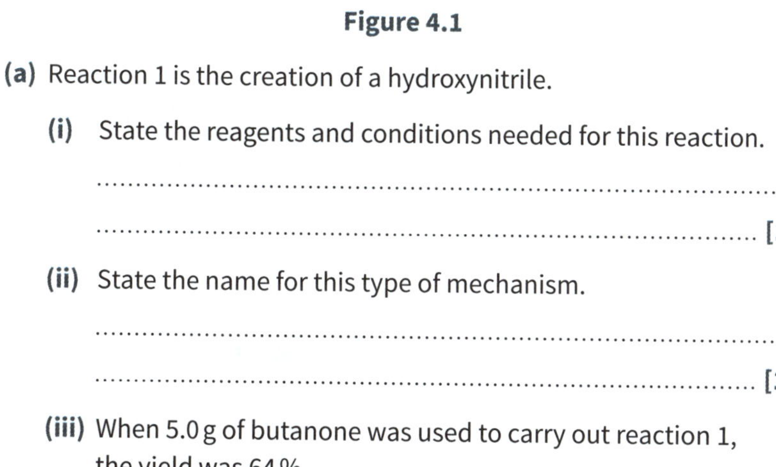


**Figure 4.1**

**(a)** Reaction 1 is the creation of a hydroxynitrile.

**(i)** State the reagents and conditions needed for this reaction.

..........................................................................................

.......................................................................................... **[1]**

**(ii)** State the name for this type of mechanism.

..........................................................................................

.......................................................................................... **[1]**

**(iii)** When 5.0 g of butanone was used to carry out reaction 1, the yield was 64 %.

Calculate the mass of compound **X** actually formed.

mass of compound **X** = ..................... g **[4]**

**(b)** Reaction 2 is a reduction reaction.

**(i)** State a suitable reagent for reaction 2.

..........................................................................................

.......................................................................................... **[1]**

**(ii)** Product **X** is found to be optically inactive. Suggest why this might be.

..........................................................................................

.......................................................................................... **[1]**

(c) The addition of excess concentrated sulfuric acid to compound **Y** followed by heating leads to the formation of a mixture of but-1-ene and but-2-ene.

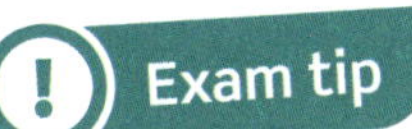

**Exam tip**

This is as simple as it sounds!

(i) Give the displayed formula of but-2-ene. [1]

(ii) Suggest why this is called a dehydration reaction.

..........................................................................................

.......................................................................................... [1]

**5** **Figure 5.1** shows the synthesis of 1-phenylpropene.

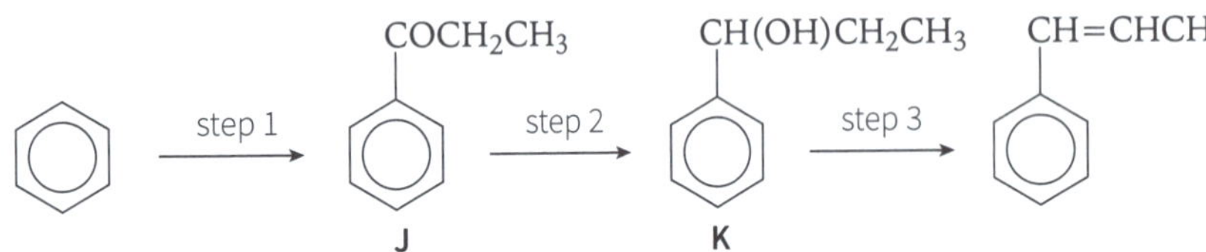


**Figure 5.1**

**Synoptic links**

6.1.1 4.1.3

(a) Step 1 is an electrophilic substitution reaction.

Suggest the mechanism for the attack of the electrophile on the benzene ring.

Represent the electrophile as $[E^+]$.

**Exam tip**

Don't forget to regenerate the aromaticity of the benzene ring in your last step.

[3]

(b) Step 2 is the formation of an alcohol.

(i) Give the name of the mechanism for step 2.

..........................................................................................

.......................................................................................... [1]

(ii) State the reagents and conditions needed for step 2.

..........................................................................................

.......................................................................................... [1]

(iii) State, in terms of their structure, why molecules of **Q** show optical isomerism.

..........................................................................................

.......................................................................................... [1]

(c) The final step in this synthesis is a dehydration reaction.

(i) Suggest a suitable reagent for step 3.

..........................................................................................

.......................................................................................... [1]

**(ii)** Give the name for the type of stereoisomerism shown by the product of this reaction.

.......................................................................................

....................................................................................... **[1]**

**(iii)** State what is required in the structure of molecules to allow them to show this type of stereoisomerism.

.......................................................................................

....................................................................................... **[1]**

**6** The conversion of compound **P** into compound **R** can be achieved in two steps, as shown in **Figure 6.1**.

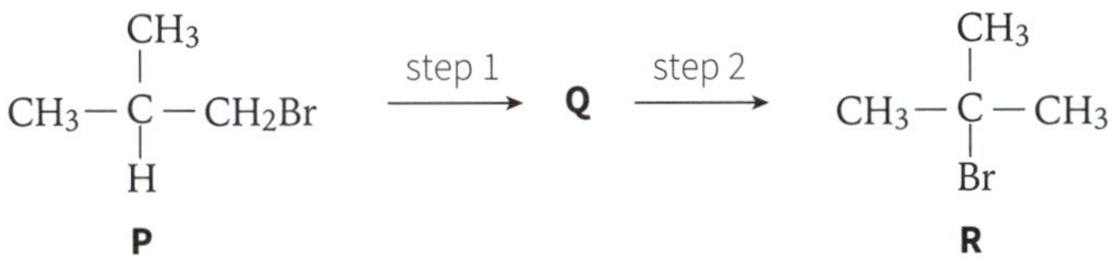



**Figure 6.1**

4.1.3 6.3.2

**(a)** The intermediate compound **Q** has an absorption peak at 1650 $cm^{-1}$ in its infrared spectrum.

Identify the structure of compound **Q** and explain your answer.

The second mark here is for the explanation.

Explanation:.......................................................................

....................................................................................... **[2]**

**(b)** Step 2 is an addition reaction.

**(i)** State the reagents and conditions needed for step 2.

.......................................................................................

....................................................................................... **[1]**

**(ii)** Outline the mechanism for the reaction in step 2.

**[1]**

**(iii)** Describe how the number of peaks in their $^{1}H$ NMR spectra would enable you to distinguish between compounds **P** and **R**.

.......................................................................................

.......................................................................................

....................................................................................... **[2]**

# Knowledge

## 26 Chromatography and spectroscopy (NMR)

### Thin layer chromatography (TLC)

Thin layer chromatography (TLC) can be used in the separation of components within a mixture. This depends on the affinity of the compound for the mobile and stationary phases. How far a sample moves depends on its solubility in the mobile phase and how easily it can move through the stationary phase. The more soluble a sample is in the solvent, the faster it will run; this can depend on things such as polarity. The stationary phase will slow down compounds that are attracted to it, or large compounds that have trouble moving through the stationary phase.

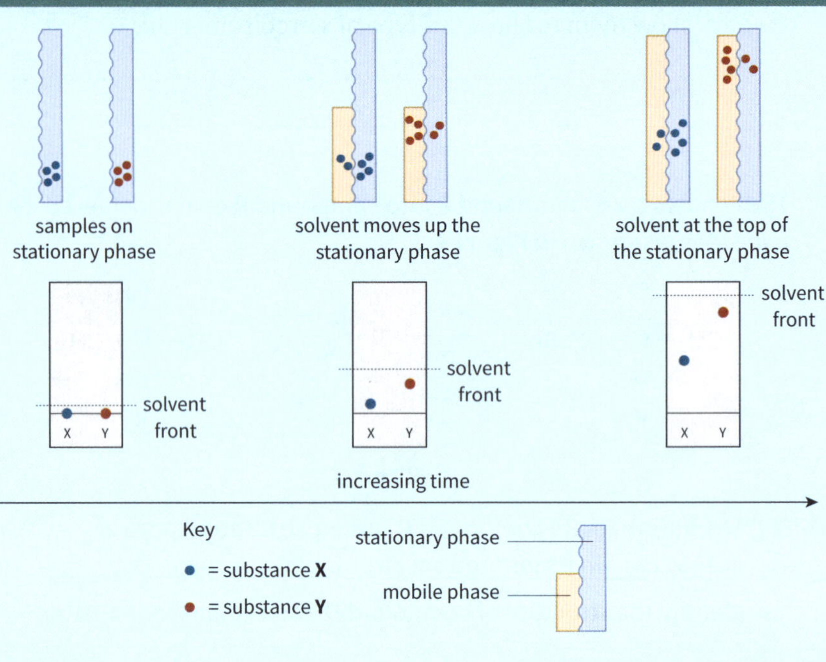


### Interpretation of chromatograms

Samples can be identified based on **retention times** (the time it takes to pass through a GC column) or **$R_f$ values**.

On a TLC plate $R_f$ values can be found using the equation:

$$R_f = \frac{\text{distance moved by spot}}{\text{distance moved by solvent front}}$$

Each component in the sample can be identified by its $R_f$ value, using the same absorbent and solvent.

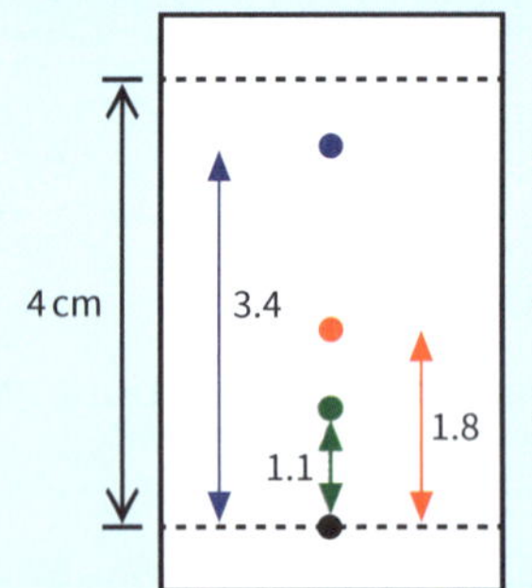


blue spot : $Rf = \frac{3.4}{4.0} = 0.85$

red spot : $Rf = \frac{1.8}{4.0} = 0.45$

green spot : $Rf = \frac{1.1}{4.0} = 0.23$

### Gas chromatography

Gas chromatography (GC) has a column packed with a solid. This solid could be coated by a liquid. A gas is passed along the column at high pressure and temperature.

Retention time can be used to identify compounds in samples by comparing them to known reference samples.

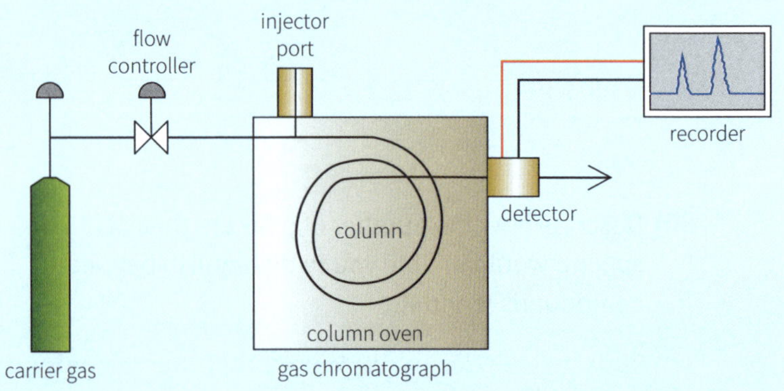

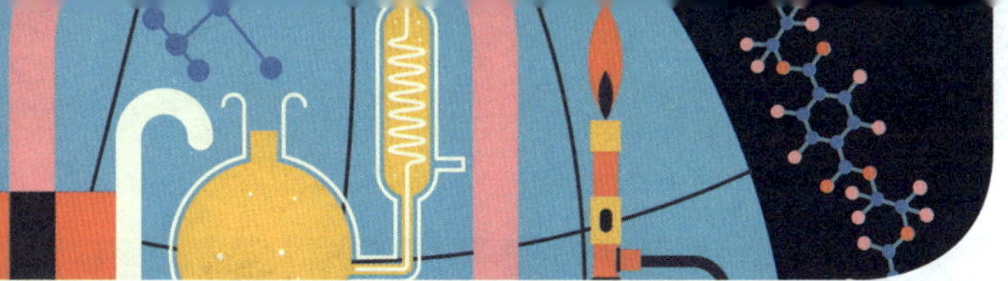



## Gas chromatography-mass spectrometry

After samples have gone through gas chromatography they can be run on a mass spectrometer.

Gas Chromatography–Mass Spectrometry (GCMS) can be used to analyse components that have been separated. This can be done automatically as each sample comes off the end of a column. The retention time can be combined with the mass and fragmentation pattern to analyse the sample.

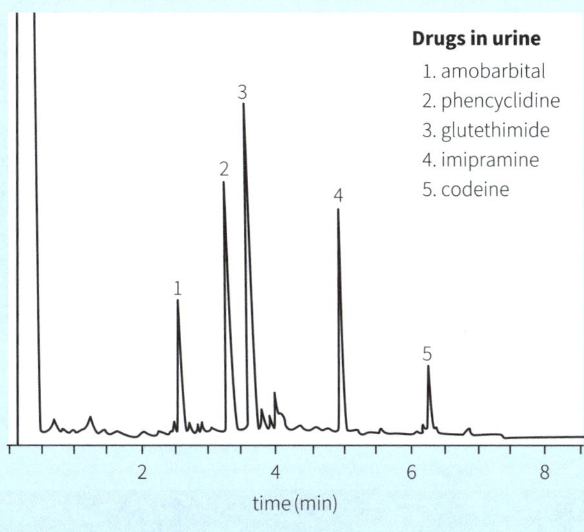


The concentration of a substance in a given sample can be found by comparing the **peak integration values** (area under the peak) with values from a standard solution.

A combination of retention times and peak integration can be used to determine the amounts and proportions of components within a mixture.

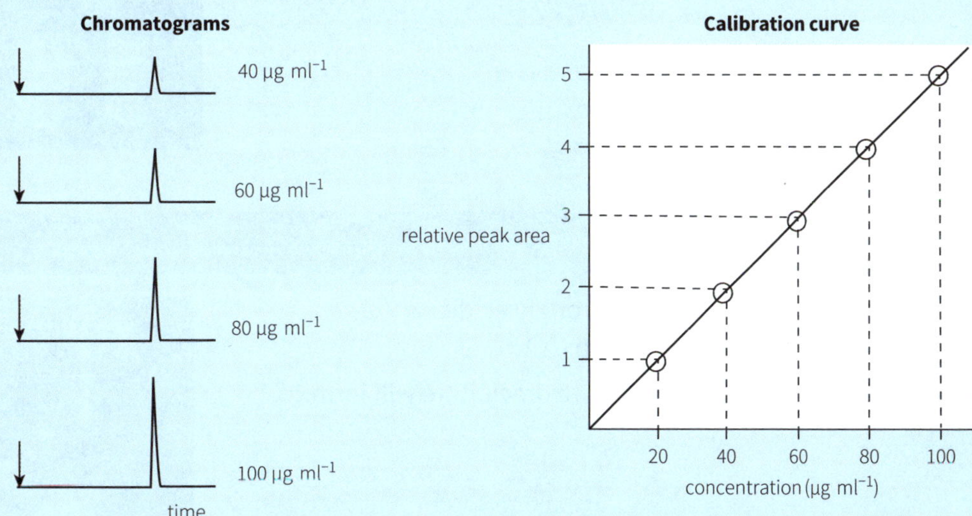

## Test for organic functional groups

There is a range of qualitative tests that can be done in a test tube to analyse functional groups in organic compounds.

## Test for alkenes

Alkenes can be identified by the reaction with bromine or bromine water. Orange/brown bromine water is added to a solution and gently shaken. The bromine water will decolourise giving a positive result for an alkene.

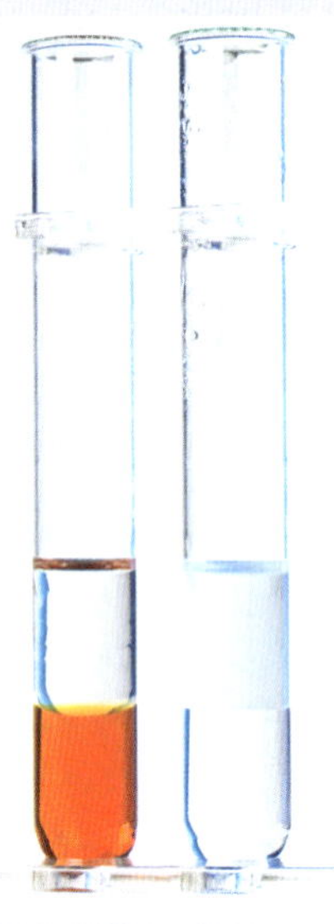

## Test for haloalkanes

The different haloalkanes can be identified and differentiated by warming with aqueous silver nitrate in ethanol. Each haloalkane produces a different coloured precipitate. Chloride ions give a white precipitate, bromide ions will give a cream precipitate, and iodide ions will give a yellow precipitate.

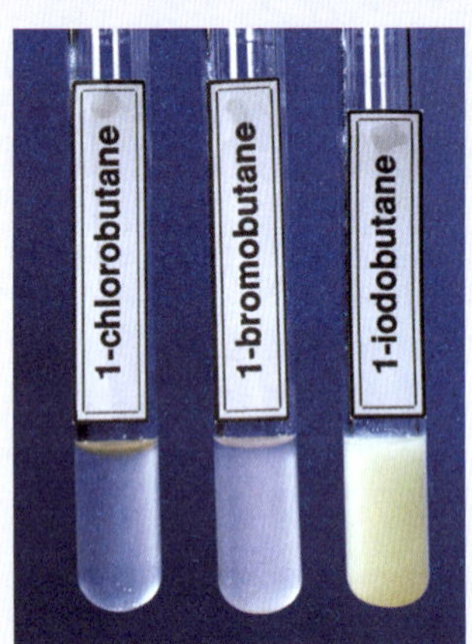


## Test for phenols

Phenols are weak acids, they are not as strongly acidic as carboxylic acids and show no reaction with $CO_3^{2-}$.

When bromine is added to a solution a white precipitate will form.

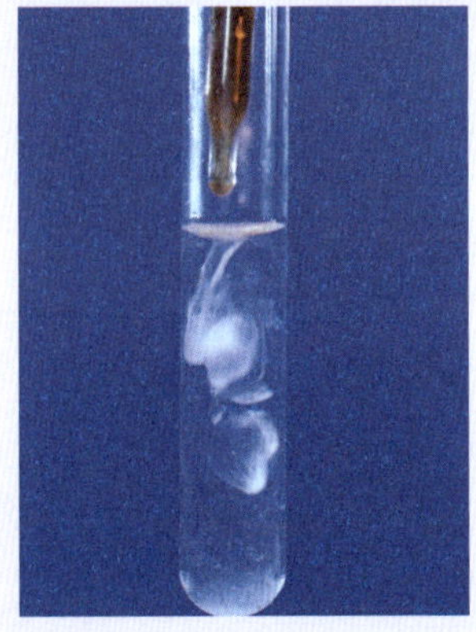

## Test for carbonyl groups

A compound with carbonyl groups can be identified by its reaction with 2,4–DNP: an orange precipitate will be formed.

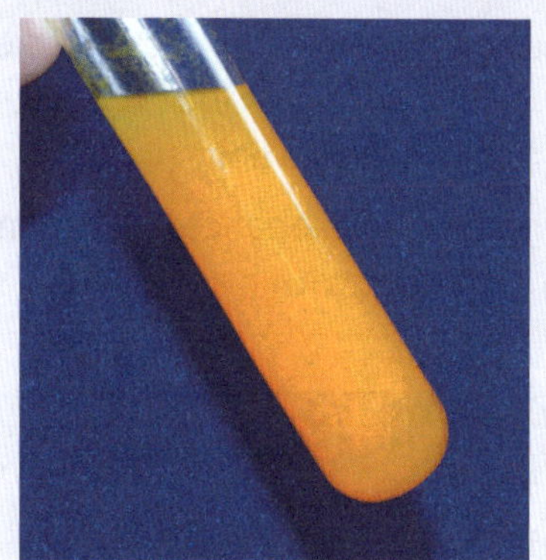

## Test for aldehydes

Aldehydes can be differentiated from ketones by their reaction with Tollens' reagent: aldehydes will form a 'silver mirror' on the inside of a test tube.

## Test for primary and secondary alcohols

Primary and secondary alcohols, and aldehydes, can be identified by their reaction with orange acidified potassium dichromate(VII) to form a dark green precipitate.

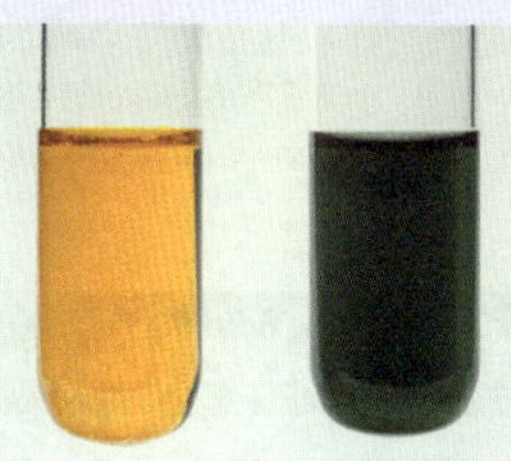

## Test for carboxylic acids

Carboxylic acid groups can be identified by the production of effervescence after reaction with $CO_3^{2-}$.

## Nuclear magnetic resonance

To identify an unknown compound scientists can use a range of analytical techniques. These can include the tests of anions and cations covered previously or be more sophisticated.

Nuclear magnetic resonance (NMR) can be used to determine the position of $^{13}C$ or $^{1}H$ atoms within a molecule.

## Chemical shift

The *x*-axis of the graph shows **chemical shift**, δ, measured in parts per million (ppm). This can be used to determine the types of bond present in a sample.

This data will be given to you in the *Data Sheet*, you do not need to recall these values.

## Tetramethylsilane

**Tetramethylsilane (TMS)** is used as a standard in NMR investigations. It is inert, non toxic, and easy to remove from the sample. All of the carbons and protons in TMS are in identical environments, so give a single sharp peak. This shift is set as 0 (δ = 0 ppm) and all other peaks are measured relative to this.

## $^{13}C$ NMR chemical shifts relative to TMS

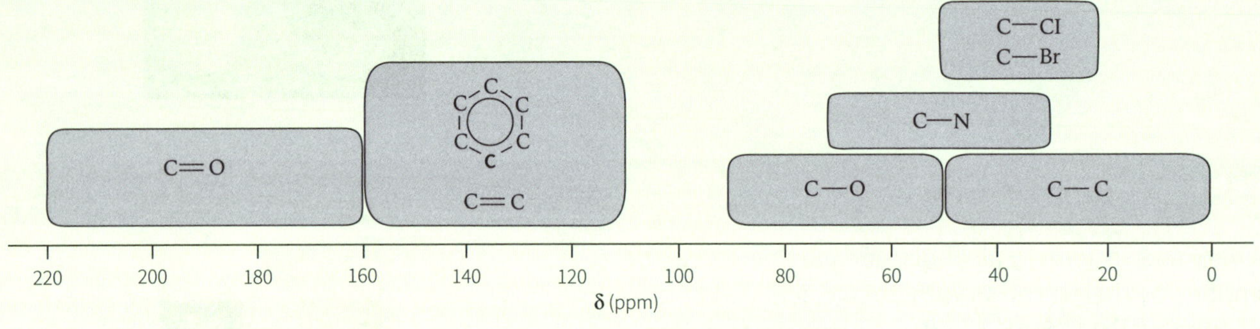


## Deuterated solvents

Deuterated solvents (where the hydrogen atoms have been replaced with deuterium, $^{2}_{1}H$) such as $CDCl_3$, are used when running an NMR spectrum to avoid the signal from $^{1}_{1}H$ in solvents being confused with a signal from the sample.

Proton exchange using $D_2O$ can be used to identify –OH and –NH protons. After an initial NMR spectrum is run, $D_2O$ is added and another NMR is run. Deuterium replaces the protons in –OH and –NH and the peak disappears.

## Integration trace

The peak on a $^{1}H$ -NMR spectra not only gives the type of environment that a proton is in, but also the number of protons in that environment. The area under each peak is the **integration trace**: this gives the *number of protons* in that particular environment.

## Carbon-13 NMR spectroscopy

Propanone and propanal are isomers with the same molecular formula but different arrangements of atoms. The spectra are dependent on the environment of the carbons, so propanone and propanal show different patterns.

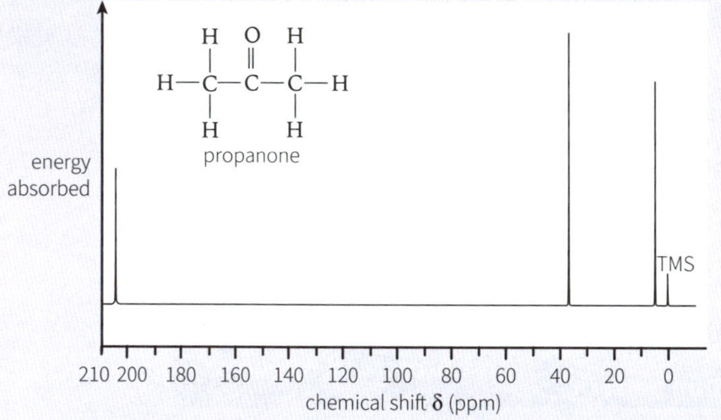


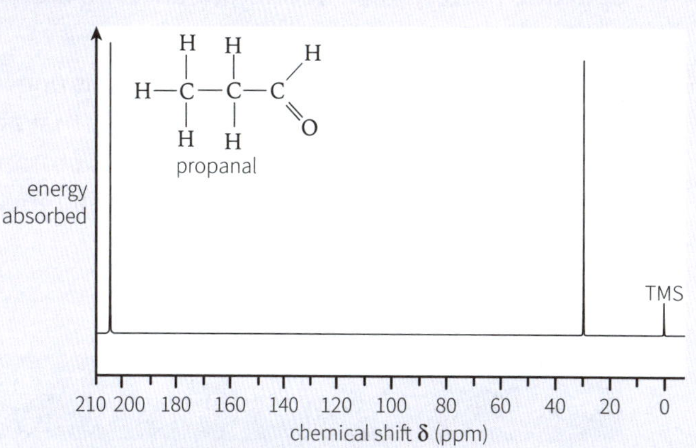


Propanal also has 3 carbons, but is not symmetrical, so it has three different environments and the chemical shift shows three different peaks.

Propanone is symmetrical, it has 3 carbons, with the carbons in two different environments.

## Proton NMR spectroscopy

Proton, $^1H$ NMR spectroscopy looks at the proton environments. These environments are not just the atom that the proton is directly connected to, but the adjacent atoms as well.

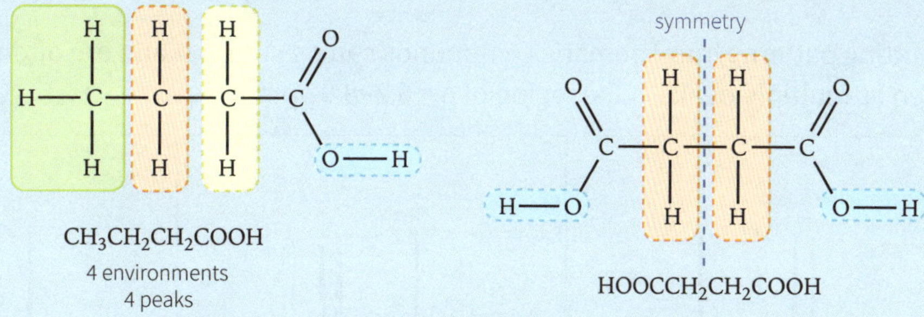


Butanoic acid and butanedioic acid have the same number of carbons, but different numbers of proton environments. Butanedioic acid has a line of symmetry down the middle, so it has two different proton environments, and produces two peaks. Butanoic acid produces four different environments, so produces four peaks.

## $^1H$ NMR chemical shifts relative to TMS

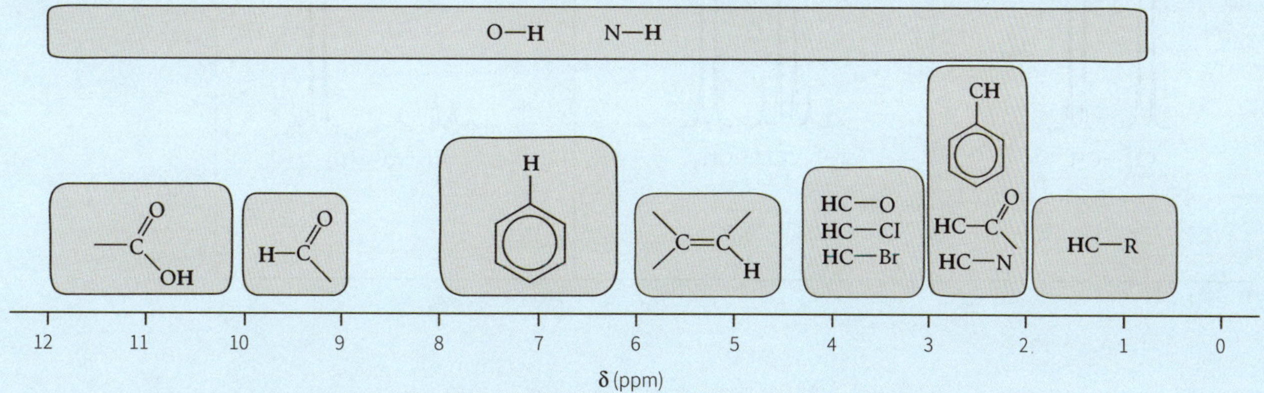

## Spin–spin splitting patterns

A closer look at the peaks on a $^1$H-NMR spectrum reveals that each peak is split. The **spin–spin splitting patterns** can be interpreted using the ***n* + 1 rule**.

The number of peaks in a splitting pattern is one more than the number of adjacent protons that are causing the splitting, so for a proton with $n$ protons attached to the adjacent carbon atom the number of peaks seen in the splitting pattern is $n + 1$.

For a proton that is attached to a carbon with no adjacent hydrogen atoms, a **singlet** pattern is produced. A proton attached to a carbon with an adjacent –CH shows a **doublet** pattern.

| $n$ | $n + 1$ | Splitting pattern | Relative peak areas within pattern | Structural features |
|---|---|---|---|---|
| 0 | 1 | singlet | 1 | no H on adjacent atoms |
| 1 | 2 | doublet | 1 : 1 | adjacent –CH |
| 2 | 3 | triplet | 1 : 2 : 1 | adjacent $-CH_2$ |
| 3 | 4 | quartet | 1 : 3 : 3 : 1 | adjacent $-CH_3$ |

Splitting patterns from aromatic compounds can be complex and are often seen as multiple peaks in the region of δ = 6.2–8.0 ppm.

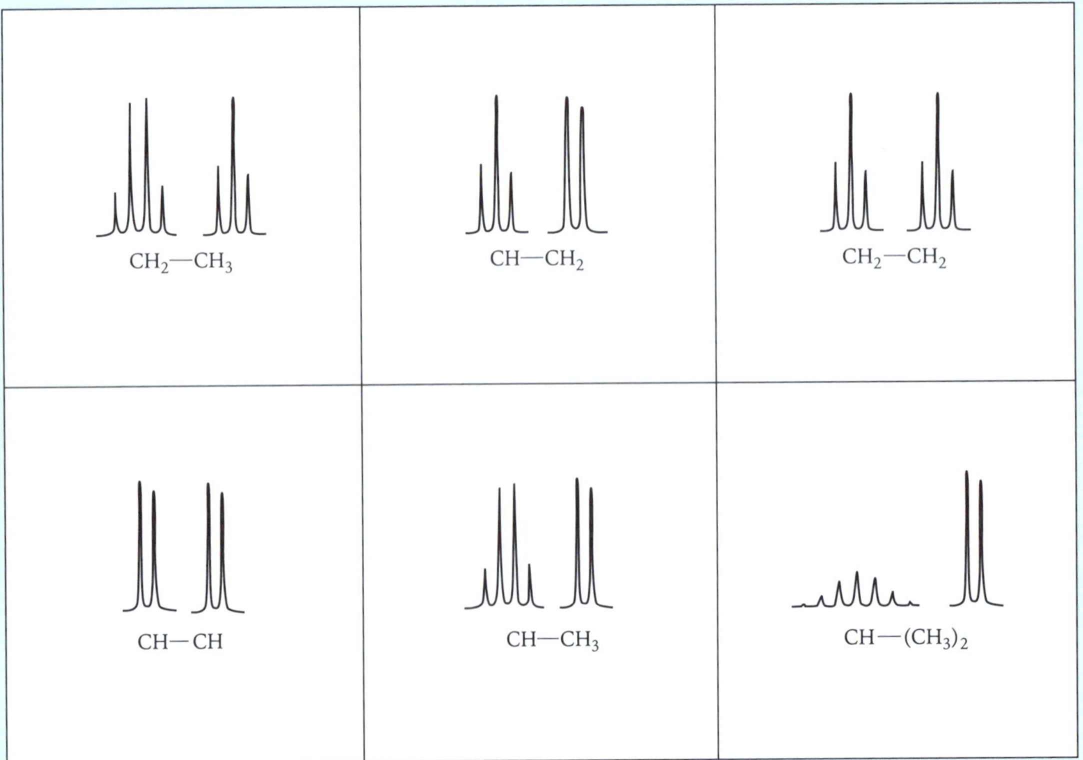

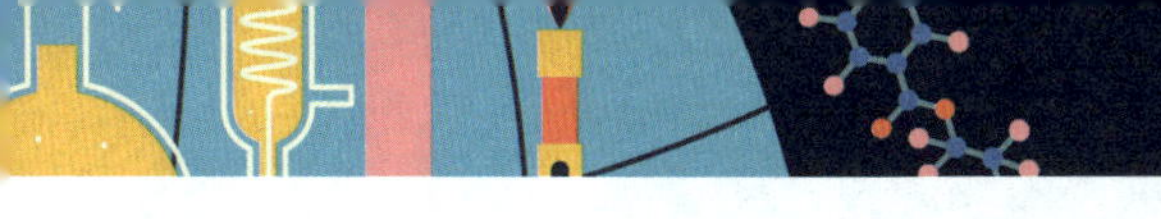

## Example of a spin-spin splitting pattern

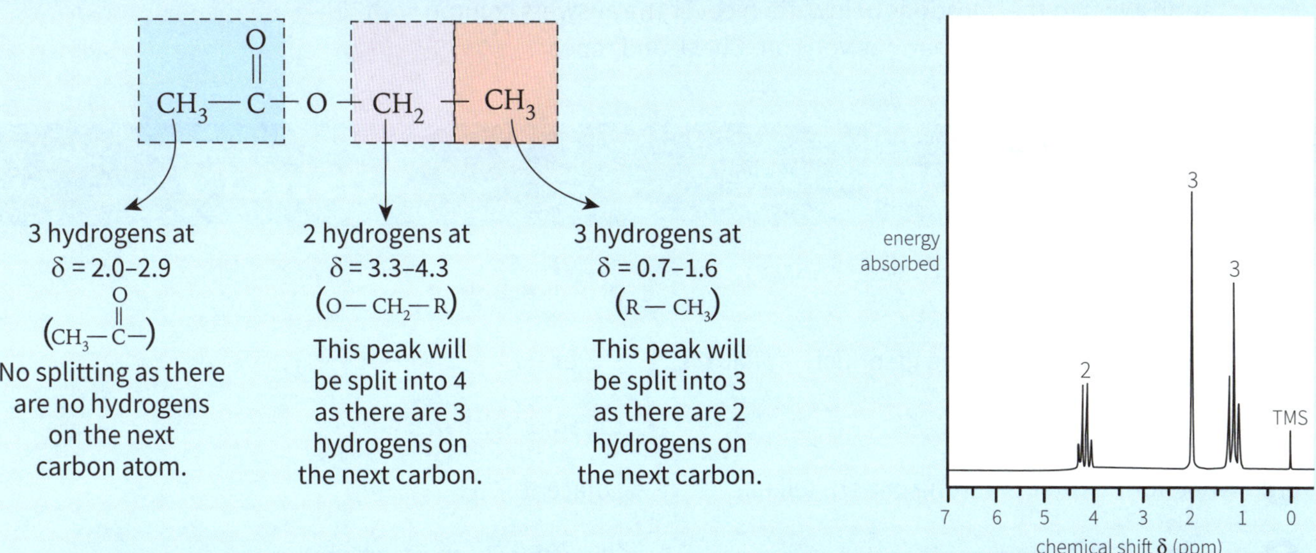


## Combined techniques

The structure of molecules in samples can be determined by a range of combined techniques. **Elemental analysis** determines the percentage composition by mass of each element which gives the empirical formulae (e.g. $C_5H_6O$ ), this can be combined with **mass spectra** data to determine the molecular formula.

The **IR spectra** shows what functional groups are in a compound and the NMR spectra gives information on the type of proton and carbon environments.

All this information can be combined to give the potential structure for a molecule.

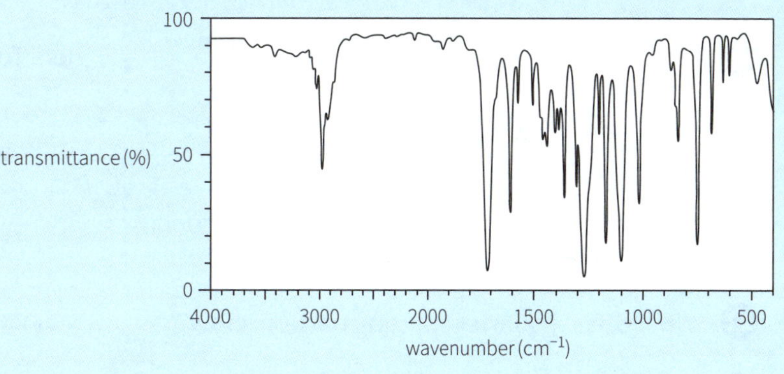


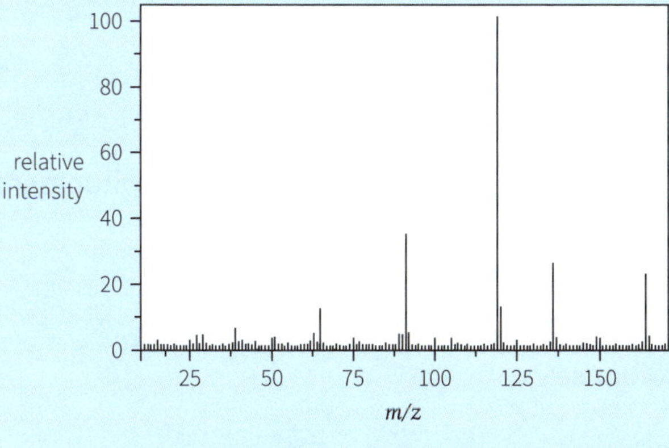


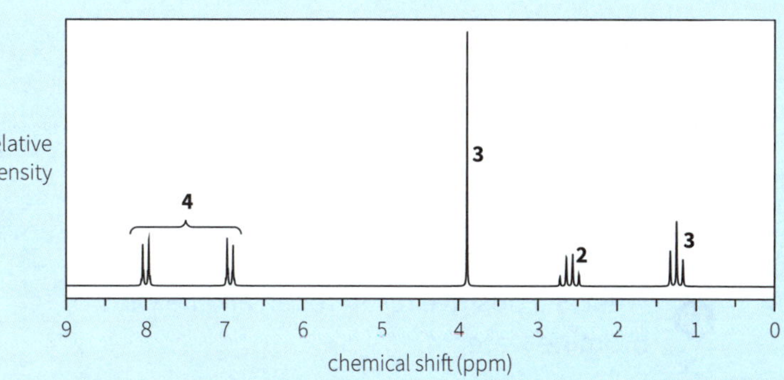

# Retrieval

Learn the answers to the questions below, then cover the answers column with a piece of paper and write as many as you can. Check and repeat.

| | Questions | | Answers |
|---|---|---|---|
| 1 | What is the area under each peak in a gas chromatogram called? | Put paper here | peak integration |
| 2 | Which two isotopes can be identified in NMR? | | $^{13}C$ or $^{1}H$ |
| 3 | What does TMS stand for? | | tetramethylsilane |
| 4 | What does a triplet splitting pattern tell us? | Put paper here | adjacent $-CH_2$ |
| 5 | What does TLC stand for? | | thin layer chromatography |
| 6 | Which rule can be used to interpret spin–spin splitting patterns? | | $n + 1$ |
| 7 | Which two ways can samples be identified from chromatography? | Put paper here | retention times or $R_f$ values |
| 8 | How can aldehydes be differentiated from ketones? | | use Tollens' silver mirror test |
| 9 | What does a positive result for a carbonyl tested with 2,4-DNP look like? | | orange precipitate |
| 10 | How many hydrogens are on adjacent carbon atoms for a 1 : 3 : 3 : 1 pattern? | Put paper here | 3 |
| 11 | What does a quartet splitting pattern show? | | adjacent $-CH_3$ |
| 12 | What does NMR stand for? | | nuclear magnetic resonance |
| 13 | Why is TMS used? | Put paper here | all protons and carbon atoms are in the same environment |
| 14 | What colour is potassium dichromate(VII)? | | orange |
| 15 | How are $R_f$ values calculated? | | $R_f = \frac{\text{distance moved by spot}}{\text{distance moved by solvent front}}$ |
| 16 | What does the integration trace show? | Put paper here | the number of protons in that particular environment |
| 17 | What does each peak represent on a carbon NMR spectrum? | | the carbon environment |
| 18 | What is a positive result for an alkene with bromine water? | Put paper here | decolourises |
| 19 | What are the units on the $x$-axis of NMR spectra? | | δ (ppm) |
| 20 | What is the chemical shift for TMS? | | δ = 0 ppm |

# Maths skills

Practise your maths skills using the worked example and practice questions below.

## Logarithms

Logarithms are basically the opposite of powers. For example:

$$10^x = 45$$

To solve for $x$, undo the 'ten to the power of', take $\log_{10}$ of each side.

$$\log_{10}10^x = \log_{10}45$$

$\log_{10}$ is the opposite of ten to the power of, they cancel, so,

$$x = \log_{10}45 = 1.65 \text{ (to 3 s.f.)}.$$

Sometimes $\log_{10}$ is written as log or lg.

The natural logarithm, $\log_e x$ is written as $\ln x$. The mathematical constant, e, is an important number in mathematics. In chemistry, e, appears in the Arrhenius equation:

$$k = Ae^{-\frac{E_a}{RT}}$$

To rearrange this equation you need to use ln.

## Worked example

**Questions**

Solve the following equations.

**1** $10^a = 90$

**2** $10^{b+5} = 500\,000$

**3** $4e^c = 20$

**Answers**

**1** Take log of both sides to undo the 'ten to the power of'.

$$a = \log_{10}90 = 1.95$$

**2** Take logs, then subtract 5.

$$b + 5 = \log_{10}500\,000 = 5.699$$

$$b = 5.699 - 5 = 0.699$$

**3** First divide by 4, then undo the 'e to the power of' by taking ln of both sides.

$$e^c = \frac{20}{4} = 5$$

$$c = \ln 5 = 1.61$$

## Practice

**1** Calculate the following expressions:

**a** $\log_{10}150$

**b** $\ln 86$

**c** $\log_{10} 1.34 \times 10^4$

**d** $\ln 1.30 \times 10^{-14}$

**e** $\log_{10}20$

**f** $\log_{10}3$

**g** $\log_{10}60$

**2** Solve the following equations:

**a** $10^{-w} = 3.2 \times 10^{-13}$

**b** $50e^x = 1250$

**c** $10^{y-3} = 316$

**d** $7.50e^{-\frac{1000}{z}} = 1.37 \times 10^{-1}$

For answers and more practice questions visit www.oxfordrevise.com/scienceanswers

Even more practice and interactive revision quizzes are available on kerboodle

# Practice

## Exam-style questions

1 The isomeric compounds in **Figure 1.1** were believed to be present in a mixture in a reagent bottle.

**Synoptic links**
2.2.2k 6.1.1

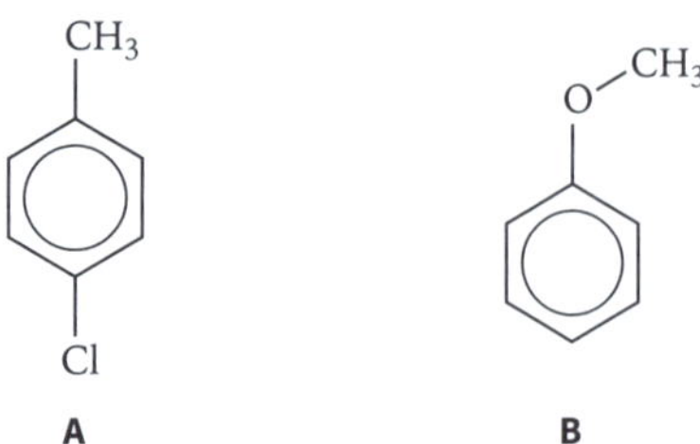

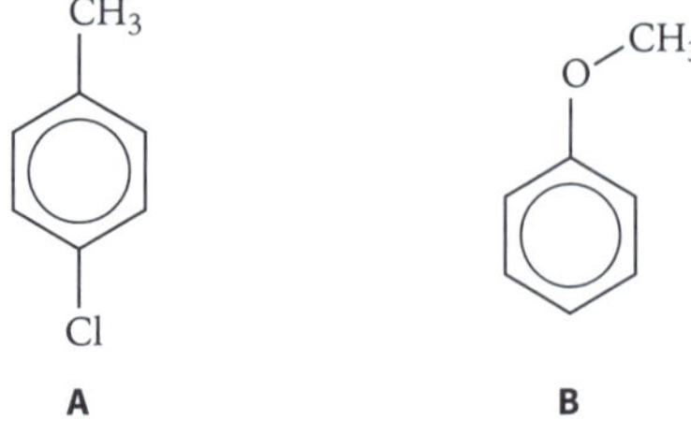

**Figure 1.1**

**(a)** For each compound, describe the main type of intermolecular force between the molecules of the compound.

**A** ..........................................................................................

**B** .......................................................................................... **[2]**

**(b)** Explain how **A** can be distinguished from **B** by a simple chemical test.

**(i)** Describe the test used.

..........................................................................................

.......................................................................................... **[1]**

**(ii)** Describe the observations.

..........................................................................................

.......................................................................................... **[1]**

**(iii)** Give the equation for the reaction.

..........................................................................................

.......................................................................................... **[1]**

**(c)** A student was asked to separate the mixture using gas–liquid chromatography. A polar stationary phase was used.

The chromatogram obtained is shown in **Figure 1.2**.

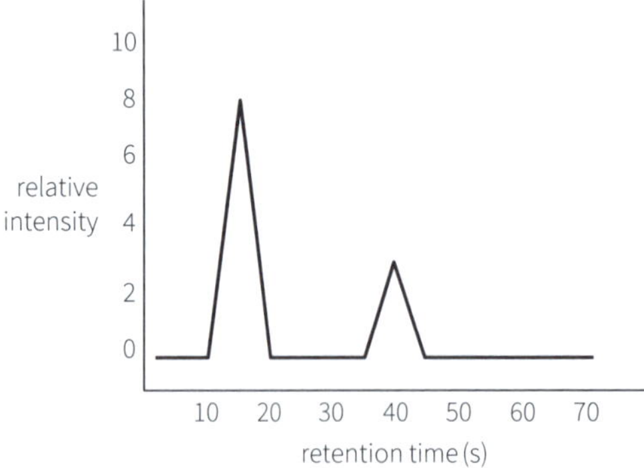


**Figure 1.2**

Give the retention times for the two compounds and explain how you arrived at your answer.

**A** ..........................................................................................

**B** ..........................................................................................

Explanation: .......................................................................................... **[2]**

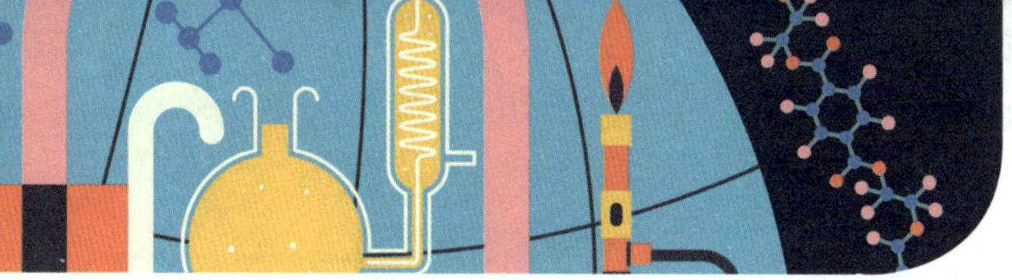



**(d)** Calculate the percentage composition of the mixture.

**[2]**

**Exam tip**

area of triangle $= \frac{1}{2} \times \text{base} \times \text{height}$

**2** A group of students was given the task of analysing data on some isomeric unbranched carbon compounds **C**, **D**, and **E**.

They received the following information:

- The percentage composition of the three compounds was carbon 58.88%; hydrogen 9.80%; oxygen 31.37%.
- The molecular ion peak on the mass spectra was at $m/z = 102$.
- All three compounds were esters but none of them were esters of methanoic acid.
- The proton NMR spectra of the three compounds are shown in **Figure 2**.

**Synoptic links**

2.1.3 3.3.9.1
4.2.4 6.1.3

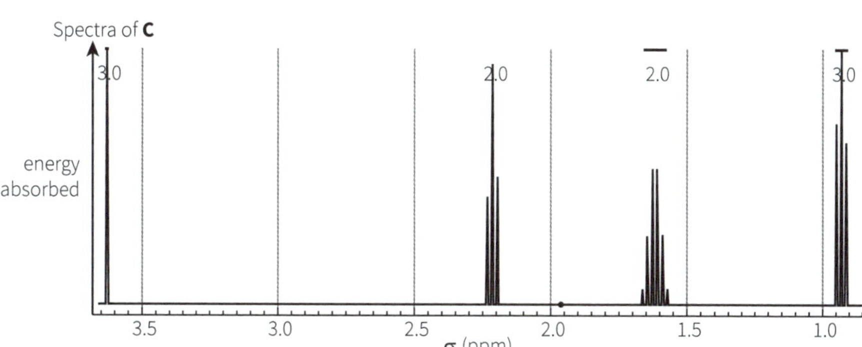


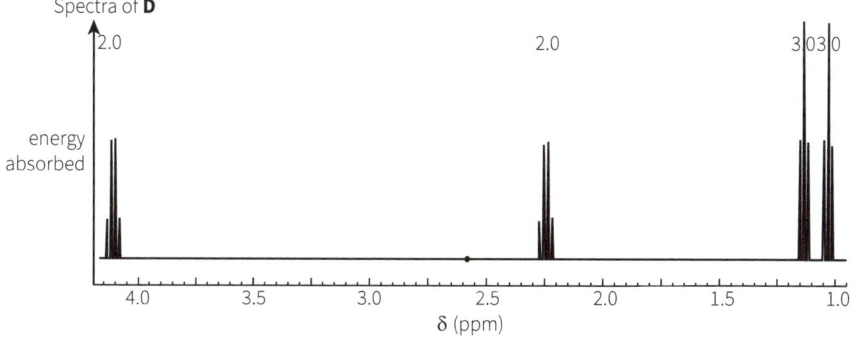


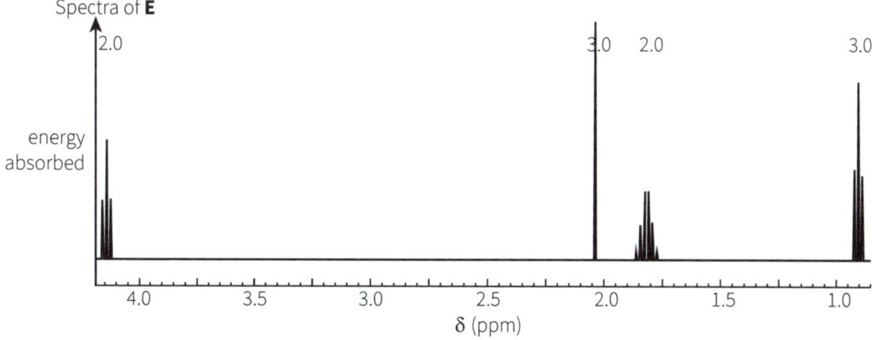


**Figure 2**

**(a)** Calculate the empirical and molecular formulae of the esters. **[2]**

**(b)** Give the structural formulae of the three esters that fit the data given. **[1]**

**(c)** Using the NMR spectra, including the relative numbers of atoms, chemical shifts, and splitting patterns, work out the structures of the three compounds. **[9]**

**Exam tip**

Work out the number of carbons, hydrogens, and oxygens and then try out some different structures of esters that fit the data. Also look for any singlets and their chemical shift.

**3** Methylbenzene can undergo substitution in both the benzene ring and in the methyl side chain (see **Figure 3.1**).

Substitution into the benzene ring occurs by electrophilic substitution.

**Synoptic link**

6.1.1

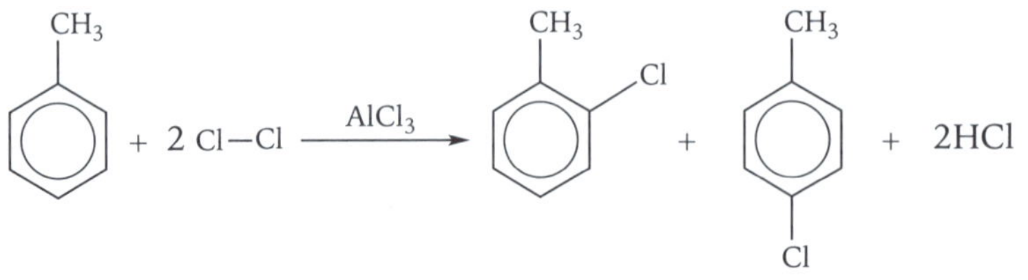

$C_6H_5CH_3 + 2\ Cl{-}Cl \xrightarrow{AlCl_3}$ 2-chloromethylbenzene + 4-chloromethylbenzene + 2HCl

**Figure 3.1**

**(a)** Explain how $^{13}C$ NMR can be used to distinguish between the two organic products. **[2]**

**(b)** Benzene can be nitrated using a mixture of concentrated acids to give nitrobenzene.

**(i)** Name the acids used and for each one give its function in the reaction. **[2]**

**(ii)** Give the balanced symbol equation for the reaction. **[1]**

**(c)** Nitrobenzene can undergo further nitration to give dinitrobenzene.

**(i)** Give the structures of the three isomers of dinitrobenzene. **[3]**

**(ii)** How can $^{13}C$ NMR be used to establish which of the three isomers is formed? **[2]**

**(iii)** Suggest which isomer is formed and how $^{13}C$ NMR can confirm that it is formed. **[3]**

**Exam tip**

Consider the different numbers of environments in each molecule.

**4** This question concerns unknown compounds **F** and **G**.

- **F** is a branched chain compound with the molecular formula $C_5H_{12}O$.
- **F** exhibits optical isomerism.
- The infrared spectrum of **F** is shown in **Figure 4.1**.

**Synoptic links**

4.1.3 6.2.2 6.3.2

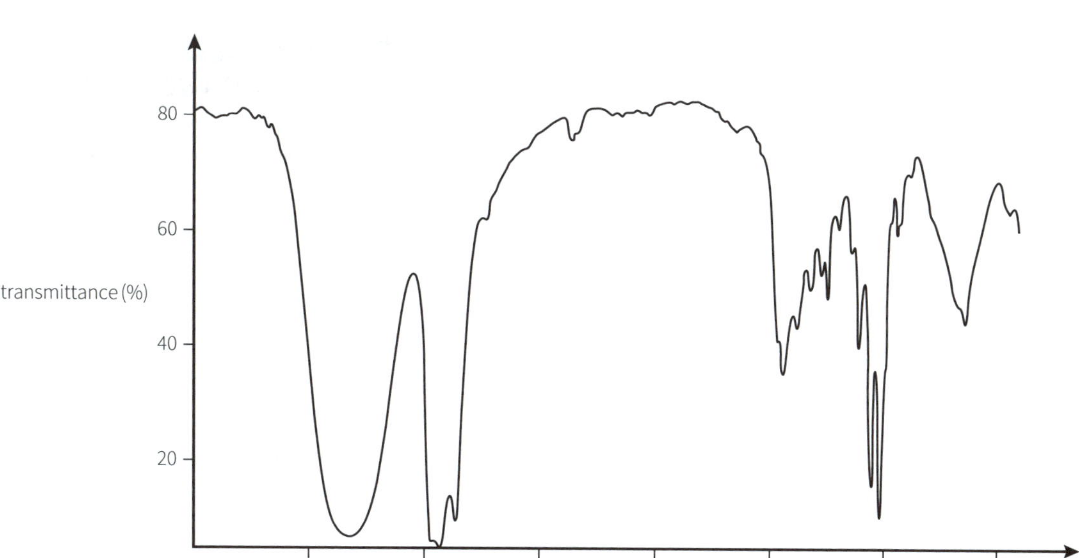


**Figure 4.1**

- When **F** is refluxed with acidified dichromate solution, **G** is formed. The infrared spectrum of the product of oxidation is given in **Figure 4.2**.

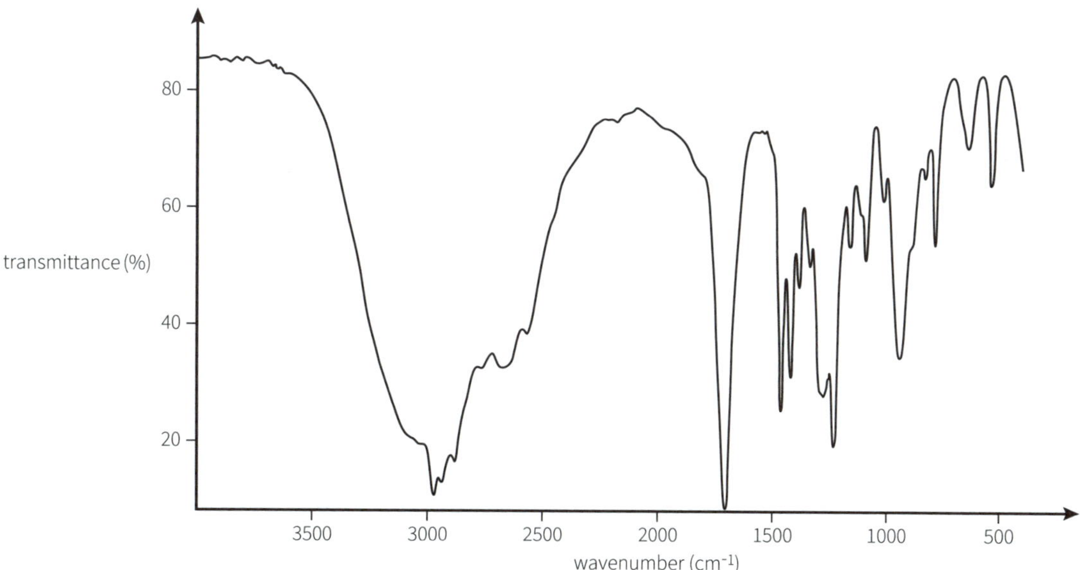


**Figure 4.2**

**(a)** Identify **F**. Explain your answer. **[3]**

**(b)** Predict the $^1$H NMR spectrum of **G** as it would appear after treatment with $D_2O$.

Your prediction should include the following:

- the structure of **G** so that you can refer clearly to the different protons
- the number of peaks
- the number of protons responsible for each peak
- the splitting pattern for each peak and the location of the proton responsible each peak. **[7]**

**(c)** Give the two optical isomers of **F**. **[2]**

**Exam tip**

Sketch out the possible structures of **F** and see which one fits the data.

**5** Leucine and isoleucine are two isomeric amino acids. Their structures are shown in **Figure 5.1**.

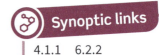

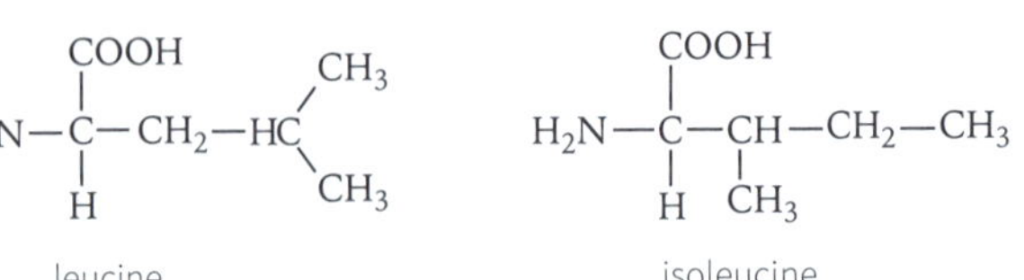


**Figure 5.1**

**(a)** Give the systematic name for leucine. **[1]**

**(b)** State how $^{13}C$ NMR spectroscopy distinguishes between the two amino acids. **[1]**

**(c)** The $^{1}H$ NMR spectra of both compounds are shown in **Figure 5.2**. These spectra do not include the peaks for the labile protons of the $NH_2$ and COOH groups.

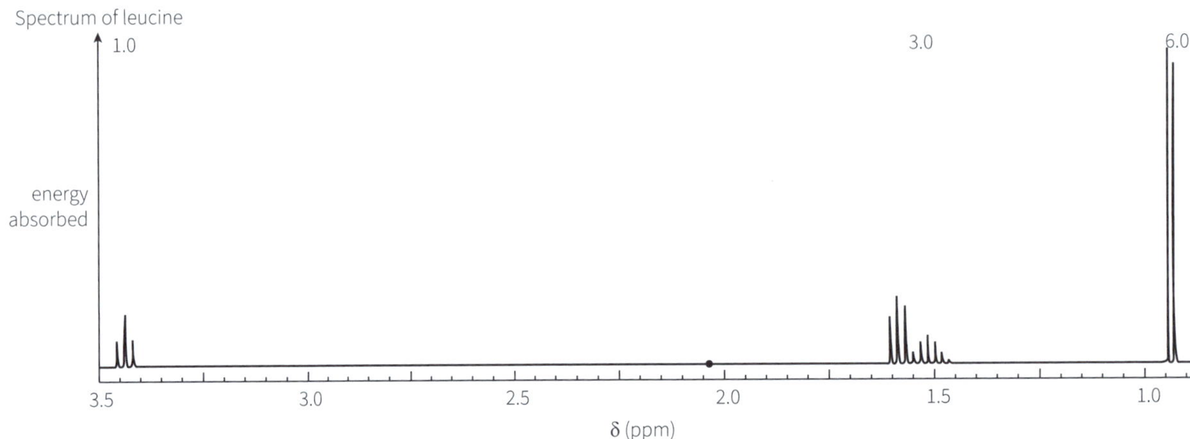


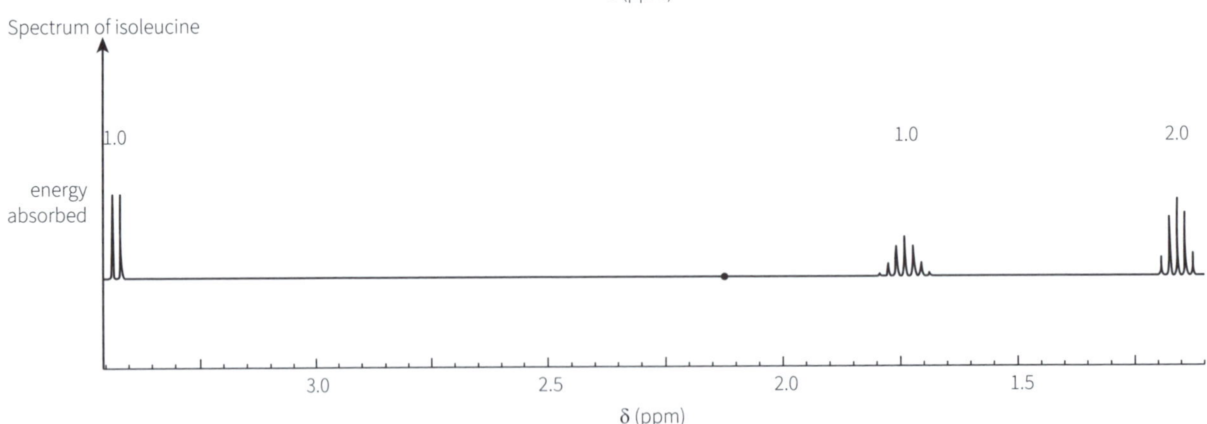


**Figure 5.2**

Using the structures of leucine and isoleucine, explain the proton NMR spectra splitting patterns listed below.

In your answers, make it clear to which proton(s) you are referring.

**(i)** Leucine

**I** the triplet at $\delta = 3.44$ ppm

**II** the doublet at $\delta = 0.93$ ppm. **[6]**

**(ii)** Isoleucine

the doublet at $\delta = 3.43$ ppm. **[2]**

**(d)** Explain why it is difficult to separate leucine and isoleucine using chromatography. **[2]**

**Synoptic links**

4.1.1 6.2.2

**Exam tip**

Count up the number of different environments in each molecule – do not be vague, use numbers.

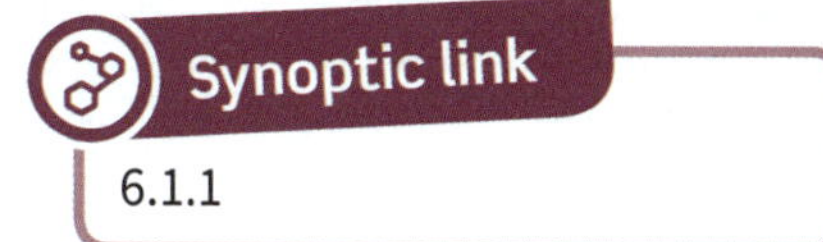


6.1.1

**6** Ibuprofen is an anti-inflammatory drug that has the structure shown in **Figure 6.1**.

**Figure 6.1**

A group of students decided to synthesise ibuprofen from benzene. Their first step was to prepare the compound in **Figure 6.2**.

**Figure 6.2**

The students deduced that in order to synthesise the compound they required benzene, 2-chloropropanoic acid, and aluminium chloride.

**(a)** Give the structure of the electrophile produced in the reaction and the equation for its formation from the 2-chloropropanoic acid and aluminium chloride. **[2]**

**(b)** The proton NMR spectrum of Ibuprofen is shown in **Figure 6.3**. Note that the spectrum is without the peak due to the COOH proton and the peaks due to the benzene protons.

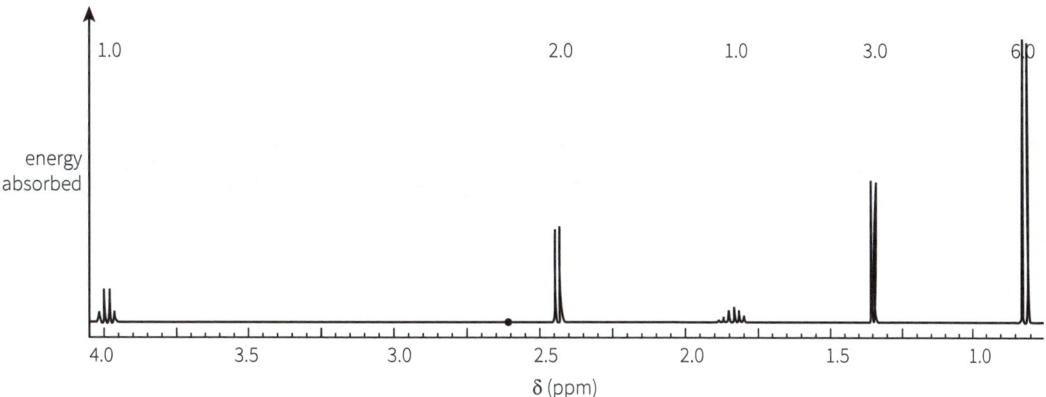


**Figure 6.3**

Using the notation shown on the structure of ibuprofen above, identify the protons responsible for the two peaks identified below and explain your choices.

The peaks at:

**(i)** $\delta = 4.0$ ppm **[3]**

**(ii)** $\delta = 2.45$ ppm. **[3]**

# Knowledge

## Unifying concepts

### What is synoptic assessment?

A synoptic question assesses your knowledge and skills from across the whole of A level chemistry. Synoptic questions are likely to span across the three traditional branches of chemistry (physical, inorganic, and organic) but may be connected by a common theme.

They are also likely to include problem solving (which may include calculations) or applying one area of chemistry to a seemingly unrelated areas; for example, reaction mechanisms to inorganic chemistry.

When you are studying and revising, look back through your notes and try to pick out connections between different areas of the specification.

### Energy

Energy changes take place during chemical reactions when bonds are made or broken. You can predict the enthalpy change of a reaction using Hess' law; which is a consequence of the conservation of energy.

Only molecules with an energy higher than the activation energy can react to form new products. This kinetic energy is related to the temperature of substances. Energy is not disturbed evenly across molecules, but its distribution can be modelled with the Maxwell–Boltzmann distribution. This has consequences for the rates of chemical reactions.

Thermodynamics can also be used to predict if reactions are theoretically possible. The total entropy of the universe must increase, and as a consequence of this a chemical reaction will only happen when its Gibb's free energy is less than zero. $\Delta G$ depends on the enthalpy change, the entropy change, and the temperature in kelvin. You can also predict reactions with $E^{\ominus}$ values.

### Patterns and trends in chemical behaviour

There are patterns throughout all of chemistry. You can use this to make predictions about chemical properties and reactions. The Periodic Table is a classic example of a series of patterns.

Using your knowledge of Group 2 and Group 7, you can predict the chemical and physical properties of radium and astatine (even though they are both very radioactive and you are very unlikely to ever come across them).

### Electronegativity and bonding

The electrostatic forces between atoms or ions and/or electrons determines the chemical and physical behaviour of all matter and influences how substances react. Bonding fundamentally depends on the electronegatively of atoms. Two atoms bonded together with different electronegativities could result in a polar molecule.

Permanent dipoles form stronger intermolecular bonds than induced dipoles. Hydrogen bonding is even stronger than permanent dipole–dipole interactions.

Atoms with very different electronegativities are likely to form ionic structures. Atoms with similar but high electronegativities bond to form covalent structures. Similar, but low electronegativities result in metallic structures.

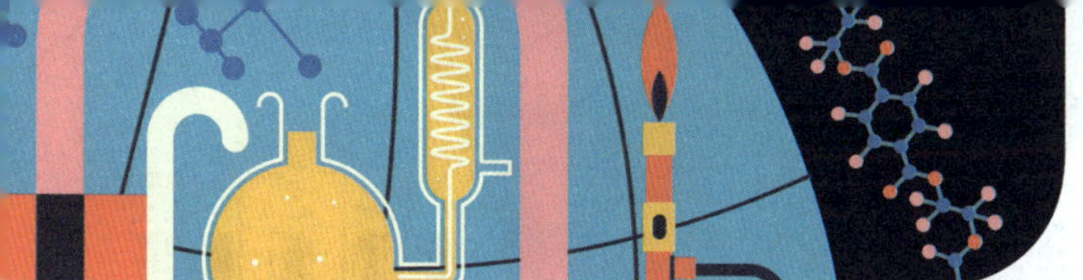



## Example question

This question is about concentrated sulfuric acid, ethanol, and sodium chloride.

Sulfuric acid can react with ethanol and sodium chloride.

## Part 1

**Question**

For the reaction with ethanol, identify the organic product and the role of sulfuric acid. **[2 marks]**

**Answer**

*This is asking you to recall AS organic chemistry. This is the 'warm up' part of the question.*

Organic product: ethene

Role: dehydrating agent

## Part 2

**Question**

Write the equation for the reaction with sodium chloride, state the role of sulfuric acid, and state any observations. **[3 marks]**

**Answer**

*This is asking you to recall AS inorganic chemistry. This is also a 'warm up' part of the question.*

Equation: $NaCl + H_2SO_4 \rightarrow HCl + Na_2SO_4$

Observations: white or misty fumes (these are the fumes of HCl gas)

Role of sulfuric acid: proton donor or acid (Remember, acids are proton donors)

## Part 3

**Question**

Concentrated sulfuric acid can react with 8.00 g of copper metal. A blue solution is formed, and a gas that occupies 3.09 dm$^3$ at 101 kPa and 298 K and has a mass of 8.06 g.

Identify the gas, the blue solution, the role of sulphuric acid, and write an equation for the reaction. **[6 marks]**

**Answer**

*This question is asking you to do a lot, so take it one step at a time.*

*You have some information which you can use $pV = nRT$ with, so start there.*

If we use $pV = nRT$ we find that $M_r$ is 64.

This means the gas cannot be $H_2$ which you might expect from normal reactions of acids with metals.

Think of common gases that contain the elements H, O, and S. $H_2O$ and $H_2S$ have too low a mass.

Try $SO_2$, 32 + 16 × 2 = 64.

The gas is $SO_2$.

*The blue solution is likely to contain Cu because you know from transition elements that copper forms blue solutions.*

The blue solution is likely to be $CuSO_4(aq)$.

*So far we have: $Cu + H_2SO_4 \rightarrow SO_2 + CuSO_4$*
*Hydrogens are missing. We know a solution forms, so water has to be made. It has to be made because concentrated $H_2SO_4$ has no water associated with it. Add water to the products then balance:*

$Cu + 2H_2SO_4 \rightarrow SO_2 + CuSO_4 + 2H_2O$

# Practice

## Multiple-choice questions

Only **one** answer per question is allowed.

For each answer completely fill in the circle alongside the appropriate answer.

CORRECT METHOD ● WRONG METHODS

If you want to change your answer you must cross out your original answer as shown.

If you wish to return to an answer previously crossed out, ring the answer you now wish to select as shown.

**1** Which of these contains the greatest number of atoms?

**A** 770 mg of bromine ○

**B** $1.54 \times 10^{-3}$ kg of silicon ○

**C** 89.5 mg of carbon dioxide ○

**D** $1.70 \times 10^{-3}$ kg of ammonia ○

**2** 125 cm$^3$ of oxygen and 45 cm$^3$ of nitrogen, each at 298 K and 100 kPa, were placed into an evacuated flask of volume 0.75 dm$^3$.

What is the pressure of the gas mixture in the flask at 298 K?

**A** 22.7 kPa ○

**B** 3.27 kPa ○

**C** 16.6 kPa ○

**D** 230 kPa ○

**3** A 25.0 cm$^3$ sample of hydrobromic acid, HBr, of an unknown concentration, was titrated against 0.125 mol dm$^{-3}$ solution of sodium hydroxide, NaOH. The average titre was 28.35 cm$^3$.

What was the concentration of the hydrobromic acid?

**A** 0.110 mol dm$^{-3}$ ○

**B** 0.143 mol dm$^{-3}$ ○

**C** 0.125 mol dm$^{-3}$ ○

**D** 1.40 mol dm$^{-3}$ ○

**4** Which of these species contains an element with an oxidation state of +6?

**A** $V_2O_5$ ○

**B** $HSO_4^-$ ○

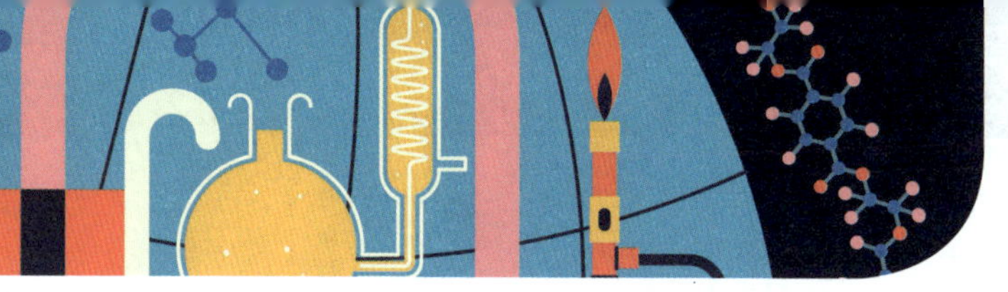

C $ClO_3^-$

D $P_4O_{10}$

5 Which of these molecules does not have a permanent dipole?

A $CH_2Br_2$

B $CCl_4$

C $NH_3$

D $CH_3I$

6 Which of the following molecules has the highest boiling point?

A butanal

B butanone

C butan-1-ol

D 2-methylpropan-2-ol

7 Which element has the lowest first ionisation energy?

A carbon

B fluorine

C neon

D oxygen

8 Which of these elements has the highest second ionisation energy?

A Na

B Mg

C Ne

D Ar

9 Which of these decreases down Group 2?

A atomic radius

B solubility of hydroxide salts

C first ionisation energy

D reactivity with water

10 Which species is the best reducing agent?

A $Br_2$

B $Br^-$

C $Cl_2$

D $Cl^-$

**11** Which one of the following solutions would not give a white precipitate when added to barium chloride solution?

**A** silver nitrate solution

**B** dilute sulfuric acid

**C** sodium sulfate solution

**D** sodium nitrate solution

**12** Which of the following compounds will give no precipitate with silver nitrate or barium chloride?

**A** NaF 

**B** $Na_2SO_4$

**C** NaBr

**D** NaI

**13** Which of the following equations illustrates the reaction equivalent to the enthalpy of formation of lithium chloride?

**A** $Li(s) + Cl(g) \rightarrow LiCl(s)$

**B** $Li(s) + \frac{1}{2}Cl_2(g) \rightarrow LiCl(s)$

**C** $Li^+(g) + Cl^-(g) \rightarrow LiCl(s)$

**D** $Li^+(aq) + Cl_2(g) \rightarrow LiCl(s)$

**14** When 0.30 g of pentane was burnt the quantity of heat evolved was 14.0 kJ. What is the enthalpy of combustion of propane in kJ $mol^{-1}$?

**A** −46

**B** −1960

**C** −3800

**D** −3400

**15** For this reaction at equilibrium, which combination of temperature and pressure would give the greatest equilibrium yield of products?

$$\mathbf{W}(g) + \mathbf{X}(g) \rightleftharpoons 2\mathbf{Y}(g) + \mathbf{Z}(g) \qquad \Delta H = +47\ kJ\ mol^{-1}$$

**A** Increase the pressure and reduce the temperature. 

**B** Decrease the pressure and increase the temperature.

**C** Increase the temperature and pressure.

**D** Decrease the temperature and pressure.

**16** Which of the following is *not* true about the Boltzmann distribution?

**A** The most probable energy is less than the average energy.

**B** The shape is independent of the number of particles.

**C** At low temperatures no particles have an energy above the activation energy.

**D** The activation energy of each reaction is a different value.

**17** The skeletal formula in **Figure 17.1** shows a hydrocarbon, **M**.

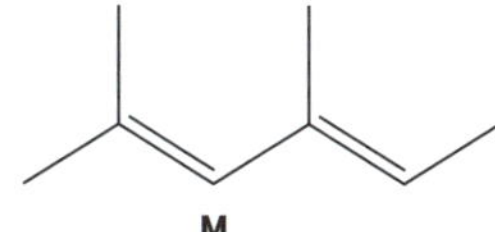


**Figure 17.1**

Which of the following is the correct molecular formula for **M**?

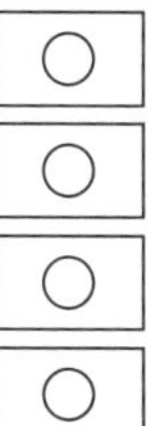

**A** $C_7H_{10}$

**B** $C_7H_{12}$

**C** $C_8H_{16}$

**D** $C_8H_{14}$

**18** The compound in **Figure 18.1** shows the displayed formula of a compound with three functional groups.

2.2.2h 3.1.3.5 MS4.1

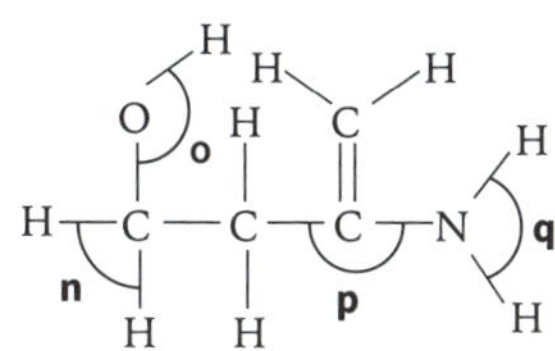


**Figure 18.1**

Which row in **Table 18.1** gives the correct values for the angles **n** to **q**?

| | n | o | p | q |
|---|---|---|---|---|
| **A** | 104.5 | 107.0 | 109.5 | 120.0 |
| **B** | 109.5 | 104.5 | 120.0 | 107.0 |
| **C** | 107.0 | 120.0 | 104.5 | 109.5 |
| **D** | 120 | 109.5 | 107.0 | 104.5 |

**Table 18.1**

**19** The symbol equation below shows the reactants and products for the incomplete combustion of the alkane heptane.

2.1.2b

$$C_7H_{16}(l) + rO_2(g) \rightarrow sC(s) + tH_2O(l)$$

Which row in **Table 19.1** gives the correct values for **r**, **s**, and **t**?

| | r | s | t |
|---|---|---|---|
| **A** | 2 | 4 | 6 |
| **B** | 3 | 5 | 7 |
| **C** | 4 | 7 | 8 |
| **D** | 5 | 7 | 9 |

**Table 19.1**

**20** A student reacted pent-1-ene with hydrogen bromide.

Which of the following is the main compound formed?

**A** 1,2-dibromopentane ○

**B** 1-bromopentane ○

**C** 2-bromopentane ○

**D** 3-bromopentane ○

**21** This question concerns the three isomeric alcohols **X**, **Y**, and **Z**, shown in **Figure 21.1**.

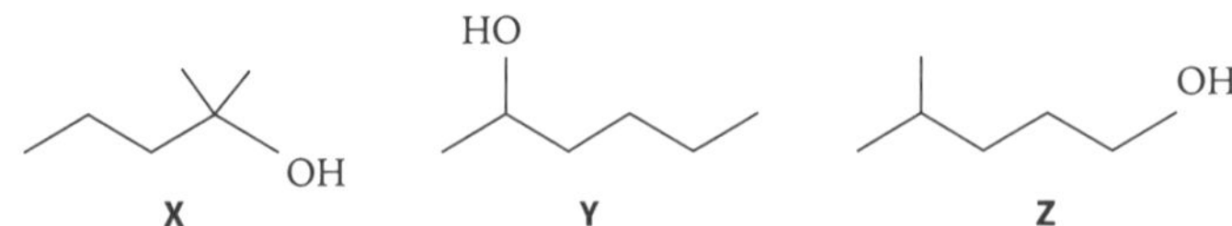


**Figure 21.1**

Which of the following statements about **X**, **Y**, and **Z** is *not* true?

**A** **X** will not change the colour of acidified dichromate solution from orange to blue/green. ○

**B** **X** can be dehydrated. ○

**C** After reaction with acidified dichromate solution, **Y** gives just one organic product. ○

**D** **Z** can give two different organic products with acidified dichromate solution depending on the conditions. ○

**22** In the upper atmosphere, chlorofluorocarbons (CFCs) undergo homolytic fission to give free radicals. A typical CFC is $CHF_2Cl$.

Which of the following equations represents the homolytic fission of $CHF_2Cl_2$?

**A** $CHF_2Cl \rightarrow CF_2Cl\bullet + H\bullet$ ○

**B** $CHF_2Cl \rightarrow CHFCl\bullet + F\bullet$ ○

**C** $CHF_2Cl \rightarrow CHF_2^+ + Cl^-$ ○

**D** $CHF_2Cl \rightarrow CHF_2\bullet + Cl\bullet$ ○

**23** **Figure 23.1** shows a synthetic pathway with some of the conditions and intermediates missing.

**Figure 23.1**

Which of the alternative combinations in **Table 23.1** gives the correct identities of **I**, **II**, and **III**?

| | I | II | III | |
|---|---|---|---|---|
| **A** | HBr | $CH_3CH_2BrCH_2CH_2CH_3$ | reflux with NaOH(aq) | ○ |
| **B** | $Br_2$ | $CH_3CH_2CHBrCHBrCH_3$ | distil with NaOH(aq) | ○ |
| **C** | $Br_2$ | $CH_3CH_2CH\ BrCHBrCH_3$ | reflux with NaOH(aq) | ○ |
| **D** | HBr | $CH_3CH_2CHBrCHBrCH_3$ | distil with NaOH(aq) | ○ |

**Table 23.1**

**24** The scheme below shows a series of reactions that could be used to prepare an addition polymer.

Which of the four pathways gives a correct sequence of compounds?

**A** $(CH_3)_3COH + (CH_3)_2C{=}CH_2 \longrightarrow \left[ -C(H_3C)(CH_3)-C(H)(H)- \right]$ ○

**B** $(CH_3)_2CHCH_2OH + (CH_3)_2C{=}CH_2 \longrightarrow \left[ -C(H)(CH_3)-C(CH_3)(H)- \right]$ ○

**C** $CH_3CH(OH)CH_2CH_3 + CH_3CH{=}CHCH_3 \longrightarrow \left[ -C(H_3C)(CH_3)-C(H)(H)- \right]$ ○

**D** $CH_3CH_2CH_3CH_2OH + CH_3CH_2CH{=}CH_2 \longrightarrow \left[ -C(H)(CH_3)-C(CH_3)(H)- \right]$ ○

**25** A compound **V** gave a pale cream precipitate when warmed with alcoholic silver nitrate solution. When **V** was warmed with acidified potassium dichromate solution there was no reaction.

Which of the following compounds is most likely to be **V**?

**A**
Br $CH_2$ OH
CH $CH_2$
$CH_3$

**B**
Cl OH
C $CH_3$
$H_3C$ $CH_2$

**C**
Br OH
C $CH_3$
$H_3C$ $CH_2$

**D**
Br
CH $CH_3$
$H_3C$ CH
OH

**26** The data below describe the properties of compound **W**.

The substance **W** has the infrared spectrum shown in **Figure 26.1**.

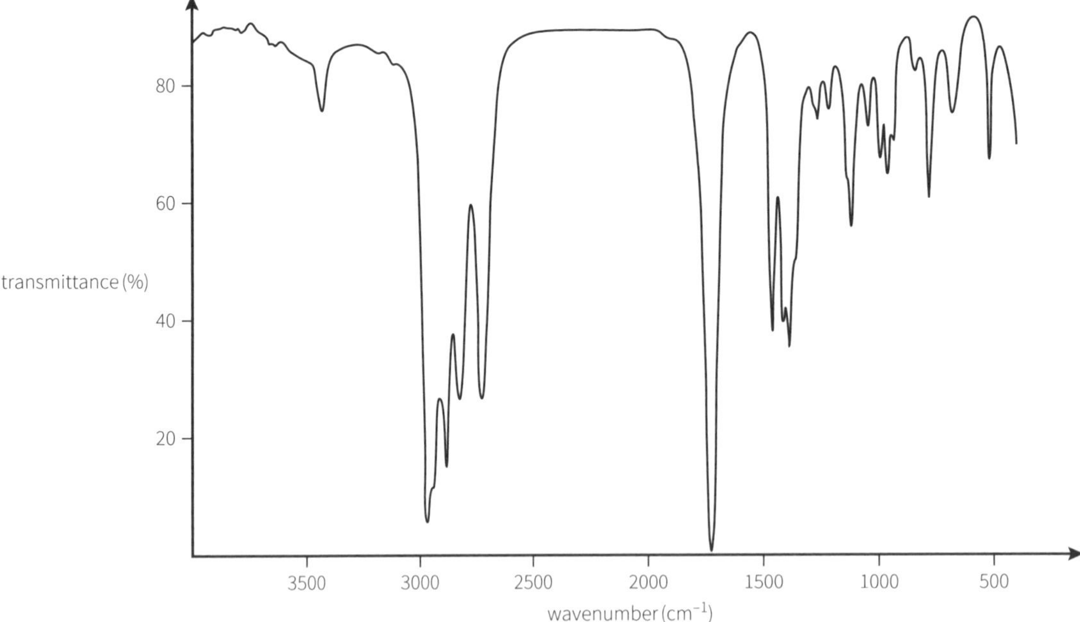


**Figure 26.1**

When warmed with acidified potassium dichromate solution, the colour of the solution changes from orange to green.

**W** has the composition C 62.1 %, H 10.3 %, and O 27.6 %.

Which of the following compounds is most likely to be **W**?

**A** $CH_3CH_2CH_2OH$

**B** $CH_3COCH_3$

**C** $CH_3CH_2COOH$

**D** $CH_3CH_2CHO$

**27** **Figure 27.1** shows how varying the concentration of **A** affects the rate of the reaction:

$$A + 2B \rightarrow C + 3D$$

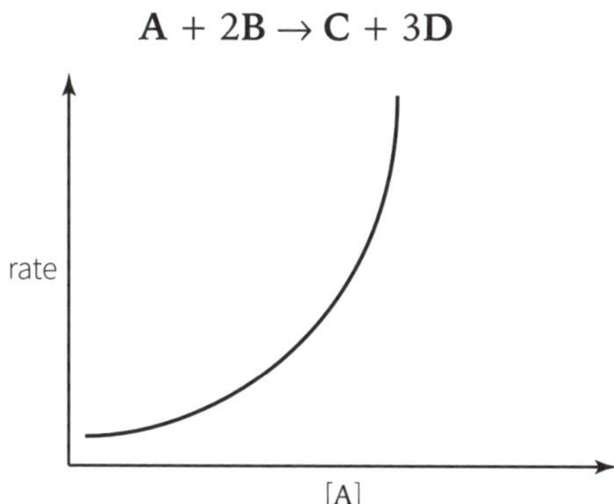


**Figure 27.1**

Which of the following statements is true?

**A** The reaction is first order with respect to **A** and second order with respect to **B**. ○

**B** The reaction is first order with respect to **A** and unknown with respect to **B**. ○

**C** The reaction is second order with respect to **A** and unknown with respect to **B**. ○

**D** The reaction is second order with respect to **A** and second order with respect to **B**. ○

**28** Which row in **Table 28.1** will cause an increase in the value of $K_p$?

| | Equation | Change | |
|---|---|---|---|
| **A** | $C_2H_4(g) + H_2O(g) \rightleftharpoons C_2H_5OH(l)$ | total pressure is halved | ○ |
| **B** | $3Fe(s) + 4H_2O(g) \rightleftharpoons Fe_3O_4(s) + 4H_2(g)$ | total pressure is doubled | ○ |
| **C** | $2CO(g) + O_2(g) \rightleftharpoons 2CO_2(g)$ | total pressure is tripled | ○ |
| **D** | $H_2(g) + I_2(g) \rightleftharpoons 2HI(g)$ | total pressure is reduced to $\frac{1}{10}$ | ○ |

**Table 28.1**

**29** Calculate the pH of a 0.0451 mol dm$^{-3}$ solution of potassium hydroxide.
Ionic product of water, $K_w = 1 \times 10^{-14}$

**A** 12.70 ○

**B** 15.35 ○

**C** 12.65 ○

**D** 1.35 ○

**30** **Table 30.1** the $pK_a$ values of two weak acids.

| Name of acid | $pK_a$ |
|---|---|
| propanoic acid | 4.87 |
| butanoic acid | 4.82 |

**Table 30.1**

Which of the following statements is correct?

**A** Propanoic acid is the stronger of the two acids. ○

**B** The $K_a$ of propanoic acid is $1.51\times10^{-5}$. ○

**C** The $K_a$ of butanoic acid is $1.35\times10^{-5}$. ○

**D** Butanoic acid is the stronger of the two acids. ○

**31** Which equation correctly illustrates the process of the standard enthalpy of atomisation of bromine?

**A** $Br_2(l) \rightarrow 2Br(g)$ ○

**B** $\frac{1}{2}Br_2(g) \rightarrow Br(g)$ ○

**C** $\frac{1}{2}Br_2(l) \rightarrow Br(g)$ ○

**D** $Br_2(g) \rightarrow 2Br(g)$ ○

**32** Which of the following equations will be feasible at all temperatures?

**A** $A(g) + B(g) \rightarrow C(s)$ exothermic ○

**B** $2G(g) + J(g) \rightarrow 2X(g)$ endothermic ○

**C** $S(g) \rightarrow 2Y(g)$ exothermic ○

**D** $3L(s) + 2Q(g) \rightarrow 2R(g)$ endothermic ○

**33** In this question consider the data in **Table 33.1**.

| Electrode half-equation | $E^\ominus$ (V) |
|---|---|
| $Ag^+(aq) + e^- \rightleftharpoons Ag(s)$ | +0.80 |
| $2H^+(aq) + 2e^- \rightleftharpoons H_2(g)$ | 0.00 |
| $Pb^{2+}(aq) + 2e^- \rightleftharpoons Pb(s)$ | −0.13 |

**Figure 33.1**

The EMF of the cell $Pb(s)\,|\,Pb^{2+}(aq)\,||\,Ag^+(aq)\,|\,Ag(s)$ is:

**A** 0.93 V ○

**B** 0.67 V ○

**C** −0.67 V ○

**D** −0.97 V ○

**34** The following cell has an EMF of +0.46 V.

$$Cu|Cu^{2+}||Ag^{+}|Ag$$

Which statement is *not* correct about the operation of the cell?

**A** Electrons flow from the copper electrode to the silver electrode. ○

**B** Silver is the cathode. ○

**C** The silver electrode will dissolve when a current flows. ○

**D** Metallic copper will be oxidised by silver ions. ○

**35** Which of the following statements is *not* true about transition metals?

**A** Complex metal ions involve ligand bonding via coordinate bonds. ○

**B** Transition metals form coloured compounds. ○

**C** Transition metals can form a variety of oxidation states. ○

**D** To be a transition metal, all its ions must have a partially filled d-orbital. ○

**36** Which of the following *cannot* act as a ligand?

**A** $F^-$ ○

**B** $CH_3OH$ ○

**C** $CH_3NH_2$ ○

**D** $CH_2CH_2$ ○

**37** Which one of the following does *not* contain any delocalised electrons?

**A** poly(propene) ○

**B** benzene ○

**C** graphite ○

**D** sodium ○

**38** What is the relative molecular mass, $M_r$, of benzene-1,4-dicarboxylic acid.

**A** 164 ○

**B** 166 ○

**C** 168 ○

**D** 170 ○

**39** Which of these structures contains a secondary alcohol?

**A** $CH_3CHO$

**B** $HOCH_2CH_2CH_2COH$

**C** $CH_3CH(OH)CH_3$

**D** $CH(CH_3)_2COOH$

**49** Which one of the following statements about but-2-enal, $CH_3CH{=}CHCHO$, is *not* true?

**A** It can be dehydrated by concentrated sulfuric acid.

**B** It has stereoisomers.

**C** It shows a strong peak in the IR spectrum at 1700 $cm^{-1}$.

**D** It will turn an acidified solution of potassium dichromate (VI) green.

**50** Butan-1-ol was converted into butyl propanoate by reaction with excess propanoic acid. In the reaction, 6.0 g of the alcohol gave 7.4 g of the ester. The percentage yield of ester was:

**A** 57 %

**B** 70 %

**C** 75 %

**D** 81 %

**51** Which one of the following would not react with aqueous silver nitrate to produce a precipitate that is soluble in concentrated aqueous ammonia?

**A** $CaBr_2$

**B** $[CoCl_4]^{2-}$

**C** $(CH_3)_4N^+I^-$

**D** $CH_3COCl$

**52** In a reaction that gave a 27.0 % yield, 5.00 g of methylbenzene was converted into the explosive 2,4,6-trinitromethylbenzene (TNT) ($M_r$ = 227.0).

Calculate the mass of TNT formed.

**A** 1.35 g

**B** 3.65 g

**C** 12.34 g

**D** 3.33 g

**53** Butylamine has a higher boiling point than propylamine. Which of the statements below gives the reason for this?

**A** The hydrogen bonds of butylamine are stronger than the hydrogen bonds of propylamine. ○

**B** The London/van der Waals forces of butylamine are stronger than the hydrogen bonds of propylamine. ○

**C** The London/van der Waals forces of butylamine are stronger than the London/van der Waals forces of propylamine. ○

**D** The covalent C—H bonds of butylamine are stronger than the covalent C—H bonds of propylamine. ○

**54** Which of the following compounds (**1–4**) can undergo addition polymerisation. Select from **A**, **B**, **C**, or **D**.

**1** ethyne ○

**2** 1,4-butandioic acid ○

**3** 1,6-hexanediamine ○

**4** ethene ○

**A** only compound **1** ○

**B** only compounds **1** and **4** ○

**C** only compounds **2** and **3** ○

**D** none of the compounds ○

**55** Select the IUPAC name for the polymer formed from the monomer below:

$$H_2C{=}C(CH_3)CH_2CH_3$$

**Figure 55.1**

**A** poly(2-methylbut-1-ene) ○

**B** poly(pentane) ○

**C** poly(2-methylbut-1-ane) ○

**D** poly(1-methyl-1-ethylethene) ○

**56** Select which one of the following pairs of reagents would react to form an organic product that shows only two peaks in its [1]H NMR spectrum.

**A** butan-2-ol and acidified potassium dichromate (VI) ◯

**B** ethanoyl chloride and methanol ◯

**C** propanoic acid and ethanol in the presence of concentrated sulfuric acid ◯

**D** ethene and hydrogen in the presence of nickel ◯

**57** Select which one of the following types of reaction mechanism is *not* involved in the sequence below:

$$CH_3CH_2CH_3 \rightarrow (CH_3)_2CHCl \rightarrow (CH_3)_2CHCN \rightarrow (CH_3)_2CHCH_2NH_2 \rightarrow (CH_3)_2CHCH_2NHCOCH_3$$

**Figure 57.1**

**A** free-radical substitution ◯

**B** nucleophilic substitution ◯

**C** elimination ◯

**D** nucleophilic addition–elimination ◯

**58** Compound **X** has the molecular formula $C_4H_{10}O$.

**X** undergoes dehydration to give an alkene.

The NMR spectrum for **X** in $D_2O$ is shown in **Figure 58.1**.

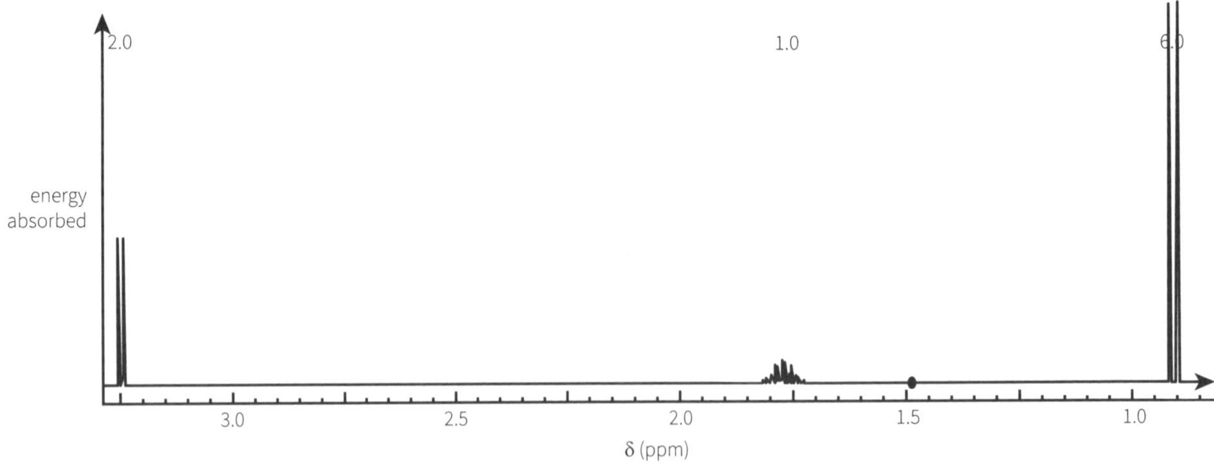


**Figure 58.1**

Which of the following compounds is **Z**?

**A** $CH_3CH_2OCH_2CH_3$ ◯

**B** $(CH_3)_3COH$ ◯

**C** $(CH_3)_2CHCH_2OH$ ◯

**D** $CH_3CH(OH)CH_2CH_3$ ◯

**59** A mixture of trihalogenomethanes ($CHF_3$, $CHCl_3$, $CHBr_3$, and $CHI_3$) was placed on a gas–liquid chromatography column with a polar stationary phase. The chromatogram in **Figure 59.1** was obtained:

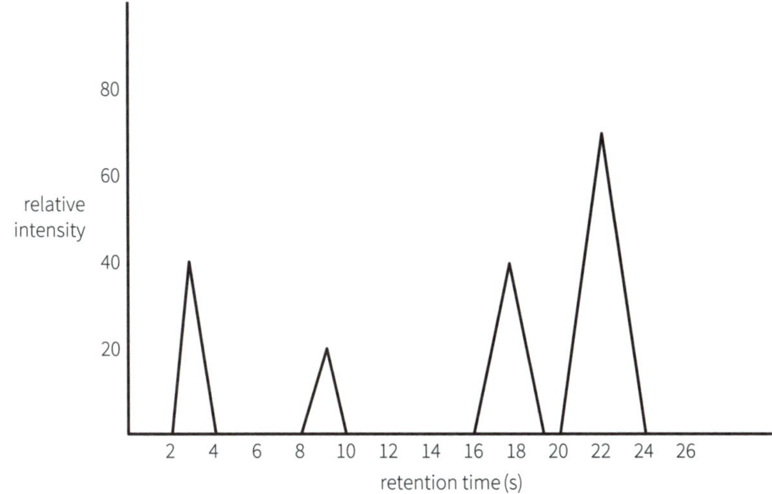


**Figure 59.1**

Which row in **Table 59.1** gives the percentage composition of the mixture?

| | $CHF_3$ | $CHCl_3$ | $CHBr_3$ | $CHI_3$ | |
|---|---|---|---|---|---|
| **A** | 21.4 | 7.2 | 21.4 | 50 | ○ |
| **B** | 50 | 21.4 | 7.2 | 21.4 | ○ |
| **C** | 7.2 | 21.4 | 50 | 21.4 | ○ |
| **D** | 7.2 | 21.4 | 21.4 | 50 | ○ |

**Table 59.1**

# Data booklet

## Table A - Infrared absorption data

| Bond | Wavenumber ($cm^{-1}$) |
|---|---|
| N—H (amines, amides) | 3300–3500 |
| O—H (alcohols, phenols) | 3200–3600 |
| C—H | 2850–3100 |
| O—H (acids) | 2500–3300 (broad) |
| C≡N | 2220–2260 |
| C═O | 1630–1820 |
| C═C (alkenes) | 1620–1680 |
| C—O | 1000–1300 |
| C—C | 750–1100 |
| C—X (X═Cl, Br, I) | 500–800 |
| C—F | 1000–1350 |
| C ═ C (arenes) | Several peaks in range 1450–1650 (variable) |

## General information

Molar gas volume = 24.0 $dm^3\ mol^{-1}$ at room temperature and pressure, RTP

Avogadro constant, $N_A$ = 6.02x$10^{23}$ $mol^{-1}$

Specific heat capacity of water, $c$ = 4.18 $J\,g^{-1}K^{-1}$

Ionic product of water, $K_w$ = 1.00x$10^{-14}$ $mol^2dm^{-6}$ at 298 K

1 tonne = $10^6$ g

Arrhenius equation: $k = Ae^{-Ea/RT}$ or $\ln k = -E_a/RT + \ln A$

Gas constant, $R$ = 8.314 $J\,mol^{-1}K^{-1}$

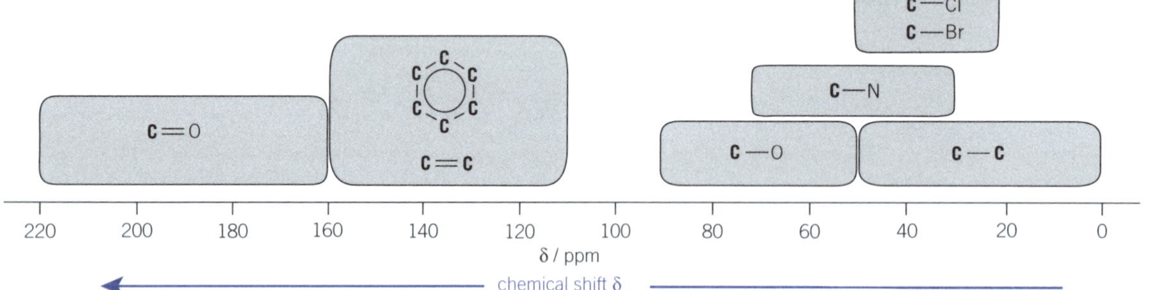


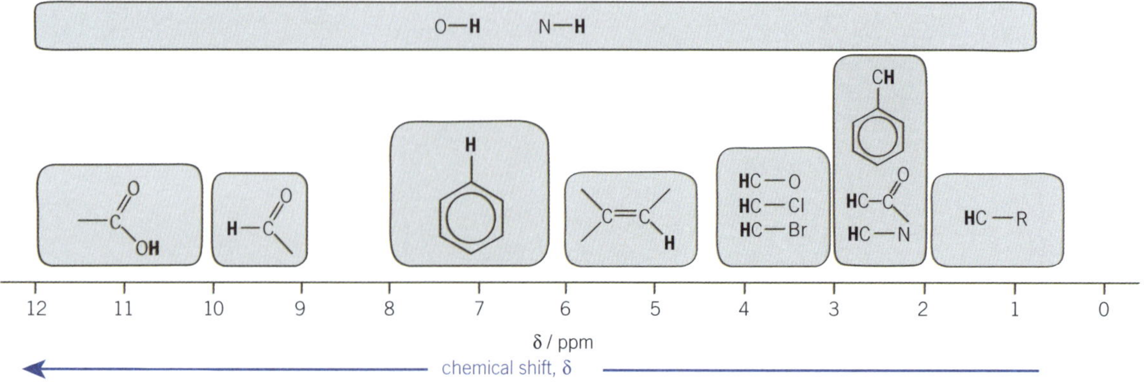

# Periodic Table

**Key**

| relative atomic mass<br>**atomic symbol**<br>name<br>atomic (proton) number |
|---|

| 1 | 2 | | | | | | | | | | | 3 | 4 | 5 | 6 | 7 | 0 |
|---|---|---|---|---|---|---|---|---|---|---|---|---|---|---|---|---|---|
| | | | | | | | 1.0<br>**H**<br>hydrogen<br>1 | | | | | | | | | | 4.0<br>**He**<br>helium<br>2 |
| 6.9<br>**Li**<br>lithium<br>3 | 9.0<br>**Be**<br>beryllium<br>4 | | | | | | | | | | | 10.8<br>**B**<br>boron<br>5 | 12.0<br>**C**<br>carbon<br>6 | 14.0<br>**N**<br>nitrogen<br>7 | 16.0<br>**O**<br>oxygen<br>8 | 19.0<br>**F**<br>fluorine<br>9 | 20.2<br>**Ne**<br>neon<br>10 |
| 23.0<br>**Na**<br>sodium<br>11 | 24.3<br>**Mg**<br>magnesium<br>12 | | | | | | | | | | | 27.0<br>**Al**<br>aluminium<br>13 | 28.1<br>**Si**<br>silicon<br>14 | 31.0<br>**P**<br>phosphorus<br>15 | 32.1<br>**S**<br>sulfur<br>16 | 35.5<br>**Cl**<br>chlorine<br>17 | 39.9<br>**Ar**<br>argon<br>18 |
| 39.1<br>**K**<br>potassium<br>19 | 40.1<br>**Ca**<br>calcium<br>20 | 45.0<br>**Sc**<br>scandium<br>21 | 47.9<br>**Ti**<br>titanium<br>22 | 50.9<br>**V**<br>vanadium<br>23 | 52.0<br>**Cr**<br>chromium<br>24 | 54.9<br>**Mn**<br>manganese<br>25 | 55.8<br>**Fe**<br>iron<br>26 | 58.9<br>**Co**<br>cobalt<br>27 | 58.7<br>**Ni**<br>nickel<br>28 | 63.5<br>**Cu**<br>copper<br>29 | 65.4<br>**Zn**<br>zinc<br>30 | 69.7<br>**Ga**<br>gallium<br>31 | 72.6<br>**Ge**<br>germanium<br>32 | 74.9<br>**As**<br>arsenic<br>33 | 79.0<br>**Se**<br>selenium<br>34 | 79.9<br>**Br**<br>bromine<br>35 | 83.8<br>**Kr**<br>krypton<br>36 |
| 85.5<br>**Rb**<br>rubidium<br>37 | 87.6<br>**Sr**<br>strontium<br>38 | 88.9<br>**Y**<br>yttrium<br>39 | 91.2<br>**Zr**<br>zirconium<br>40 | 92.9<br>**Nb**<br>niobium<br>41 | 95.9<br>**Mo**<br>molybdenum<br>42 | [98]<br>**Tc**<br>technetium<br>43 | 101.1<br>**Ru**<br>ruthenium<br>44 | 102.9<br>**Rh**<br>rhodium<br>45 | 106.4<br>**Pd**<br>palladium<br>46 | 107.9<br>**Ag**<br>silver<br>47 | 112.4<br>**Cd**<br>cadmium<br>48 | 114.8<br>**In**<br>indium<br>49 | 118.7<br>**Sn**<br>tin<br>50 | 121.8<br>**Sb**<br>antimony<br>51 | 127.6<br>**Te**<br>tellurium<br>52 | 126.9<br>**I**<br>iodine<br>53 | 131.3<br>**Xe**<br>xenon<br>54 |
| 132.9<br>**Cs**<br>caesium<br>55 | 137.3<br>**Ba**<br>barium<br>56 | 138.9<br>**La***<br>lanthanum<br>57 | 178.5<br>**Hf**<br>hafnium<br>72 | 180.9<br>**Ta**<br>tantalum<br>73 | 183.8<br>**W**<br>tungsten<br>74 | 186.2<br>**Re**<br>rhenium<br>75 | 190.2<br>**Os**<br>osmium<br>76 | 192.2<br>**Ir**<br>iridium<br>77 | 195.1<br>**Pt**<br>platinum<br>78 | 197.0<br>**Au**<br>gold<br>79 | 200.6<br>**Hg**<br>mercury<br>80 | 204.4<br>**Tl**<br>thallium<br>81 | 207.2<br>**Pb**<br>lead<br>82 | 209.0<br>**Bi**<br>bismuth<br>83 | [209]<br>**Po**<br>polonium<br>84 | [210]<br>**At**<br>astatine<br>85 | [222]<br>**Rn**<br>radon<br>86 |
| [223]<br>**Fr**<br>francium<br>87 | [226]<br>**Ra**<br>radium<br>88 | [227]<br>**Ac**†<br>actinium<br>89 | [261]<br>**Rf**<br>rutherfordium<br>104 | [262]<br>**Db**<br>dubnium<br>105 | [266]<br>**Sg**<br>seaborgium<br>106 | [264]<br>**Bh**<br>bohrium<br>107 | [277]<br>**Hs**<br>hassium<br>108 | [268]<br>**Mt**<br>meitnerium<br>109 | [271]<br>**Ds**<br>darmstadtium<br>110 | [272]<br>**Rg**<br>roentgenium<br>111 | [285]<br>**Cn**<br>copernicium<br>112 | [286]<br>**Nh**<br>nihonium<br>113 | [289]<br>**Fl**<br>flerovium<br>114 | [289]<br>**Mc**<br>moscovium<br>115 | [293]<br>**Lv**<br>livermorium<br>116 | [294]<br>**Ts**<br>tennessine<br>117 | [294]<br>**Og**<br>oganesson<br>118 |

| | | | | | | | | | | | | | | | |
|---|---|---|---|---|---|---|---|---|---|---|---|---|---|---|---|
| *** 58 – 71** Lanthanides | | 140.1<br>**Ce**<br>cerium<br>58 | 140.9<br>**Pr**<br>praseodymium<br>59 | 144.2<br>**Nd**<br>neodymium<br>60 | 144.9<br>**Pm**<br>promethium<br>61 | 150.4<br>**Sm**<br>samarium<br>62 | 152.0<br>**Eu**<br>europium<br>63 | 157.3<br>**Gd**<br>gadolinium<br>64 | 158.9<br>**Tb**<br>terbium<br>65 | 162.5<br>**Dy**<br>dysprosium<br>66 | 164.9<br>**Ho**<br>holmium<br>67 | 167.3<br>**Er**<br>erbium<br>68 | 168.9<br>**Tm**<br>thulium<br>69 | 173.0<br>**Yb**<br>ytterbium<br>70 | 175.0<br>**Lu**<br>lutetium<br>71 |
| **† 90 – 103** Actinides | | 232.0<br>**Th**<br>thorium<br>90 | 231.0<br>**Pa**<br>protactinium<br>91 | 238.0<br>**U**<br>uranium<br>92 | 237.0<br>**Np**<br>neptunium<br>93 | 239.1<br>**Pu**<br>plutonium<br>94 | 243.1<br>**Am**<br>americium<br>95 | 247.1<br>**Cm**<br>curium<br>96 | 247.1<br>**Bk**<br>berkelium<br>97 | 252.1<br>**Cf**<br>californium<br>98 | [252]<br>**Es**<br>einsteinium<br>99 | [257]<br>**Fm**<br>fermium<br>100 | [258]<br>**Md**<br>mendelevium<br>101 | [259]<br>**No**<br>nobelium<br>102 | [260]<br>**Lr**<br>lawrencium<br>103 |

# Index

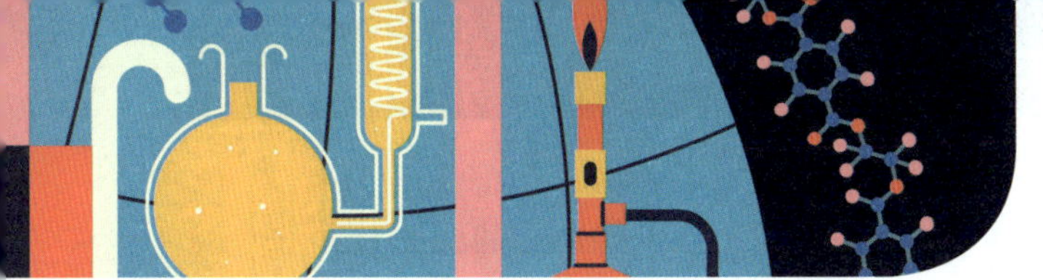

Great Clarendon Street, Oxford, OX2 6DP, United Kingdom

Oxford University Press is a department of the University of Oxford.

It furthers the University's objective of excellence in research, scholarship, and education by publishing worldwide. Oxford is a registered trade mark of Oxford University Press in the UK and in certain other countries



First published in 2020



British Library Cataloguing in Publication Data

Data available

978-1-38-200866-2

10 9 8 7 6 5 4 3 2 1

Paper used in the production of this book is a natural, recyclable product made from wood grown in sustainable forests.

The manufacturing process conforms to the environmental regulations of the country of origin.

Printed in Great Britain by Bell and Bain Ltd. Glasgow

**Acknowledgements**

**p72**: Andrew Lambert Photography/Science Photo Library; **p294t**: Martyn F. Chillmaid/Science Photo Library; **p294m**: Andrew Lambert Photography/Science Photo Library; **p294b**: Andrew Lambert Photography/Science Photo Library; **p295t &b**: Courtesy of the author; **p295tm**: Andrew Lambert Photography/Science Photo Library; **p295bm**: Andrew Lambert Photography/Science Photo Library.

Cover artwork by: John Devolle

All other artwork by QBS Learning

Every effort has been made to contact copyright holders of material reproduced in this book. Any omissions will be rectified in subsequent printings if notice is given to the publisher.

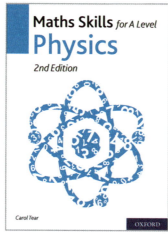

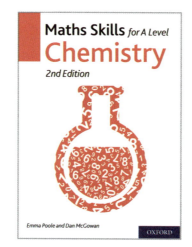

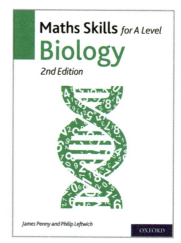